Biogeography and Biodiversity

IGU Commission Contribution to International Year of Planet Earth

Biogeography and Biodiversity

Editor

R.B. Singh

RAWAT PUBLICATIONS

Jaipur • New Delhi • Bangalore • Guwahati • Kolkata

ISBN 978-81-316-0248-5

Published by
Prem Rawat for **Rawat Publications**
Satyam Apts., Sector 3, Jawahar Nagar, Jaipur 302 004 (India)
Phone: 0141 265 1748/7006 Fax: 0141 265 1748
E-mail: info@rawatbooks.com
Website: www.rawatbooks.com

New Delhi Office
4858/24, Ansari Road, Daryaganj, New Delhi 110 002
Phone: 011 2326 3290

Also at *Bangalore, Guwahati* and *Kolkata*

Typeset by Rawat Computers, Jaipur
Printed at Chaman Enterprises, New Delhi

Contents

Preface

Forests and biodiversity provide the basis for ecosystem services in which people, particularly the poor, depend. The loss of biodiversity restricts our future development options. There has been substantial concentration of Green House Gases (GHGs) in the atmosphere over the last several decades by anthropogenic processes and changes in land use, particularly deforestation. Demand from the forests has grown rapidly over the last 50 years. Subsequently, it causes the loss of biodiversity. Therefore, the full value of goods and services provided by biodiversity is being threatened and puts constraints for achieving sustainable development as described in the report of the World Commission on Environment and Development (Brundtland Commission Report). The report identifies following key issues:

- Biodiversity plays a critical role in providing livelihood security for people.
- From the use of genetic resources to harnessing other ecosystem services, agriculture throughout the world is dependent on biodiversity.
- Many of the factors leading to the accelerating loss of biodiversity are linked to the increasing use of energy by society.
- Human health is affected by changes in biodiversity and ecosystem services.
- Human societies everywhere have depended on biodiversity for cultural identity, spirituality, inspiration, aesthetic enjoyment and recreation.

Recently, biodiversity loss continues because current policies and economic systems do not incorporate the values of the biodiversity effectively in market systems, and many current national and international policies have not been fully implemented. The 1992 United Nations Conference on Environment and Development (UNCED) Chapter 15 of Agenda 21 and Earth System Science Partnership Programme-DIVERSITAS also promote biodiversity and call on countries and their national institutions to:

- develop efficient methodologies for baseline resource surveys and inventories as well as the systematic sampling and evaluation of biological resources.

- promote establishment and strengthening of the national inventory with respect to survey, data collection, sampling and evaluation; and
- produce regularly updated reports on biodiversity based on national assessments.

Keeping above issues in mind, the focus of the present research volume has been on biogeographic dimensions, biodiversity conservation and sustainable use of its components for socio-economic development. Holistic perspective of biodiversity conservation includes bio-geo monitoring and indicators, climate change, tourism, invasive and alien species. Special focus is on mountain, coastal and marine biodiversity, eco-development in protected areas, local knowledge, technology transfer, education and public awareness.

The Kyoto Protocol (1997) agreed on Clean Development Mechanism (CDM) which includes two activities, namely, afforestation and reforestation. Marrakech Accords (2001) defined afforestation as the direct human-induced conversion of land that has not been forested for a period of at least 50 years through planting, seeding and/or the human-induced promotion of natural seed sources. Reforestation is the direct human-induced conversion of non-forested land to forested land through planting, seeding and/or the human-induced promotion of natural seed sources on land that was forested but that has been converted to non-forested land. Conservation and sustainable use of biodiversity is an essential element of any strategy to adapt to climate change as per Convention on Biological Diversity and the United Nations Framework Convention on Climate Chang (UNFCC).

In a nutshell, the present volume highlights:

1. introduction to basic concepts and techniques for making biodiversity assessments at various biogeographic levels;
2. application of emerging areas of science and new techniques for making assessments at local, regional, national and global levels; and
3. planning for the conservation and sustainable use of biodiversity components.

The book comprises 26 chapters relating to conceptual and empirical case studies from both the developed as well as developing countries. The book also combines science and policy perspectives to the biogeography and biodiversity. It will benefit students, researchers and teachers of Geography, Environmental Studies, Biosciences, Ecology and Policy Science.

Finally, I would take this opportunity to express my sincere thanks and gratitude to the Chairman of the IGU Commission on Biogeography and Biodiversity, Prof. Udo Schickhoff (University of Hamburg, Germany), for supporting my initiative.

R.B. Singh

1 Forest Resources, Biogeography and Biodiversity

Indian Experience

R.B. Singh

> Rapid environmental change is all around us. This assault on the global environmental risks undermining the many advances human society has made in recent decades. It is undercutting our fight against poverty. It could even come to jeopardize international peace and security.
>
> —Ban Ki-moon

Forests and biodiversity play significant roles in maintaining and enhancing the livelihood security and well-being of the world's more than 6.7 billion people, including rich, poor, and indigenous group together with rural and urban communities. Biodiversity comprises much of the renewable natural capital on which livelihoods and development depend. However, degradation and losses in biodiversity over the past 20 years have decreased the capacity of many ecosystems to provide services, and have had profound negative impacts on opportunities for sustainable development around the earth. These impacts are prominent in the developing countries (UNEP, 2007).

Forest and bio-resources in India are dynamic in both space and time. Indian ecosystems are getting enormous pressure due to rapid population growth, underdevelopment and careless application of development technology. Emerging forest resource degradation has sought geographers for intensive research on how human beings developed a relationship with nature and natural resources. Geographers are examining the spatio-temporal dimensions of natural, social and human environment taking different critical regions of India. The major environmental issues include population growth dynamics, land utilization, food and hunger, water stress, urbanization and industrialization, socio-economic disparities, poverty, environmental refugees, gender, migration, extinction of plants and animals, deforestation, soil erosion, desertification, natural disasters, toxic and nuclear waste, climatic change, atmospheric pollution, water pollution and land use/cover change (Goel and Singh, 2006; Krishna et al., 2000; Singh, Fox and Himiyama, 2001; Singh, 1990, 1995, 2007).

Drivers of Resource Use and Environment

Major human driving forces of resource use include (1) demographic change of environment including migration; (2) land use and land cover change includes biophysical and social dimensions of rapidly changing land use, human settlement and land cover patterns; (3) urbanization and industrialization transformation highlights the linkage between environmental services and infrastructures for regulating the environment; (4) attitude and behaviour and their role in driving environmental responses and the potential role of alternative development paths; and (5) decentralized decision-making process promoting the linkages between national, regional and local skills and the constraints to the transfer of policy instruments from one region to another (Singh, 2002). In recent years, dimensions of environmental change have encompassed a full range of social sciences disciplines necessary to analyze and understand people's role as both the possible cause and target of environmental change as well as recognizing the local issues and use of local field-based geographical studies supplemented by national and regional data. This should provide an indispensable contribution to analyze the key driving forces of land use maintenance and change. These issues especially reflect the wide diversity of economic, socio-cultural and institutional practices and traditional knowledge at the local level (Singh, 2002, 2006; Jha and Singh, 2008), which has direct implications for policy development and implementation in order to develop strong links between research community, policy makers and community leaders (Singh, 1996, 1998).

Geographical Monitoring of Diverse Ecosystems

Geographical regions are units for studying and monitoring environments. As the seventh largest country in the world, India poses a challenge for researchers to plan viable use of natural resources in diverse condition. The enormous variety of ecosystems in India ranges from alpine zones in the Northern Himalaya to sandy coastal beaches at the southern peninsular India. To the west lies an arid and desert and on the east a flood-prone delta. Between the northern/southern and eastern/western extremes are fertile agricultural lands, and a myriad of dense forests. Entrusted with the monumental task of surveying the land for these natural resources are some agencies like the Botanical Survey of India (BSI), Forest Survey of India (FSI), National Remote Sensing Agency (NRSA), and Revenue Departments of the Government of India and states. Each agency has their own way of monitoring the natural resources and Indian environment (Singh, 1991).

Land Utilization Pattern

Land utilization statistics are available for 92.8 per cent of total geographical area of 328.7 million hectares. According to land use statistics available from states, area under forests had increased from 40.5 million hectares in 1950–51 to 69.4 million hectares in 2000–2001. Net sown area increased from 118.8 million hectares to 141.1 million hectares during the same period (Table 1).

Table 1

Pattern of Land Utilization in India (1950–51 to 2000–01)

('000 hectares)

Year	Total reported area	Net area sown	Area sown more than once	Total cropped area	Forests	Not available for cultivation	Permanent pastures and other grazing lands	Land under miscellaneous tree crops and groves (not included in net sown area)	Culturable waste	Fallow lands other than current fallows	Current fallows
1950–51	284,315	118,746	13,147	131,893	40,482	47,517	6,675	19,828	22,943	17,445	10,579
1960–61	298,458	133,199	19,573	152,772	54,052	59,751	13,966	4,459	19,212	11,180	11,639
1970–71	393,758	140,784	25,007	165,791	63,917	44,640	13,262	4,298	17,500	8,759	10,598
1980–81	304,167	140,270	33,054	173,324	67,421	39,619	12,008	3,494	16,728	9,817	14,810
1990–91	304,017	142,234	43,243	185,477	67,955	40,880	11,804	3,703	15,014	9,590	13,807
1995–96	304,875	142,197	45,274	187,471	68,817	41,371	11,064	3,481	14,098	10,016	13,831
1996–97	304,878	142,813	46,779	189,592	68,750	41,543	11,040	3,567	13,947	9,892	13,326
1997–98	305,786	142,083	48,487	190,570	69,012	42,136	11,046	3,616	13,880	9,748	14,265
1998–99	306,053	142,580	50,453	193,034	68,980	42,315	11,127	3,629	13,904	9,929	13,589
1999–00	306,054	141,231	48,510	189,740	69,024	42,407	11,040	3,618	13,828	10,108	14,798
2000–01	306,249	141,101	45,908	187,009	69,407	42,828	10,897	3,366	13,660	10,191	14,799

Source: Statistical Abstract of India (2004).

Broad cropping pattern indicates that though food grains have preponderance in gross cropped area as compared to non-food grains, their relative share came down from 76.7 per cent during 1950–51 to about 66 per cent at present. Land utilization in India shows a rather diversified picture. Forests in India constitutes 22 per cent, land not available for cultivation is 14 per cent, permanent pastures are only 4 per cent, whereas land under miscellaneous tree crops etc., are only 1 per cent. Total cultivable wasteland in India is 5 per cent, fallow land is 8 per cent and net area sown is 46 per cent (Statistical Abstract of India, 2004). Perhaps, one of the most valuable natural resources to a country is the land that has the ability to feed its people. Out of the 328 million hectares that constitutes India, 143 million hectares are agriculturally cultivable lands. Considering that agriculture employs 64 per cent of the labour force, it is the agriculture which is the predominant mainstay in almost every state in India. Land use study has been the subject of intensive research and monitoring in post-independence period at the national, state and district levels. But, agro-ecosystem and agro-climatic research and education are being preferred by geographers in order to provide particularly a strong base for detecting responses to global climatic changes and preferred differentiating effects of climates, pollution and land use cover. The impact of increasing CO_2 concentration in the atmosphere will have enormous effect on agriculture and vegetation (Sen, Roy and Singh, 2002; Sharma and Kumar, 2003; Rawat, 2003). It is important to understand processes relating to changes in land use, affecting the resources of the river basin and coastal zone. Subsequently, changes in sea level and climate influence coastal ecosystem. Therefore, focus on agro-ecosystem in Indian geography is needed, highlighting different agriculture ecosystems of India, agricultural input environment, eco-farming and ecologically based pest management (Jha and Singh, 2008; Mohammad, Singh and Datta, 2007; Singh and Verma, 2000; Sastry, 2000).

Bio-Resource Survey and Inventory

Resource use studies require analysis of phenomena, processes and their interaction throughout space and time (Singh and Parijat, 2001). The Resource Information System has to be considered as multidimensional, i.e., attribute, spatial and temporal dimensions. Indian geographic education offers such capabilities as they integrate multi-sector, multi-space and multi-period perspectives. Vulnerable environment is highly sensitive and responds rapidly to anthropogenic intervention. Its restoration needs a long time, especially if the vulnerable environment is degraded or stressed beyond a certain point (Singh, 1998). Study of the effects of positive and negative impact of resource use change is needed for risk mitigation, environmental preparedness and long-term planning. Observations using remote sensing and Geographical Information System (GIS) in such environmental changes allow identification of major processes. Empirical diagnostic models of environmental change can be developed from these observations. New technology applications like GIS for safeguarding the environment have increased dramatically in recent years. The development of regional and global models based on remote sensing and

GIS can simulate both the socio-economic and biophysical driving forces, interactions, feedbacks and their responses to environmental change.

Remote Sensing and GIS-Based Monitoring and Management

India's forest areas have been decreasing rapidly, and a system of continuous operational monitoring is necessary. Coloured infrared aerial photography helps assess the annual occurrence of losses, especially in the less accessible areas. Thermal scanners can detect fire. It is through remote sensing that the actual forest cover of India was known through findings of satellite data analysis as highlighted by the Forest Survey of India. Only 10.88 per cent forms the closed forests of adequate density. In the last 40 years, as much as 4.32 million hactares of forest land have been lost, 0.7 million hactares encroached upon and the rest subjected to shifting cultivation (Government of India, 2007; Singh et al., 2003). Deforestation has increased because of large-scale consumption of fuelwood. Each year, Bangalore alone consumes 0.4 million tonnes of fuelwood, which is equivalent to, 5,000 hactares of forests. The cattle carrying capacity of forests for grazing purposes is 90 million head, as against the 400 million cattle population (Mehrotra and Suri, 2002).

The first attempt to categorize forest cover types by computer analysis of LANDSAT digital data was done in 1978 for Nagaland. In this study, a colour-coded categorized map delineated the broad forest cover types. Similar studies in Mizoram and Andhra Pradesh were also conducted. In a study conducted by NRSA, the satellite digital covering of entire Periyar-Thodupuzha drainage basin was analyzed. Multi-stage approach is being adopted by the National Remote Sensing Agency (NRSA), which gives information like timber volume. Apart from forestland classification, stock mapping and volume estimation, remote sensing is also used for damage assessment and fire detection, which is a common feature of Indian forests. GIS is also used in biodiversity conservation plan (Ravan, 1999; Kushwaha et al., 2000).

Use of aerial photographs or other imageries is to complement and improve or reduce fieldwork rather than absolutely replacing it. The Forest Survey of India (FSI) prepared forest cover type and land use maps on 1: 50,000 and 1: 63,360 scale by interpreting medium to small-scale panchromatic aerial photographs for about 4,20,000 km^2 in India. The main application of remote sensing in forest management has been focused on timber harvest planning and monitoring of logging and deforestation. FSI has also suggested establishment of about 30 industries relating to pulp and paper mills, plywood, saw mills, fibre board, hard board and newsprint, based on the results of forest inventory using aerial photo interpretation. This project was applied extensively in Central India, Kerala, Andhra Pradesh and Manipur.

Alaknanda valley in the Chamoli district of Uttrakhand is considered one of the most severely degraded areas in the Central Himalayas. At the behest of Dasholi Gram Swarjya Mandal (DGSM), the Space Application Centre took up study to monitor the impact of Chipko movement using multi-temporal satellite data available since 1972. Satellite data of 1972, 1982, and 1991 were interpreted and the spatial distribution of forest type and other land use/cover have been mapped. Analysis of these results in terms of forest lost and gain showed that Chipko movement, launched in 1973, has helped forest gain of 8.2 per cent during 1972–91. The present study attempts to classify various vegetation classes using time integrated NDVI (T-NDVI) values derived from IRS-P3 WiFS data.

The study area forms the hilly northern part of East Godavari district (Eastern Ghats), Andhra Pradesh. The major land use/cover classes in the study area include moist mixed dry deciduous vegetation, degraded/scrub forest, fallow lands, water bodies and shifting cultivation areas, accounting for about 9,265 km^2. The forest of the study area corresponds to southern dry mixed deciduous forests and southern tropical forests as classified by Champion and Seth. IRS-P3 WiFS data of 15/2/98, 25/2/98, 21/3/98, 19/4/98 and 3/5/98 covering study area has been taken for analysis.

Vegetation indices are built as linear combinations of albedos or bi-directional reflectance, which depend on radiation properties of surface and can be effectively utilized for discriminating different vegetation types. NDVI is sensitive to the presence of green vegetation and has been successfully used in numerous regional and global applications for studying the state of vegetation. The typical phonology of dry deciduous and moist mixed vegetation types is clearly evident from the IRS-P3 WiFS data. The deciduous vegetation could be easily distinguished from the moist mixed vegetation with its characteristics tonal variation and also from rest of the classes, viz., shifting cultivation areas and degraded scrub forests. The results of the study indicate that discrimination of the forest types is possible using T-NDVI characteristics and this method can be effectively utilized for forest type classification.

Coarser resolution satellite data is increasingly used for land use/cover monitoring and mapping at regional scale. Several studies indicate the importance of vegetation indices in the delineation of land use/land cover classed and vegetation types. The present study suggests the utility of time integrated NDVI approach using coarse resolution IRS-P3 WiFS data for forest types and land use/land cover discrimination at broad spatial scales (Chaturvedi and Barthwal, 1999; Nagaraja, 2002; Nanoti et al., 2000).

The IRS–1B FCC imagery, FSI topographical maps and ground observations were used to generate the spatial information on the extent of forest cover in the Chilla Sanctuary of Rajaji National Park in Shivaliks. It includes Gohri and Chilla forest ranges. The study area, falling in Pauri Garhwal district of newly created Uttarakhand state, covers a geographical area of about 258 km^2. The visual interpretation of satellite imagery was carried out for stratification of various forest types and densities. Image elements like tone, texture, shape, size, shadow, location and association were used to identify various forest/vegetation cover categories. According to the study done by this author, mixed Sal forests happens to be the most dominant land cover occupying 63 per cent of the total sanctuary area. It is second largest cover category followed by pure Sal forests plantations. Remote sensing and GIS technologies provide vital information support in terms of relevant, reliable and timely information needed for conservation planning. The reliable information about forest cover in Chilla Sanctuary has been already discovered (Kushwaha, Munkhtuya and Roy, 2000).

Forest stock assessment has been done using IRS LISS-III and PAN merged data in Timli Forest Range, Dehradun. The study demonstrates the use of high resolution IRS–1C LISS-III and PAN merged data for growing stock assessment in Timli Forest Range in the west of Dehradun. The merged datasets was generated using principal component-based image fusion. The merged data had advantage of colour and high resolution from LISS-III and PAN respectively. It facilitated in differentiation and mapping of a number of forest categories in terms of type and density (Mehrotra and Suri, 2002). The study was carried out in Timli Forest Range located 45 km west of Dehradun. Timli village and Chidiya valley agricultural land fall within the forest of the study area (Singh, Das and Kushwaha, 2003).

According to the data obtained by remote sensing sources, the majority of area was found to be Sal forest (54.6%) followed by Sal estimated to be 2.08 million m^3. The present studies demonstrate the utility of IRS–1C LISS-III and PAN merged data in discrimination of forest type and canopy density and estimation of growing stock. Forest cover type map and forest canopy density generated with remote sensing data can be significant inputs for the forests management (Narayanan, 2000).

Singh, Roy and Dash have done another study using IRS-P3 MOS data in the Indo-Gangetic basin and peninsular regions. Monitoring and mapping of the Himalayan region on day-to-day basis is possible using the broadband remote sensing sensors. The spatial resolution of microwave sensors is poor compared to visible optical sensors. The visible sensors suffer from the problem of cloud cover, especially in high altitude regions, such as the Himalayan region. The spectral reflectance of snow, ice, cloud, vegetation and bare ground have distinct behaviour. However, due to the presence of cloud, the information from the snow, ice, vegetation and bare ground are marked by the high reflectance of the cloud. The effect of the cloud can be distinguished in the Short Wave Infrared (SWIR). The Institute of Space Sensor Technology of the German Aerospace Centre (DLR) has developed a Visible/Near Infrared (VIS/NIR) imaging spectrometer named Modular Optoelectronic Scanner (MOS)[2,3] launched on 21 March 1996 on-board the Indian Remote Sensing (IRS) satellite IRS-P3 (ref. 4) into a sun-synchronous polar orbit at 817 km. MOS sensors consist of two separate imaging spectro-radiometers MOS-A (4 bands) and MOS-B (13 bands) in the VIS and NIR and a CCD-line camera MOS-C (1 band) in the SWIR. The image has been classified using Maximum Likelihood Supervised Classification method from MOS-B data (Rao et al., 2003).

Roy et al. (2002) has contributed an article on map atlas on biodiversity characterization. This is a multi-institutional programme on bio-prospecting of biological wealth jointly supported by Department of Space and Department of Biotechnology. The map atlas allows identifying gaps in the conservation planning by setting priority in decision-making and at management level for conservation of biodiversity. Biosphere reserves areas are those habitats where landscape ecological conditions are favourable for natural speciation and evolutionary process. The satellite data provided the key input. Digital Elevation Model (DEM) was used to prepare terrain complexity map. The resultant map is the biological richness. The goals and scales of inventorying and monitoring programmes may change with time. The present effort to characterize vegetation cover, fragmentation, disturbance and biological richness across the landscape is organized in the form of map atlas. It allows us to identify gaps in the conservation and management areas, and can help in biodiversity conservation planning by setting priority areas. This information is immensely valued in the biodiversity hotspots of the country. It is expected that maps will be used for planning detailed ground level inventories of flora and fauna by premier institutions (Babu Rao, 2001; Bhagabati, 2001).

A creation of biosphere reserves, which is considered as restored polygon in different bio-geographic regions, is one of the important programmes of Government of India. UNESCO-MAB initiated programme as an integral part of wide spectrum of complementary and transverse scale observations from man-on-the ground to geosynchronous satellites and polar orbital platforms. Such a vast task can be largely assisted by recent advances in computer-based GIS. In the wildlife census, the fish counts, migratory bird numbering, their resting spots, etc., could be photographed through thermal infrared light imagery, this would provide a considerable management input to the protected areas and biodiversity (Sankar et al., 2003; Singh, 1994; Singh and Bortamuly, 2005).

Methodology of Monitoring Bio-Resources and Forest Changes

India is rich in flora. Currently available data place India in the tenth position in the world and fourth in Asia in plant diversity. From about 70 per cent geographical area surveyed so far, the Botanical Survey of India has described 47,000 species of plants. Biogeography, which integrates physical and ecological aspects of surrounding, can play an effective role in environmental management. In assessing environmental change and prediction of future of that environment in an integrated and holistic manner, there are a number of emerging research areas relating to environmental changes which require geographical enquiry in Indian context. The understanding of global environmental change requires integrated emphasis on some promising geographical tasks, i.e., reconstruction of landscape processes, environmental history for ecosystem, state of environment and resource use, modelling of the dynamics of present-day landscapes, analysis of causes and consequences of anthropogenic changes. This is possible with enhanced GIS supported by detailed site studies (Dasgupta, 2002; Singh and Murai, 1998; Singh, 2005; Singh and Mishra, 2005).

Data Source and Data Period

The satellite imagery (data) is used from the NRSA, Hyderabad in digital form as well as False Colour Composite (FCC) prints. The first State of Forest Report (SFR), published in 1987, was based on Landsat data of US satellite having a resolution of 80 m. From the second assessment, the resolution of the sensor improved to 30 m. The IRS satellite data has been used since fifth assessment. The data of Linear Imaging and Self-scanning Sensor (LISS) II of IRS–1B with a resolution of 36.25 m and LISS III data of IRS 1C & 1D satellites with a resolution of 23.5 m have been used in the present assessment.

Forest Cover Classes

On the basis of crown density forest types include (see Tables 2 and 3):

- *Dense forest:* All lands with tree cover of canopy density of 40 per cent and above.

Table 2

Crown Density Range of Different Forest Cover

Forest cover	*Crown density range*
Very Dense Forest (VDF)	> 70%
Moderately Dense Forest (MDF)	40–70%
Open Forest (OF)	10–40%
Non-forest cover	< 10%
Scrub	—
Non-forest	—
Water bodies	—

Source: State of Forest Repot, 2003.

Table 3

Status of Forest Cover in India

Class	*Area (km²)*	*Per cent of Geographic Area*
Forest Cover		
(a) VDF	51,285	1.56
(b) MDF	339,279	10.32
(c) Open	287,769	8.76
Total Forest Cover*	678,333	20.64
Non-forest Cover		
Scrub	—	1.23
Non-forest**	—	78.13
Total Geographic Area	—	100.00

* Including 4,461 km^2 under mangroves (0.14% of country's geographic area)
** Excludes scrubs and includes water bodies

Source: State of Forest Report, 2003.

- *Open forest:* All lands with tree cover of canopy density between 10 to 40 per cent.
- *Mangrove:* Salt tolerant forest ecosystem found mainly in tropical and sub-tropical inter-tidal regions.
- *Scrub:* All lands with poor tree growth mainly of small or stunted trees having canopy density less than 10 per cent.
- *Non-forest:* Any area not included in the above classes.

Spatial Distribution of Different Types of Forests in India

Forests have acquired increasing importance in the recent past for their role not only in meeting the material requirements but also for their ecological and environmental functions. Therefore, accurate and updated information on forest resources (Ghoshal et al., 2003) (see Table 4) is of paramount importance. The forest types in India include:

1. *Moist Tropical Forests:*
   (a) Tropical Wet Evergreen: Western Ghats (Kerala, Karnataka, Maharashtra).
   (b) Tropical Semi-Evergreen: Lower Hills of Eastern Ghats (Karnataka and Kerala).
   (c) Tropical Moist Deciduous in Dehra Dun, Mahabaleshwar, and Eastern Ghats.
   (d) Littoral and Swamp Forests spread over Sunderbans, Andamans and Bengal.
2. *Dry Tropical Forests:*
   (a) Tropical Dry Deciduous found in Madhya Pradesh and Uttar Pradesh.
   (b) Tropical Thorn Forests: Delhi, Punjab and Gujarat.
   (c) Tropical Dry Evergreen: Eastern Ghats (Andhra Pradesh, Tamil Nadu).
3. *Montane Sub-Tropical Forests:*
   (a) Sub-Tropical Broad: Shillong, Nilgiris (Tamil Nadu).
   (b) Sub-Tropical Pine Forests: Khasia (Arunachal Pradesh), Kashmir, Punjab (Western Himalaya).
   (c) Sub-Tropical Dry Evergreen: Jammu, foothills of Himalaya.
4. *Montane Temperate Forests:*
   (a) Montane Wet Temperate: Nilgris, Palni Hills (Tamil Nadu).
   (b) Himalayan Wet Temperate: Simla, Assam, Himachal Pradesh.
   (c) Himalayan Dry Temperate: North-West Kashmir.
5. *Sub-Alpine Forest:* Laddakh, Sikkim.
6. *Alpine Scrubs:*
   (a) Moist Alpine Scrub High Himalayan.
   (b) Dry Alpine Scrub: Sikkim.

Table 4
Spatial Distribution of Forest Cover in India, 2001

State/UT	*Geographical area*	*Total forest cover*	*Per cent*	*Scrub (area in sq km)*
Andhra Pradesh	275,069	44,419	16.15	9,748
Arunachal Pradesh	83,743	68,019	81.22	116
Assam	78,438	27,826	35.48	219
Bihar	94,163	5,558	5.90	150
Chhattisgarh	135,191	55,998	41.42	88
Delhi	1,483	170	11.47	1
Goa	3,702	2,156	58.24	0
Gujarat	196,022	14,946	7.62	1,743
Haryana	44	1,517	3.43	68
Himachal Pradesh	55	14,353	25.78	389
Jammu & Kashmir	222	21,267	9.57	2,947
Jharkhand	79	22,716	28.50	807
Karnataka	191	36,449	19.00	3,141
Kerala	38	15,577	40.08	72
Madhya Pradesh	308	76,429	24.79	2,378
Maharashtra	307,713	46,865	15.23	4,175
Manipur	22,327	17,219	77.12	74
Meghalaya	22,429	16,839	75.08	169
Mizoram	21,081	18,430	87.42	274
Nagaland	16,579	13,609	82.09	231
Orissa	155,707	48,366	31.06	5,346
Punjab	50,362	1,580	3.14	22
Rajasthan	342,239	15,826	4.62	4,564
Sikkim	7,096	3,262	45.97	360
Tamil Nadu	130,058	22,643	17.41	2,040
Tripura	10,486	8,093	77.18	1
Uttar Pradesh	240,928	14,118	5.86	749
Uttarakhand	53,483	24,465	45.74	320
West Bengal	88,752	12,343	13.91	75
Andaman & Nicobar Islands	8,249	6,964	84.42	1
Chandigarh	114	15	13.16	1
Dadra & Nagar Haveli	491	225	45.82	—
Daman & Diu	112	8	7.45	—
Lakshdweep	32	23	71.88	—
Puducherry	480	40	8.33	—
Total	3,287,263	678,333	20.64	40,269

Source: State of Forest Report, 2003.

Since 1987 remote sensing technology is being used for the assessment of the forest cover. FSI also prepares thematic maps using aerial photographs, and carries out inventory of natural forests as well as Trees Outside Forests (TOF). Thematic maps describes forest types, major species composition and crown density of forest (see Table 5).

Table 5
Temporal Distribution of Forest Cover in India

(in sq km)

State/UTs	*1987*	*1993*	*2001*	*2001**
Andhra Pradesh	49,573	47,256	44,637	53,648
Arunachal Pradesh	64,132	68,661	68,045	68,523
Assam	25,160	24,508	27,714	29,656
Bihar	28,482	26,587	5,720	9,413
Jharkhand	—	—	22,637	25,331
Delhi	15	22	111	151
Goa	1,240	1,250	2,095	2,157
Gujarat	11,991	12,044	15,152	19,188
Haryana	513	513	1,754	3,280
Himachal Pradesh	12,480	12,502	14,360	14,757
Jammu & Kashmir	20,905	20,443	21,237	23,454
Karnataka	32,268	32,343	36,991	44,437
Kerala	10,292	10,336	15,560	16,706
Madhya Pradesh	130,099	135,396	77,265	83,016
Chhattisgarh	—	—	56,448	59,983
Maharashtra	45,616	43,859	47,482	55,751
Manipur	17,475	17,621	16,926	17,021
Meghalaya	16,466	15,769	15,584	15,724
Mizoram	19,084	18,697	17,494	17,589
Nagaland	14,394	14,348	13,345	13,415
Orissa	53,253	47,145	48,838	53,202
Punjab	943	1,343	2,432	4,066
Rajasthan	12,758	13,099	16,367	21,653
Sikkim	2,756	3,119	3,193	3,207
Tamil Nadu	17,472	17,005	21,482	27,536
Tripura	5,953	5,538	7,065	7,133
Uttar Pradesh	31,226	33,961	13,746	21,291
Uttarakhand	—	—	23,938	24,386
West Bengal	8,432	8,186	10,693	13,957
Andaman & Nicobar Islands	7,601	7,624	6,930	7,013
Chandigarh	2	5	9	11
Dadra & Nagar Haveli	238	206	219	246
Daman & Diu	0	0	6	10
Lakshdweep	0	0	27	27
Puducherry	0	0	36	71
Grand Total	640,819	639,386	675,538	757,009
Per cent	19.49	19.45	20.55	23.03

*forest and tree cover

Source: State of Forest Report, 2003.

The states of Andhra Pradesh, Arunachal Pradesh, Delhi, Gujarat, Haryana, Himachal Pradesh, Jammu & Kashmir, Karnataka, Madhya Pradesh, Maharashtra, Orissa, Punjab, Rajasthan, Tamil Nadu, Tripura, Uttar Pradesh and West Bengal have registered increase in the forest cover. There is decrease in forest cover in states like Assam, Bihar, Goa, Kerala, Manipur, Meghalaya, Mizoram, Nagaland, Sikkim, Andaman & Nicobar Islands, and Dadra & Nagar Haveli. In Chandigarh and Daman & Diu, no change has been observed.

Species Composition

There are a number of species found in India due to varied agro-climatic conditions. *Eucalyptus spp., Tectona grandis* and *Acacia spp.* are the main species having larger area under the plantations compared to other species. Among eucalyptus, *E. globulus, E. grandis, E. tereticornis* are most common. Among acacias, *A. auriculiformis, A. catechu, A. mearnsii, A. nilotica* and *A. tortalis* are common. Other broad-leaved species are *Albizia spp., Azadirachta indica, Dalbergia sissoo, Gmelina arborea, Populus spp., Prosopis spp.* and *Terminalia spp.* Among conifers, *Cedrus deodara* and *Pinus roxburghii* and *Pinu kesiya ocupy* major area. *Pinus patula* and *P. carribaea* have been planted in a limited area.

Shifting Cultivation in North-Eastern Region

Shifting cultivation (*jhum*) is a common land use in many states, particularly in the north-eastern region of the country. According to a Task Force on shifting cultivation (1983), the area affected by shifting cultivation in the north-eastern region was reported to be 3.81 million hectares. According to North-Eastern Council (1975), it is 2.80 million hectares (see Table 6). The extent of area under shifting cultivation is maximum (0.39 m ha) in Nagaland, followed by Mizoram (0.38 m ha) and Manipur (0.36 m ha). These states together account for about 65 per cent of the total area under shifting cultivation in the north-eastern region (Government of India, 2007).

Table 6

Loss of Forest Cover due to Shifting Cultivation in North-Eastern States

(area in km^2)

State	*Dense Forest*	*Open Forest*	*Total*
Assam	272	337	609
Arunachal Pradesh	663	262	925
Manipur	125	730	855
Meghalaya	141	543	684
Mizoram	351	336	687
Nagaland	321	1,011	1,332
Tripura	221	163	384
Total	2,094	3,382	5,476

Source: State of Forest Report, 2003.

Mangroves

India has a wealth of wetland ecosystems distributed in different geographical regions from the cold arid zone of Ladakh in the north to the wet humid climate of Imphal in the east, the warm arid zone of Rajasthan in the west to the tropical monsoon Central India and the wet and humid zone of Southern Peninsula. Most of the wetlands in India are directly or indirectly linked with major river systems such as Ganga, Brahmaputra, Narmada, Tapti, Godavari, Krishna, Cauveri, etc. (see Table 7).

Table 7

Temporal Change in Mangrove Cover

(area in km^2)

State/UT	1987	1989	1991	1993	1995	1997	1999	2001
Andhra Pradesh	495	405	399	378	383	383	397	333
Goa	0	3	3	3	3	5	5	5
Gujarat	427	412	397	419	689	901	1031	911
Karnataka	0	0	0	0	2	3	3	2
Maharashtra	140	114	113	155	155	124	108	118
Orissa	199	192	195	195	195	211	215	219
Tamil Nadu	23	47	47	21	21	21	21	23
West Bengal	2,076	2,109	2,119	2,119	2,119	2,123	2,125	2,081
Andaman & Nicobar	686	973	971	966	966	966	966	789
Puducherry	0	0	0	0	0	0	0	1
Total	4,046	4,255	4,244	4,256	4,533	4,737	4,871	4,482

Source: State of Forest Report, 2003.

The Ministry of Forests & Environment has been implementing a programme on conservation of wetlands since 1987 and a total number of 20 wetlands in 13 states have been covered under this programme. Several initiatives have been taken up with the main focus on biological methods of conservation rather than adopting engineering options. Management Action Plans prepared for 18 wetlands are in operation and two Action Plans for wetlands of Kerala have been finalized. A directory on wetlands in India already published gives information on location, area and ecological categorization of wetlands. India is signatory to the Convention on Wetlands of International Importance, especially as Waterfowl Habitat (Ramsar Convention).

Biodiversity in India

Biodiversity is the variety of life on earth. It includes diversity of the genetic level, such as, between individuals in a population or between plant varieties, the diversity of species and the diversity of ecosystems and habitats (UNEP,

2007). The World Commission on Environment and Development (WCED) identified following roles and issues of the biodiversity:

1. Biodiversity plays a critical role in providing livelihood security for people.
2. From the use of genetic resources to harnessing other ecosystem services, agriculture throughout the world is dependent on biodiversity.
3. Many of the factors leading to the accelerating loss of biodiversity are linked to the increasing use of energy by society.
4. Human health is affected by changes in biodiversity and ecosystem services.
5. Human societies everywhere have depended on biodiversity for cultural identity, spirituality, inspiration, aesthetic enjoyment and recreation.

India is one of the 16 mega diverse countries where 60–70 per cent of the world's biodiversity is found. India has a large network of protected areas. At macro level, there are 14 biosphere reserves including Sunderban, Gulf of Mannar and Nilgiris, which are included in the world network of biosphere reserves. At meso level there are about 100 national parks together with 500 wildlife sanctuaries at micro level.

The number of plants (flowering and non-flowering) species in India is estimated to be over 47,000 representing about 12 per cent of world's flora. These are categorized in different taxonomic divisions including over 1,500 flowering plants. Estimates for lower plants are: 64 gymnosperms, 2,843 bryophytes, 1,012 pteridophytes, 1,940 lichens, 12,480 algae and 23,000 fungi. Around 5,150 species of flowering plants are endemic to the country. Among the endemic species, 2,532 species are found in Himalaya (Singh and Haigh, 1995) and surrounding areas, followed by 1,782 species in peninsular India. About 1,500 endemic species are facing varying degree of threats. Forests are rich in floral and faunal diversity and regarded as home of the biodiversity.

Conservation and Management of Forests and Biodiversity

In recent years, India is becoming one of the important countries for involving non-governmental initiatives for resource management, environmental and developmental capacity building in order to achieve sustainable development. Many government programmes are facilitating linkages between Environmental Non-Governmental Organizations (ENGOs) and Developmental Non-governmental Organization (DNGOs) and development decision-making. This attracted a greater attention after 73rd and 74th amendment to the Indian Constitution, encouraging participatory decision-making and empowerment of the people. Such organizations respond to social problem fast with more focused approach and often with better success. Several NGOs have been involved in programmes and service delivery for environmental decision-making. It has been suggested that with the help of the mass media, traditional knowledge and community education programmes a greater awareness and sensitivity needs to be build in the public and the communities. Such awareness, it is hoped, will contribute to a greater participation by people in programmes and services (Prasad et al., 1999; Siddique, 2001; Phutego and Chanda, 2004, Kunwar and Kachhawah, 2001).

The Indian Forest Act, 1927, The Wildlife Act, 1972, The National Forest Policy, 1988, The Environment Protection Act, 1986 and Biodiversity Act, 2002 provided legal basis for conservation and management. Biodiversity Act, 2002 and Biodiversity Rules, 2004 focus on conservation and sustainable use of components of biodiversity and fair and equitable sharing of benefits. A National Biodiversity Authority has been set up at Chennai under the Biological Diversity Act, 2002. These Acts initiated establishment of Biodiversity Management Committees (BMC) at local village level, State Biodiversity Boards at state level and a National Biodiversity Authority. Various types of protected areas are included (Government of India, 2007).

Biosphere Reserves

- Biosphere reserve have been set up to protect representative ecosystems and also serve as laboratories for evolving alternative models of development.
- Three biosphere reserves from India are now included in the World Network of Biosphere Reserves. These are Sundarbans (West Bengal), Gulf of Mannar (Tamil Nadu) and Nilgiri (Kerala, Karnataka and Tamil Nadu).

Wetlands, Mangroves and Coral Reefs

- For 22 Wetlands Management Action Plans have been prepared.
- Calimer from Tamil Nadu and East Kolkata from West Bengal have been added to the wetland list of the country.
- Nineteen sites have already been declared as Ramsar sites of international importance in India.
- A Directory of Wetlands, covering 2,107 natural and 65,253 man-made wetlands, occupying an area of 4.1 million hectares and information on the status of 183 wetlands of national/international importance, was prepared.
- Twenty-four wetlands, 35 mangrove and four coral reef areas in the country have been identified by the Government of India for conservation and management (Munyati et al., 1999).

National Afforestation and Eco-Development Board

- A total of 515 projects in 23 states have been operationalized for treating an area of 0.8 million hectares for the afforestation programme with people's involvement for the sustainable management of the country's forests.
- The NAEB has seven regional centres located in universities/national level institutions. During the year, these regional centres have conducted a number of training programmes on Joint Forest Management (JFM),

interactive workshop on forestry programmes, micro-planning exercise, etc. (see Table 8).

Table 8

Total Population Covered under Joint Forest Management (JFM), 2004

State	*Population covered*
Andhra Pradesh	1,822,344
Arunachal Pradesh	142,217
Assam	321,103
Bihar	1,370,714
Chhattisgarh	29,886,000
Goa	1,680
Gujarat	1,005,609
Haryana	827,000
Himachal Pradesh	1,088,589
Jammu & Kashmir	339,220
Jharkhand	2,863,487
Karnataka	1,216,785
Kerala	165,475
Madhya Pradesh	3,391,000
Maharashtra	4,849,875
Manipur	134,575
Mizoram	191,464
Nagaland	416,831
Orissa	—
Punjab	181,624
Rajasthan	1,501,475
Sikkim	16,340
Tamil Nadu	532,479
Tripura	111,391
Uttar Pradesh	3,761,325
Uttarakhand	3,999,900
West Bengal	2,326,975
Total	62,393,477

Source: Government of India, 2007.

In India, Environmental Impact Assessment (EIA) is emerging as a measure tool for ensuring that environmental quality is fully taken into account in the decision-making processes of any developmental programme (Singh and Parijat, 2001; Singh, 2002). This technique will help the development planners in to identify environmental impacts and minimize degradation of environment as well. Geographers may evolve consensus and resolve conflict coming from different interest group. Different stages of environmental assessment, i.e., objectives, scoping, policy alternatives, prediction, significance assessment, evaluation, public participation, plant implementation, mitigation and monitoring will be accurately understood by the geographical communities. With different domains such as environmental, economic, social, developmental, structural-functional, institutional, organizational, regional and geo-political, the geography can bridge the gap between physical and social sciences. Under ecological change research, environmental and social approaches should be linked and established for better understanding of Indian environment (Singh, 2000).

The National Environment Policy, 2006 also focuses to conserve the biodiversity, inter-generational and intra-generational equity, sustainable utilization of biodiversity, integration of environmental, social and economic concerns and principles of good environmental governance.

Conclusion

As biophysical and the social processes are integrated in an ecosystem, there is an interactive linkage and connection between human impacts on the environment and environmental impact on human. Geographers will better understand such location-specific and time-specific interaction. They can describe and manage accurately such processes of environmental change. Interacting area between forests, biogeography and biodiversity is a multidisciplinary science that involves assimilation and interaction between basic services on matters pertaining to the environment integrating conventional and modern techniques and approaches. To have a better understanding of vulnerable environment, Indian geographers should focus on how physical and human ecosystems operate and interact. However, there is a lack of integrated techniques and approaches to vulnerable environment. In recent years, integrated watershed management is emerging as an important tool for resource management. Understanding complex resource-environmental interactions within space provides an important base for sustainable environmental planning and management.

References

Babu Rao, C. (2001), *Image Analysis in Materials and Life Sciences,* Oxford & IBH, New Delhi.

Bhagabati, Abani Kumar (2001), "Biodiversity and Associated Problems in the Islands of the Brahmaputra, Assam", *Geographical Review of India*, Vol. 63, No. 1–4, pp. 330–43.

Chaturvedi, Arun and Barthwal, A.K. (1999), "Land Cover Mapping of Chindwara District Using Remote Sensing Techniques", *Annals, NAGI*, Vol. XIX, No. 1–2, pp. 118–25.

Dasgupta, Malabika (2002), "Jhuming: The Traditional Mode of Cultivation of the Tribal of the Tripura", *Geographical Review of India*, Vol. 64, No. 4, pp. 346–53.

Ghoshal, Choudhuri, S., Pramanik, S.C. and Dinesh, R. (2003), "Effects of Land Use on Tropical Moist Evergreen Forest Soils of Little Andaman", *Indian Journal of Soil Conservation*, Vol. 31, No. 1, pp. 92–94.

Goel, A. and Singh, R.B. (2006), "Sustainable Forestry in Mega-cities of India for Mitigating Carbon Sequestration: Case Study of Delhi", *Advances in Earth Science*, Vol. 21, No. 2, February, pp. 144–50.

Government of India (2004), *Statistical Abstract of India*, New Delhi.

—— (2005), The State of Forest Report, 2003, Forest Survey of India, Dehradun.

—— (2007) *India Year Book*, Ministry of Information & Broadcasting, Publication Division, New Delhi.

Jha, M.M. and Singh, R.B. (eds.) (2008), *Land Use: Reflection on Spatial Informatics, Agriculture and Development*, Concept Publishing Company, New Delhi.

Krishna, N.D.R., Krishna Murthy, Y.V.N., Rao, B.S.P. and Srinivas, C.V. (2000), "Geoinformatics for Ecological Economic Zoning Towards Land Use Planning in Yerrakalava Catchment, Andhra Pradesh", *Agropedology*, Vol. 10, No. 2, pp. 88–100.

Kunwar, P. and Kachhawaha, T.S. (2001), "Transformation of Landuse/Landcover Using Remote Sensing and GIS Techniques", *National Geographical Journal of India*, Vol. 47, 1–4, March-December, 2001.

Kushwaha, S.P.S., Munkhtuya, A. and Roy, P.S. (2000), "Geospatial Modelling for Coral Habitat Evaluation", *Journal of the Indian Society of Remote Sensing*, Vol. 28, December, pp. 277–80.

Mehrotra, A. and Suri, R.K. (eds.) (2002), *Remote Sensing for Environment and Forest Management*, Indus Books, New Delhi.

Mohammad, N., Singh, R.B. and Datta, A. (eds.) (2007), *Spatial Information Technology for Natural Resources Management*, Concept, New Delhi.

Mohammad, Noor (2000), "Geo-Economics of Management of Common Land Resources in Sonipat District, Haryana", *National Geographical Journal of India*, Vol. 46, No. 1–4, pp. 25–32.

Munyati, Christopher Evaristo, Kapungwe, M. and Singh, R.Y., (1999), "The Role of Remote Sensing in Wetland Habitat, Carrying Capacity Change Studies for Endemic Lechwe (Kobus lache Kafuensis) on the Kafu, Flats, Zambia", *Annals, NAGI*, Vol. XIX, No. 1–2, pp. 126–135.

Nagaraja, R. (2002), "Land Use/Land Cover Studies Using Remote Sensing Data: Indian Experience", In proceedings of ISPRS & SIS, Vol. 34, Part 7, *Resource and Environmental Monitoring*, Hyderabad.

Nanoti, M.V., Roy, R.R., and Andey, S.P. (2000), "Role of Simulation in EIA of Water Resources Projects", In R.S. Goel (ed.), *Environmental Management in Hydropower and River Valley Projects*, Oxford and IBH, New Delhi, pp. 393–407.

Narayan, L.R.A., (2000), "Remote Sensing for Crop Yield Predictions", *The Hindu Survey of Indian Agriculture*, Chennai, pp. 193–195.

Narayanan, M.P. (2000), "Application of GIS and Remote Sensing for Assessment and Management of Environmental Impacts", In R.S. Goel (ed.), *Environmental Management in Hydropower and River Valley Projects*, Oxford and IBH, New Delhi, pp. 377–382.

Panda, Damodar (2000), "Landuse/Landcover Mapping of the Rukshikulya Basin: A Remote Sensing Approach", *The Deccan Geographer*, Vol. 38, No. 1–2, pp. 1–16.

Phutego, T.C. and Chanda, R. (2004), "Traditional Ecological Knowledge and Community Based Natural Resource Management: Lesson from a Botswana Wildlife Management Area", *Applied Geography*, Vol. 24, No. 1, pp. 57–76.

Prasad, V.K., Yogeshkant and Badrinath, K.V.S. (1999), "Vegetation Discrimination Using IRS-P3 WiFS Temporal Dataset: A Case Study from Rampa Forests, Eastern Ghats, A.P", *Journal of the Indian Society of Remote Sensing*, Vol. 28, December, pp. 277–280.

Rao, M.J., Kumar, J.S., Prakasa Rao, B.S. and Srinivasa Rao, P. (2003), "Geomorphology and Land Use Pattern of Vishakhapatnam Urban Industrial Area", *Journal of the Indian Society of Remote Sensing*, Vol. 31, No. 2, pp. 119–127.

Ravan, S.A. (1999), "GIS in Biodiversity Conservation", *GIS Development*, Vol. 3, No. 5 pp. 31–32.

Rawat, V. (2003), "Climate Change and Its Impact on Forest Biodiversity", *The Indian Forester*, Vol. 129, June, No. 6, pp. 787–98.

Roy, et al. (2002), "Map Atlas on Biodiversity Characterization", *GIS@ Development*, November, Vol. VI, Issue II, CSDMS, Noida, Thompson Press, pp. 40–42.

Sankar, Dipak, Sahoo, A.K., Paul, P.P. and Gajbhiya, K.S. (2003), "Natural Resources of Tripura for Sustainable Agriculture", *Geographical Review of India*, Vol. 65, No. 2, pp. 162–68.

Sastry, G.S. (2000), "Land Resource System: Problems, Patterns and Policy", *Indian Journal of Regional Science*, Vol. XXXII, No. 2, pp. 49–60.

Sen Roy, S. and Singh, R.B. (2002), *Climate Variability, Extreme Events and Agricultural Productivity in Mountain Regions*, Oxford & IBH, New Delhi.

Sharma, V.K. and Kumar, Anil (2003), "Sustainable land Use Planning for Upper Maul Khad Catchment in Mid Hills Zone of Himachal Pradesh", *Indian Journal of Soil Conservation*, Vol. 31, No. 2, pp. 139–47.

Siddique, Giyasuddin (2001), "Forests and Tribes of Southwestern West Bengal", *Geographical Review of India*, Vol. 63, No. 1–4, pp. 161–176.

Singh, I.J., Das, K.K. and Kushwaha, S.P.S. (2003), "Forest Stock Assessment Using IRS LISS III and PAN Merged Data in Timili Forest Range", Dehradun, *Journal of Indian Society of Remote Sensing*, Vol. 31, No. 1, March, pp. 49–55.

Singh, R.B. (ed.) (1990) *Environmental Geography, Contribution to Indian Geography*, Vol. XI, Heritage, New Delhi.

Singh, R.B. (ed.), (1996), *Research in Geography*, Vol. I, *Disaster and Environment: Monitoring and Forecasting*, Vol. 2, *Land Use Changes and Sustainable Development*, APH, New Delhi.

Singh, R.B. (ed.), (1998), *Ecological Techniques and Approaches to Vulnerable Environment: Hydrosphere and Geosphere Interaction*, Oxford and IBH, New Delhi.

Singh, R.B. (1991), *Dynamics of Mountain Geosystems*, Ashish, New Delhi.

—— (1991), *Environmental Monitoring: Application of Remote Sensing and GIS*, Hong Kong: Geocarto Int. Centre.

—— (1994), "Space Technology for Disaster Monitoring and Mitigation in India," INCEDE, Institute of Industrial Science, University of Tokyo.

—— (1995), *Global Environmental Change: Perspectives on Remote Sensing and GIS*, (ed.) Oxford & IBH, New Delhi.

—— (1996), *Disasters, Environment and Development*, Oxford & IBH, New Delhi.

—— (1996), *Environmental Law: Issues and Responses*, in R.B. Singh and S. Misra (eds.) Concept, New Delhi.

—— (1998), *Sustainable Development of Mountain Environment of India and Canada*, Oxford & IBH, New Delhi.

—— (2002), *Human Dimensions of Sustainable Development*, Rawat Publications, Jaipur.

—— (2002), "Environmental Degradation and its Impact on Land Use/Land Cover Changes in Agriculturally Developed Regions of India", Proceedings of the IGU-LUCC International Conference, (eds.) I. Blicik et al., Charles University, Prague, pp. 67–74.

—— (2002), "Himalayan Environment, Land Use Cover Changes and Highland Lowland Interactive System with Indus-Ganga-Brahmaputra Plains", In *Land Use-Historical Perspectives-Focus on Indo-Gangetic Plains*, (eds.) Y.P. Abrol et al., Allied, New Delhi, pp. 475–98.

—— (2005), "Risk Assessment and Vulnerability Analysis", IGNOU PG Diploma in *Disaster Management*-MPA-003, New Delhi.

—— (2006), *Natural Hazards and Disaster Management*, Rawat Publications, Jaipur.

—— (2007), Land Use/Cover Change, Environment and Climate Change in Delhi Metropolitan Region: Towards Promoting Sustainable City. In Proceedings of the International Symposium on *Sustainable Urban Environment*, 2007, Tokyo Metropolitan University, Tokyo, pp. 73–79.

Singh, R.B. and Bortamuly, M. (2005), "Impact of Floods on Biodiversity in the Brahmaputra Valley". In Proceedings of the International Conference on *Monitoring, Prediction and Mitigation of Water-Related Disasters*. In Takara, K. et al., Kyoto University, Kyoto, pp. 599–606.

Singh, R.B. and Haigh, M.J. (eds.) (1995), *Sustainable Reconstruction of Highland and Headwater Regions*, Oxford & IBH, New Delhi.

Singh, R.B. and Mishra, P. (2005), "Forest-Cover Mapping of Uttaranchal State, India", in *Understanding Land-Use and Land-Cover Change in Global and Regional Context*, edited by Milanova, E. et al., Science Publisher, Inc Enfield (NH), USA. pp. 249–58. 2005.

Singh, R.B. and Murai, S. (eds.) (1998), *Space Technology for Sustainable Development*, Oxford & IBH, New Delhi.

Singh, R.B. and Parijat, R. (2001), "Ecological Impact of Land Use Change in Delhi Ridge: Anthropogenic Stress and Spatial Realities", in *Urban Sustainability in the Context of Global Change*, (ed.) R.B. Singh, Science Pub. Inc., Enfield (NH), USA, pp.189–201.

Singh, R.B., Fox, J. and Himiyama, Y. (eds.) (2001), *Land Use and Cover Change*, Science Pub. Inc., Enfield (NH), USA and Oxford & IBH, New Delhi.

Singh, R.B. and Verma, Mukesh (2000), "Optimum Utilization of Land and its Management: A Case Study of Mursan Block, Hathras District, U.P.", *National Geographical Journal of India*, Vol. 46, No. 1–4, pp. 59–68.

United Nations Environment Programme (2007), *Global Environment Outlook*, GEO 4, Nairobi, p. 160.

www.gisdevelopment.net/education.

2 Evaluation of Remotely Sensed Vegetation Indices in Asian Mountainous Region to Analyze Dynamics of Vegetation Activities

Roshan Shrestha, Toshiharu Kojima, Yasuto Tachikawa and Kaoru Takara

During the period of last three decades, the satellite remote sensing has been proven to be a very useful tool for monitoring environmental features, analyzing changes in the environment such as land use, deforestation, reforestation, forest damage, planning and managements of natural resources such as quantitative and qualitative water resources assessments and disaster prevention planning. A considerable number of publications produced in this period shows actual and potential application of remote sensing information to a wide range of scientific fields in earth sciences (Rango and Ritchie, 1996). The study of small to large scale vegetation processes is one of the major parts of remote sensing studies that are carried on with the goal of understanding how the Earth functions as a system. For this, a non-dimensional index called as the vegetation index is widely used which is obtained through the remotely sensed radiance field subjected into spectral transformation of two or more bands detecting the photosynthetic activity. There are different methods to obtain the so called vegetation indices that are effectively used in land cover classification, (Townshend et al., 1991; Roy, 1997), vegetation mapping (Bakker and Kootwijk, 1993) land use changes, and understanding biophysical properties of the vegetation activities such as fractional vegetation cover, biomass, leaf area index, fraction of absorbed photosynthetic active radiation and phenology (Sellers et al., 1994). Vegetation indices also have been used to estimate net productivity (Tucker and Sellers, 1986) and investigating the biogeochemical cycles of vegetation and CO_2 sources and sinks (Fung et al., 1987; Tans et al., 1990).

The distribution of vegetation index in space and time is considered as important information on spatial/temporal variations of biophysical and canopy structural properties of the Earth surface due to vegetation activities or more precisely photosynthetic activities. Spatial and temporal distribution of vegetation indices may be subjected to scale effects (for example Davidson and Csillag, 2001). A range of multiple resolutions of data is available in this regard to investigate various aspects of the spatial and temporal distribution of vegetation indices. However, it needs a prior judgment of a proper index, which is able to represent a gross figure of photosynthetic activities such as it truly represents the vegetation index. Normalized Difference Vegetation Index (NDVI) is one of such index that is used for this purpose since earlier stage of remote sensing development, as it detects photosynthesis active green vegetations. The Enhanced Vegetation Index (EVI) is another index proposed by Huete et al. (1994, 1997). The EVI additionally includes blue band information and some coefficients to improve known disadvantages of NDVI such as the effects of noises and scaling problems.

In this study, we evaluate the temporal dynamics of vegetation indices using approximately 16 days interval MODIS/TERRA remotely sensed data (250 m resolution). Our hypothesis is that the dynamics of vegetation activity should reflect in the temporal dynamics of the remotely sensed vegetation indices. We produce a field-based vegetation dynamic mapping via relative temporal differences of the crop-based indices from homogenous agricultural lands for the period of November 2002 to October 2003. We conducted extensive field surveying of vegetation activities in an annual cycle within the study region to produce the field based vegetation dynamic. We compare the dynamics of field-based vegetation activity with the temporal dynamics of remotely sensed vegetation indices. The study is conducted in agricultural lands of Indrawati River Catchment, which extends from the Himalayan region (6,000 m above mean sea level) to temperate Siwalik region (700 m above mean sea level) at the southern face of the Himalayan ridges. The agricultural lands inside the study region are spread out from nearly 3,000 m high elevation to 700 m above mean sea level. This region is more prone to disasters, difficult to access and underdeveloped (Shrestha et al., 2003), where the remotely sensed data is more important than any other region for a variety of purposes.

Overview/Objective

The MODIS/TERRA is configured to provide consistent spatial and temporal comparisons of vegetation condition that can be used to monitor photosynthetic activity (Justice et al., 1998), which it aims through two vegetation indices, the NDVI and the EVI. The MODIS NDVI is often referred to as the 'continuity index' to NOAA-AVHRR NDVI time series of past two decades as it extends the hope of using longer term data record in operational monitoring studies.

The NDVI has been widely used in analyses and operational applications. It has served as an important parameter to various kinds of modelling tasks from local to global scales (Townshend et al., 1991; Rango and Ritchie, 1996 and so many). The NDVI is a normalized ratio of the Near Infra Red (NIR) and red bands, which is given by:

$$\text{NDVI} = \frac{\rho_{NIR} - \rho_{red}}{\rho_{NIR} + \rho_{red}} \qquad \text{...(1)}$$

where, is the surface bidirectional reflectance of y band (for y = NIR and red). The NDVI is simple to calculate and is successful to detect the presence of chlorophyll, which absorbs the red light but strongly reflects infrared radiation by mesophyll tissues.

The NDVI performs poor when there is a background noise due to various reasons. Huete et al. (2002) argued for a need of more reliable and robust vegetation index for operational monitoring of the Earth's vegetation, and proposed the EVI. The EVI is developed as an improved vegetation index that performs better than NDVI and that optimizes the vegetation signal with improved sensitivity in high biomass region and with improved vegetation monitoring through a decoupling of the canopy background signal and reduction in atmosphere influences. Unlike the chlorophyll sensitive NDVI, the EVI is argued as more responsive index to canopy structural variations, including leaf area index, canopy type, plant physiognomy and canopy architecture (Gao et al., 2000). The EVI is given by:

$$\text{EVI} = G\frac{\rho_{NIR} - \rho_{red}}{\rho_{NIR} + C_1\rho_{red} - C_2\rho_{blue} + L} \qquad \text{...(2)}$$

where P_y is the surface bidirectional reflectance of y band (for y = NIR, red and blue); G is gain factor; L is canopy background adjustments; C_1 and C_2 are coefficients of the aerosol resistance term that uses blue band to correct the aerosol influences in the red band. Inclusion of these additional variables is assumed to reduce the noise and errors in the measured reflectance advancing nearer to the fundamental objective of isolating the photosynthesis active signals from the spatially and temporally variable mixed pixels, which allows meaningful spatial and temporal inter-comparisons of vegetation activity. The coefficients adopted in the EVI algorithm are L = 1, C_1 = 6, C_2 = 7.5 and G = 2.5 (Huete et al., 1994, 1997).

Global scale comparative studies are reported and published in the process of developing the EVI and setting its parameters, which mostly use the alternate sources of remotely sensed data such as Landsat-TM, NOAA-AVHRR, and SPOT4-VEGETATION. However, the comparative analysis of field based data and remotely sensed data are not so common. Our approach here is to rely on field-based vegetation activity to use in comparing the performance of the vegetation indices instead of using a different sensor. Because it is hard to get field-based vegetation activity covering the entire globe; we go for a pilot study in a small area in the mountainous region and perform comparative evaluation between the field-based and derived vegetation information from the remotely sensed data over different topographic and climatic zones. With this, we investigate performance of the NDVI and EVI in order to test our hypothesis.

Data and Methodology

The study area lacks systematic installation of data observation network. We conducted an extensive survey to gather the field-based information on the vegetation activity. A total of 50 locations within the area are chosen such that to obtain sufficient details of the agricultural activity in space and time. We communicated with 3 to 7 groups of farmers at each location to collect details of their farming activities throughout the year. Each group of farmers included 4 to 5 farmers. The survey was conducted during May, June, July and August of the year 2000. The survey results show that the cropping trends are significantly different on various topographic altitudes and availability of irrigation facility. The data is further analyzed to identify regions having tentatively homogenous vegetation activities as reflected via the crop information, farming schedule and topographic altitude and then classifying them into four different categories of agricultural lands having distinct vegetation activities (Figure 1). They are designated as the crop zone 1, 2, 3 and 4, respectively. A detailed description of the crop-zone-categorization procedure is omitted in this paper. Generally, the crop zone 1 is well irrigated agricultural lands at lower altitudes mostly below 1,200 m msl. The crop zone 2 also encompasses mostly lower altitude agricultural lands, but their vegetation activity is

Figure 1

Classified crop zone coverage, which is based on crop information, altitude, and farming schedule in the Indrawati catchment; the well irrigated (z = 1) crop zone is scattered within low land (z = 2) crop zone.

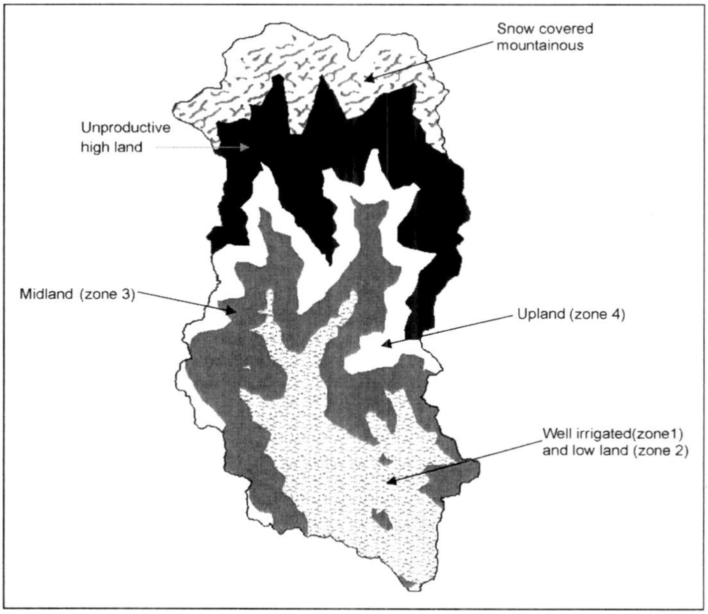

different due to restricted irrigation facility. The zone 3 covers mostly agricultural lands of mid-altitude ranging from 1,200 m msl to 1,800 m msl and the lands from 1,800 m msl to 2,600 m msl are in zone 4. There are some portions of agricultural lands that do not meet the criteria to fall into any of the above-mentioned zones due to mismatches in cropping condition, which are omitted in the analysis.

Depending on the cropping information and farming schedule, the vegetation activity of each pixel of lands is estimated via observation of photosynthetic activity. The photosynthetic vegetation activity is assumed to vary based on fraction of individual crop coverage and photosynthesis-active stage of that crop. At a particular time, it may be represented by:

$$Q_t \alpha \sum_{i=1}^{C} k_i P_i \qquad \ldots(3)$$

where, Q_t is the photosynthetic vegetation activity at time t; k_i is a coefficient representing fractional coverage of the crop i within the same crop zone; c is the total number of crops; and P_i is the index of photosynthetic activity of the crop i. The P_i is dependent on the stage of crop growth, hence can be related with the crop growth curve. Figure 2 presents a simplified version of the replicated crop growth curve that also depicts the crop coefficients for its initial, development, mid and late season of growths. We use the standard crop coefficient values K_c referring the FAO manual of irrigation and drainage paper No. 56 (Allen et al., 1998) for all the crops. We link the values of K_c as linearly proportional variable to obtain the crop growth curve.

Figure 2

Generalized crop coefficient K_c curve for the single crop coefficient approach (Allen et al.*, 1998).*

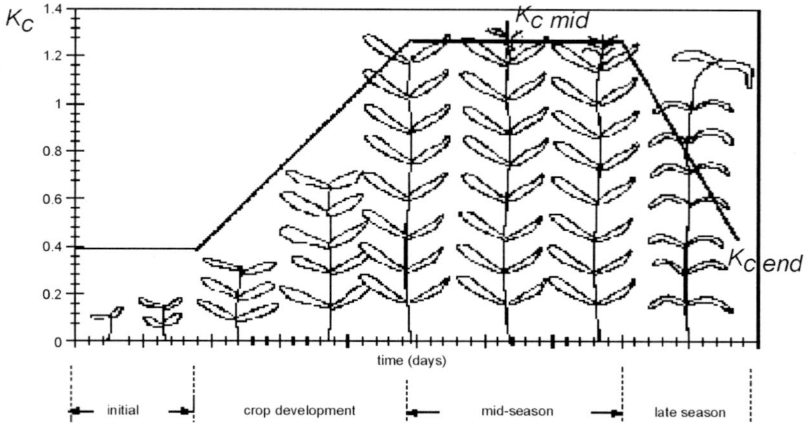

The crop calendars vary for each crop at each crop zone. We develop different crop calendars for each crop at each crop zone based on the field survey. With this, we estimate corresponding Q for each zone at 10 days interval approximately. From the series of Q, the indices of vegetation activity is given by:

$$V_{t,z} = \frac{Q_t}{\max Q_t} \qquad \text{for } z = 1, 2, 3 \text{ and } 4 \quad ...(4)$$

where, $V_{t,z}$ is the index of temporal vegetation activity at the zone z, which is considered as field based temporal vegetation activity information.

The MODIS Vegetation Indices (NDVI and EVI) is considered as robust spectral measures of the amount of vegetation present on the ground. They involve transformations of the red (620–670 nanometres), near infrared (841–876 nanometres), and blue (459–479 nanometres) bands designed to enhance the 'vegetation signal' and allow for precise inter-comparisons of spatial and temporal variations in terrestrial photosynthetic activity. Both NDVI and EVI are directly referred from the online data product set MOD13Q1 (MODIS/TERRA vegetation indices 16-DAY L3 GLOBAL 250M SIN GRID V004) for the period from November 2002 to October 2003, which is provided as operation ready data product from atmospherically corrected bidirectional surface reflectance that have been masked for water, clouds, and cloud shadows. The vegetation index products (MOD13Q1) also includes vegetation index quality information in addition to the composite surface reflectance bands 1–3 and 7 (i.e., the red, NIR, blue, and MIR bands, 2,105–2,155 nanometres). The data is validated, meaning that product uncertainties are well defined over a range of representative conditions. The EVI uses the blue band to remove residual atmospheric contamination due to smoke and sub-pixel thin clouds. The EVI also uses feedback adjustment to minimize canopy background variations and to enhance vegetation sensitivity from sparse to dense vegetation conditions.

The mean and standard deviation of NDVI and EVI is calculated within each zone separately. The standard deviation is checked for observing the extent of spatial variation of both of the indices at the moment. The zonal mean values of vegetation indices are considered as the photosynthetic vegetation activity at a particular time. The temporal vegetation activity from the remotely sensed data are given by:

$$N_{t,z} = \frac{(\text{NDVI})_t}{\max (\text{NDVI})_t} \qquad \text{for } z = 1,2,3 \text{ and } 4 \quad ...(5)$$

$$E_{t,z} = \frac{(\text{EVI})_t}{\max (\text{EVI})_t} \qquad \text{for } z = 1,2,3 \text{ and } 4 \quad ...(6)$$

where, $N_{t,z}$ is the index of temporal vegetation activity at the zone z given by the NDVI data; and $E_{t,z}$ is the index of temporal vegetation activity at the zone z given by the EVI data. The zonal mean values are given by:

$$(\text{NDVI})_t = \frac{1}{n} \sum_{j=1}^{n} \text{NDVI}_j \quad ...(7)$$

$$(\text{EVI})_t = \frac{1}{n} \sum_{j=1}^{n} \text{EVI}_j \qquad \ldots(8)$$

where, n is the number of pixels inside each zone z; $(\text{NDVI})_t$ is the mean zonal NDVI at time t; and $(\text{EVI})_t$ is the mean zonal EVI at time t.

Equations (4), (5) and (6) yield the normalized figures of all variables in the time series with their maximum value. We treat the maximum of the normalized index as equivalent as that of the 100 per cent coverage of the vegetation activity reflected in that index. This provides a basis for comparing the performances of the indices among each other reflecting the vegetation activity. This does not necessarily mean that a 100 per cent value of a particular index represents the full coverage of agricultural land with the most active vegetation stage, but it represents the state of maximum photosynthetic activity.

Results and Discussions

Figure 3, shows the time series of vegetation activity obtained from the remotely sensed vegetation indices and field based data for all the crop zones. The $N_{t,z}$ series obtained from the NDVI is found quite similar to the variation of $V_{t,z}$ series in three zones: well-irrigated (z = 1), low land (z = 2) and mid land (z = 3) crop zones. However, the $N_{t,z}$ series does not match with $V_{t,z}$ series in up land (z = 4) crop zone. Notice that the $E_{t,z}$ series seems more close to the $V_{t,z}$ series in the zone 4. This may stand as the evidence of better representation of vegetation activity by the EVI in the high altitude region. In general, the $E_{t,z}$ series is found less similar to $V_{t,z}$ series compared to the $N_{t,z}$ series. The $E_{t,z}$ series obtained from the EVI performs better during higher photosynthetic active period of June to October than the lower photosynthetic active period of November to May.

Similar patterns of $N_{t,z}$ series and $V_{t,z}$ series shows the NDVI is a better index than the EVI to represent vegetation activity in three zones 1, 2 and 3. A less similar result in zone 4 manifests that the NDVI is not always better. We reason following three possibilities behind this inconsistent observation. First, it may be due to inaccuracies in the field based information that is used to infer field based vegetation activity. The field-survey was conducted during rainy season. During this period, several active landslides and floods occurred in the study area that imposed some practical-limitations in the field-survey such as delays in reaching the survey-spots, particularly in zone 4, and less attendance of local farmers who were able to communicate with survey-team. Second, the remotely sensed vegetation indices may not be representative for high altitude region where cloud is present for elongated period. Presence of cloud may change the entire statistics of the reflected radiance causing the NDVI perform much worse in high altitude regions. Third, it may be due to erroneous MODIS data that awaits proper radiometric corrections in the high altitude region before producing the vegetation indices.

Figure 3

Comparison of temporal vegetation activity from MODIS NDVI data ($N_{t,z}$ series) and MODIS EVI data ($E_{t,z}$ series) with field-based data ($V_{t,z}$ series) in different crop zones; (a) Well irrigated (z = 1) crop zone (b) Low land (z = 2) crop zone (c) Midland (z = 3) crop zone (d) Upland (z = 4) crop zone.

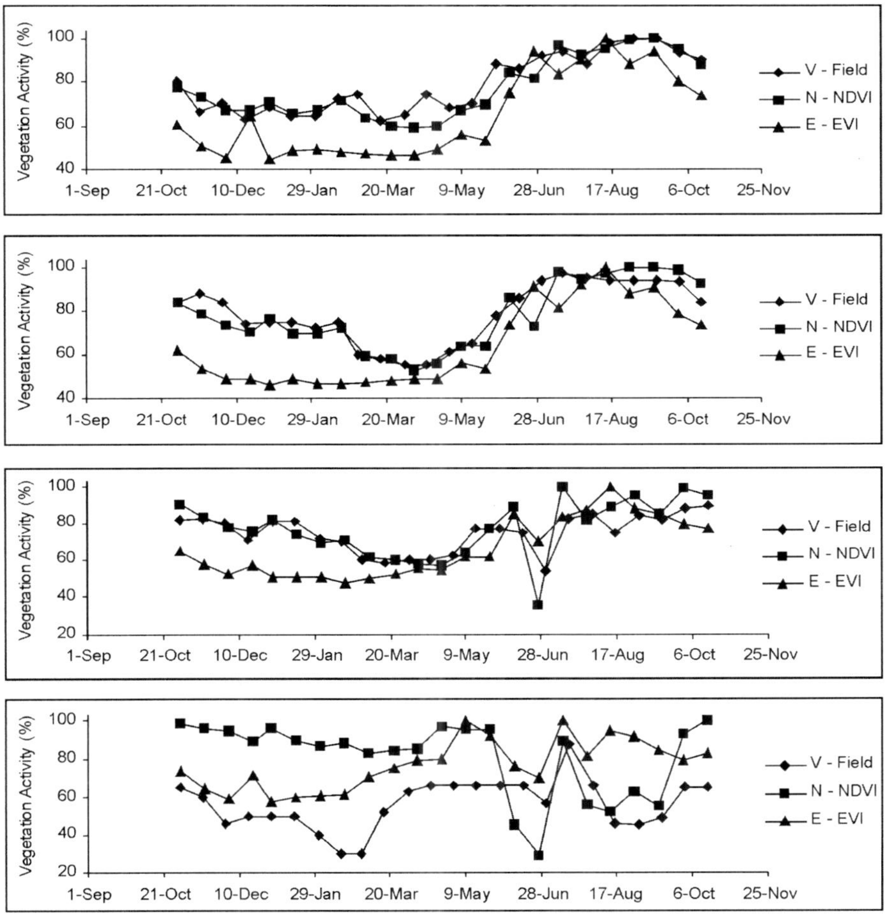

By utilizing more atmospheric sensitive blue band to correct the red band for atmospheric resistances such as aerosol effects (Kaufman and Tanre, 1992), the EVI is reported to perform better in the heavy aerosol, biomass burning conditions in Brazil (Miura et al., 1998). This additional measure may be insignificant if the aerosol presence and biomass burning condition is not abundant. To our best knowledge, there is no study done yet on aerosol presence and biomass burning condition in the Indrawati catchment of Nepal. There is a small scale burning of firewood and biomass as the domestic fuel resources and foggy weather at high altitudes with moist air. But there is no industrial activities escalating aerosol presence and biomass burning condition. The EVI also intends to remove the background influence of open canopy that exerts some effect on the canopy reflectance. The background influence of open canopy exerts obvious noise in the remote sensing of Earth's terrestrial surface. In agricultural lands, the open canopies are usually removed by farmers for better cultivation. Hence, the correction applied uniformly throughout a large domain or the entire globe, as it was done for the case of EVI calculation, is likely to err the reflectance of agricultural lands where the correction may not be necessary. We argue this as the primary cause that might have affected the performance of EVI adversely in the above presented result because our study is confined within the agricultural lands measuring the dynamics of vegetation activity via information of the crops where back-scattered reflectance from open canopy is likely to be absent.

In general, the NDVI outperforms the EVI in being a measure of temporal vegetation activity in the Asian Mountainous region and also is sufficiently stable to permit meaningful comparisons with field-based vegetation activity. However, it is often claimed that the NDVI is inherently nonlinear, which preserves the influence of additive noise effects, such as the atmospheric path radiances. We also notice that the EVI is not as successful as the NDVI within different subsets of photosynthetic active period. Particularly, the EVI performs very poor during lower photosynthetic active period despite it incorporates additional corrections. At the moment, it may be too early to assert that the EVI is necessary or not in the mountainous region in Asia, however, it is clear that the normalized coefficients of existing EVI do not match with that of the NDVI and field based vegetation activity.

The weaknesses in the performance of the EVI data reveals that it is not providing better vegetation signal in agricultural land use surfaces, where there is a lot of potentials to apply the remotely sensed data for several kinds of analyses. Since the vegetation index functions as an important parameter to detect major changes on the Earth's surface, a correct estimation of vegetation index becomes very fruitful to all kinds of vegetation index derived products such as estimation of land use changes, hazard mapping, disaster management, and so on. This study does not intend to claim the NDVI as a better vegetation index since the NDVI has some known disadvantages like asymptotic signals over high biomass condition and high sensitivity to canopy background variations. At the same time the obtained results indicated serious weakness of the improvements incorporated in the EVI. Perhaps by changing the parametric coefficients in a distributed fashion, the EVI formulation might have better performance. Applying a distributed parameterization of the correction coefficient to take account of different climatic zone, different topographic altitudes and different land cover types could be suggested as the issues to investigate further in near future.

Conclusion

The vegetation indices are effectively used in many applications either as a basic data or as a model parameter. Therefore, improvements and increased accuracy in estimating vegetation indices plays significant role in all kinds of secondary analyses that use the vegetation indices. In this study, two kinds of vegetation indices, the Normalized Difference Vegetation Index (NDVI) and the Enhanced Vegetation Index (EVI) from MODIS/TERRA products are used to estimate relative temporal differences of spatially averaged vegetation indices within four different categories of agricultural lands in the Indrawati River Catchment, Nepal. The land category classification is done on the basis of field data of cropping information, farming schedule and topographic altitudes such that each category is having more or less spatially homogenous vegetation activity in approximately two weeks time.

Extensive field surveying is conducted to know the annual vegetation activities in the catchment. The field based temporal vegetation activity is based on the photosynthetic active stages of individual crop inside the agricultural lands. The remotely sensed temporal vegetation activity is extracted from the temporal fluctuation of mean NDVI and EVI data in the respective zones. They are normalized to their own maximum values to make compatible for comparing each other.

From the comparison of the temporal vegetation activities, it is found that the NDVI is more close to the observed temporal vegetation activity than the EVI. Although EVI is claimed as improved vegetation index with improved sensitivity in high biomass regions, improved vegetation monitoring through a decoupling of the canopy background signal and reduction in atmospheric influences, its fundamental expression and parameters may need modification to materialize the improvement in Asian Mountainous region, which might be achieved through distributed parameterization of the product.

Acknowledgement

Free distributions of MODIS/TERRA products from the NASA Earth Observing System Data Gateway are highly acknowledged.

References

Allen, R., Pereira, L., Raes, D., and Smit, M., (1998). "Crop Evapotranspiration (Guideline for Computing Crop Water Requirements)", *FAO Irrigation and Drainage,* Paper No. 56, 290 pp. ISBN 92 5 104219 5.

Bakker, H.C. and Kootwijk, E.J., (1993). "Vegetation Mapping in a Dutch Tidal area Using a Combination of Visual and Digital Image Analysis", *Remote Sensing for Monitoring the Changing Environment of Europe, Winkler,* (eds.), AA Balkema, Rotterdam, ISBN 90 5410 311 6.

Davidson, A. and Csillag, F., (2001), "The Influence of Vegetation Index and Spatial Resolution on a Two-date Remote Sensing Derived Relation to C4 Species Coverage", *Remote Sensing of Environment,* 75 (1), 138–151.

Fung, Y., Tucker, C.J. and Prentice, K.C. (1987), "Application of Advanced Very High Resolution Radiometer Vegetation Index to Study Atmosphere-biosphere Exchange of CO_2", *J. Geoph. Res.* 92, 2999–3015.

Gao, X., Huete, A., Ni, W. and Miura, T., (2000), "Optical Biophysical Relationships of Vegetation Spectra Without Background Contamination", *Remote Sensing of Environment,* 74, 609–620.

Huete, A., Justice, C. and Liu, H., (1994), "Development of Vegetation and Soil Indices for MODIS-EOS", *Remote Sensing of Environment,* 49, 224–234.

Huete, A., Didan, K., Miura, T., Rodriguez, E.P., Gao, X., and Ferreira, L.G., (2002), "Overview of the Radiometric and Biophysical Performance of the MODIS Vegetation Indices", *Remote Sensing of Environment,* 83, 195–213.

Huete, A., Liu, H.Q., Batchily, K., and van Leeuwen, W., (1997), "A Comparison of Vegetation Indices over Global set of TM Images for EOS-MODIS", *Remote Sensing of Environment,* 59, 440–451.

Justice, D.H., Salomonson, V., Privette, J., Riggs, G., Strahler, A., Lucht, R., Mynemi, R., Knjazihhin, Y., Running, S., Nemani, R., Vermote, E., Townshend, J., Defries, R., Roy, D., Wan, Z., Huete, A., van Leeuwen, R., Wolfe, R., Giglio, L., Muller, J.P., Lewis, P., and Barnsley, M., (1998), "The Moderate Resolution Imagined Spectroradiometer (MODIS)", Land Remote Sensing for Global Change Research, *IEEE Transactions on Geoscience and Remote Sensing,* 36, 1228–1249.

Kaufman, Y., and Tanre, D., (1992), "Atmospherically Resistant Vegetation Index (ARVI) for EOS-MODIS", *IEEE Transactions on Geoscience and Remote Sensing*, 30, 261–270.

Miura, T., Huete, A.R., van Leeuwan, W. and Didan, K., (1998), "Vegetation Detection Through Smoke-filled AVIRIS Images: An Assessment Using MODIS Band Passes", *Journal of Geoph. Res.*, 103, 32001–32011.

National Aeronautics and Space Administration, (2000), MODIS data products Catalog (EOS AM platform), Greenbelt, Maryland, Goddard Space Flight Center, http://modis.gsfc.nasa.gov.

Rango, A. and Ritchie, J.C. (eds.), (1996), "Remote Sensing Applications to Hydrology", Special Issue, *Hydrol. Sci. J.*, 41 (4).

Roy, D.P., (1997), "Investigation of the Maximum Normalized Difference Vegetation Index (NDVI) and the Maximum Surface Temperature (T_s) AVHRR Compositing Procedures for the Extraction of NDVI and T_s Over Forest", *Int. J. Remote Sensing*, 18, 2383–2401.

Sellers, P.J., Tucker, C.J., and Collatz, G.J., (1994), "A Global 1°–1° NDVI Data Set for Climate Studies, Part 2, The Adjustment of the NDVI and Generation of Global Fields of Terrestrial Biophysical Parameters", *Int. J. Remote Sensing*, 15, 3519–3545.

Shrestha, R., Jha, R.N. and Takara. K., (2003), "GIS Assisted Water Resources Planning in Mountainous Catchment", Nepal, Proc. of First International Conference on Hydrology and Water Resources in Asia Pacific Region, Kyoto, Japan, 2, 939–944.

Tans, P.P., Fung, I.Y., and Takahashi, T., (1990), "Observational Constraints on the Global Atmosphere CO_2 Budget", *Science*, 247, 1431–1438.

Townshend, J.R.G., Justice, C., Li, W., Gurney, C. and McManus, J., (1991), "Global Land Cover Classification by Remote Sensing: Present Capabilities and Future Possibilities", *Remote Sensing of Environment*, 35, 243–256.

Tucker, C.J. and Sellers, P.J., (1986), "Satellite Remote Sensing of Primary Productivity", *Int. J. Remote Sensing*, 7, 1395–1416.

3 Method and Case Study of Regional Ecological Risk Assessment

Case Study of Liaohe River Delta of China

Xuegong Xu, Zaiyi Fu and Huiping Lin

Ecological Risk Assessment is a new study field that applies theories and methods of ecology, geography, environmental science and traditional risk assessment. Its objective is to estimate the possibility and magnitude that some undesired ecological events will occur, by using mathematics, probability and other risk analysis techniques. These undesired ecological events are ecological endpoints brought by uncertain disasters or environmental accidents. In fact, ecological risk assessment is a method of drawing risk concept into ecological impact assessment. Since ecological risk assessment provides the scientific foundation for risk management, its importance is now recognized by both academic researchers and environmental managers (Cao and Shen, 1991).

Regional ecological risk assessment is one branch of ecological risk assessment. It adapts to assess the possibility and magnitude of those undesired events affected on a regional scale. Compared with general ecological risk assessment, both the receptors and the sources of risk must be in a region, the exposure and effects assessment must account their uncertainty and spatial heterogeneity. So, regional ecological risk assessment may be more complex.

Through a wetland case study in the Liaohe River delta, this article inquires into the concept of regional ecological risk assessment and develops a set of assessment methods. Consulting from the methods of ecological risk assessment developed by both Chinese and foreign scientists (Yin, 1995; Bartell et al., 1992, Bolger, 1997; McDonald et al., 1997; Gaudet, 1994; Hunsaker et al., 1989), and considering the scale and spatial dimension of the region to be evaluated, we divide the method of regional ecological risk assessment into 5 steps: region background analysis, risk receptor selection, risk sources analysis, exposure and hazard analysis, and integrated risk assessment. In the next section, the 5 steps are illustrated in a case study of wetland in the Liaohe River delta in China.

Case Study of Wetland in the Liaohe River Delta

Regional Background Analysis

Regional ecological risk assessment must be based on fully understanding the research region. So, we begin our case study from the regional background of the Liaohe River delta.

The Liaohe River delta is one of the 7 largest deltas in China. It is located in the southern part of Liaoning province, China, lying on the beach of Bohai Sea with an area of 3,959 km^2. The Liaohe River delta spans from 121°25′ E to 122°3′ E, and from 40°39′ N to 41°27′ N. It is an area with a temperate, semi-humid continental monsoon climate. In its geological structure, this region belongs to a block-indentation basin of Mesozoic eras, and is an area with rich oil and natural gas deposits. It is also an alluvial delta plain with flat landform. There are 21 rivers in the region, in which Liaohe River, Daliaohe River, Raoyanghe River and Dalinghe River are larger. Expansive wetland developed at the rivers' mouths and on sea beaches. The entire area is covered mainly by wet and saline soil, with meadow, being the typical natural vegetation. Part of the delta has been designated as a national nature reserve for protecting wetland ecosystem and rare species of birds.

By cause of formation, the wetlands of the Liaohe River delta include such types as the river mouth wetland, coastal wetland, water-gathering swamp of the depressions, as well as rivers, reservoirs, pools, ditches, paddy fields and salt pans. Total area of wetland amounts to 3,022 km^2 or 76 per cent of the Liaohe River delta. Of the total wetland, natural wetland covers 1,532 km^2 (or 50.7%), and human-created wetland covers 1,490 km^2 (or 49.3%). By using remote sensing data and GIS techniques, we were able to produce a map of landscape types and a map of corridors in the Liaohe River delta. These maps are then used as the foundation for regional ecological risk assessment.

Economically, the Liaohe River delta is an important base of petroleum production as well as significant food and reed production bases. The Liaohe-Oilfield is the third largest oilfield in China. The reed-field is the largest in Asia and the second largest in the world. Regional ecological risk research in this region will therefore have both theoretical and applied significance.

Analysis of Risk Receptor

Selection and Description of Risk Receptor

'Receptor' is also known as receiver of risk. When regional ecological risk assessment is made, those sensitive and key species, populations, communities, and even the entire ecosystems all can be selected as risk receptors. We select birds and their habitats as receptor in this research. Because in the region many rare species of birds belong to important national protected species such as red-crowned crane *(Grus japonensis)*, hooded crane *(Grusmonacha Temminck)* and white stork *(Ciconia ciconia boyciana Swinhoe)*, they are sensitive to ecological risk or are designative species of a ecosystem in the region. And that habitat is a ligament linking birds with risk sources. According to habitat needed by different bird groups, forms of landscape in the Liaohe River delta can be distinguished into 5 types: Saline seepweed beach, reed marsh, fresh-water, farmland and meadow, and settlement. Each bird-habitat eco-group has similar bird communities and habitat environment. The habitats also feel impacts of human activities differently. In our case study, all 5 types of bird-habitat are studied as risk receptors.

Ecological Endpoint

'Ecological endpoint' refers to possible damage to receptors that is caused by some uncertain risk sources. These damages then result in structural and functional changes in regional ecosystems. In the Liaohe River delta, possible ecological endpoints include annihilation of rare species, birds quantity decrease, important habitat degeneration and disappearance.

Risk Sources Analysis

There are two types of ecological risk sources: natural disasters and human accidents. In the Liache River delta, natural disasters include flood, drought, strong wind, hailstone, freezing injury, earthquake, and storm tide; human accidents are mainly in the form of petroleum pollution caused by oil well blowout and oil pipe leaking. Otherwise agriculture development also impact to wetland.

Sources of regional ecological risk should be described with its probability, intensity and distribution. After analyzing their probability, intensity and endangerment of the above-mentioned risk sources one by one, flood, drought, storm tide and petroleum pollution are selected as main ecological risk sources in the region.

The characteristics of the four ecological risk resources can be summarized in Table 1. The probability of flood and rain waterlogging disaster is calculated according to records since 1840 (Panjin Chorography Office, 1998). Drought probability is calculated from data since 1870 (Panjin Chorography Office, 1998). The probability of storm tide is based on records in the station of Liaodong Bay from 1951 to 1995 (Liu et al., 1998). The probability of oil pollution accident is based on statistics for recent 15 years.

Table 1

Probability of Ecological Risk Sources in the Wetland of Liaohe River Delta

Disaster/Accident	*Probability (%)*			
	Total	*Serious*	*Heavy*	*Larger*
Flood disaster	46.6	7.3	12.0	27.3
Drought disaster	17.5	4.2	5.8	7.5
Storm tide disaster	25.0			
Oil pollution accident	0.41			

Exposure and Hazard Analysis

Ecological Harm Measurement of Each Habitat

As mentioned above, this research selects birds and their habitats as receptor, in which habitats expose directly under different risk sources, and impact directly to birds. Different types of habitats are places where relevant bird communities inhabit, scout food, propagate and migrating rest. There are different effects in maintaining biological diversity, protecting species, perfecting whole structure and function, and promoting landscape natural succession. As well, they have different capacity in resisting disturbance.

We use ecological index to denote the important degree of each habitat; use fragile index to show their anti-risk features; and use ecological damage index to denote their hazarded degree.

Ecological Index (E)

Four such indexes can be used to measure degrees of importance of each habitat:

Species Conservation Index (C):

C_i shows a habitat type's status in maintaining important bird species to be conserved or endangered.

$$C_i = \lambda_j U_{ij}/U_j \qquad \text{...(1), in which}$$

C_i : the species conservation index of habitat i;

U_{ij} : bird species number for j grade conservation in habitat i;

U_j : bird species number for j grade conservation in the region;

λ_j : weight of j grade of conserved species in the region.

Biological Diversity Index (V):

$$V_i = M/M \qquad \text{...(2), in which}$$

V_i : biological diversity index of habitat 1,

M_i : number of bird species in habitat i;

M : number of bird species in the region.

Disturbance Intensity (W) and Nature Degree (N):

$$W_i = L_i/S_i \qquad \text{...(3), in which}$$

W_j : disturbance intensity to habitat i;

L_j : total length of manual corridor such as highway, railway, dyke and canal in habitat i;

S_i : area of habitat i.

Because W is negatively correlated with the natural state of the habitat,

$$N_i = I/W_i \qquad \text{...(4)}$$

is used to show the nature degree of habitat i.

After C_i, V_i, N_i are worked out, they must be normalized because their units are different. Then the ecological index (E) is computed using the following formula:

$$E_j = aC_i + bV_i + cN_i \quad ...(5), \text{ in which}$$

E_j : ecological index of habitat i;

a, b, c: weights of each index; (a+b+c= 1).

In consideration of the region's situation, 0.7, 0.2 and 0.1 are assigned to a, b and c as their respective weights.

Fragile Index (F)

Frangibility of a habitat ecosystem varies in different phases of the natural succession process. Generally, ecosystems in the primary phase of succession, with simple food chain structure and small Vi are more fragile than others, whereas ecosystems such as farmland that can be put into minus entropy through management is more steady than others. So, the order of frangibility for the 5 types of habitats is as Saline seepweed beach, reed marsh, fresh water, farmland and meadow, and settlement. After normalization, Fi can be gained.

Ecological Damage Index (D)

$$D_i = E_iF_i \quad ...(6), \text{ in which}$$

D_i : ecological damage index of habitat i;

E_j : ecological index of habitat i;

F_j : fragile index of habitat i.

Based on statistics from 236 species of birds in the region, D values of each habitat are calculated by using the above formulas and weights.

Table 2

The Ecological Damage Index of Bird Habitats in the Liaohe River Delta

Habitat	C_i	V_i	N_i	E_i	F_i	D_i
Saline seepweed beach	0.679	0.418	0.767	0.636	0.333	0.212
Reed marsh	1	1	1	1	0.267	0.267
Fresh water	0.132	0.131	0.602	0.179	0.200	0.036
Farmland & meadow	0.271	0.542	0.576	0.358	0.133	0.048
Settlement	0.050	0.222	0.450	0.124	0.067	0.008

Damage Effects of Each Risk Source on Receptors

Many forms of damage are caused when risk sources affect receptors. Flood submerses reed, paddy and other plants to bring on their death, which result in food, nest and shelter of rare waterfowl cut down or destroyed. Flood is also likely to destroy oilfield and industry to make oil pollution and affect habitat indirectly.

Water is a basic element of wetland ecosystem. But drought disaster is able to make vegetation dried up, so as to block wetland nature succession or make wetland disappear. It also makes harmful effect on birds and habitats.

Storm tide badly destroy coast wetland, whose effect is same as flood harm on birds. More seriously, it will result in soil salinization and wash away vegetation into barren land. Original bird communities lose their habitat. Storm tide also harms reed field, prawn pool and salt pan, as well as some farmlands, industries and settlements.

Petroleum pollution results in bad regional environmental quality. Sometime oil well blowout shoots down and suffocates birds. The habitat polluted by oil is hardly restored, so harm from oil pollution is especially grave.

It is therefore clear that the effects of risk sources will change the structure and function of receptors in the region.

Integrated Weight of Risk Sources

Effect intension of each risk source varies on different receptors, thus posing different regional ecological risks. To evaluate the 4 major types of ecological risk sources by using the AHP method, the following weights are assigned to each risk sources: 0.133 to flood; 0.163 to drought; 0.256 to storm tide; and 0.448 to petroleum pollution. These weights lay a foundation for further integrated risk assessment.

Integrated Regional Ecological Risk Assessment

Risk Areas

Spatial heterogeneity is an important characteristic of receptors and risk sources in regional ecological risk assessment. Through GIS overlay of the 4 risk sources, the whole region is divided into 122 risk areas. Integrated action of risk sources is similar within each risk area, but different among different risk areas.

Measurement of Risk Value

The following indexes and formulas are put forward to calculate risk value of each risk area:

1. Integrated Risk Probability (P_k)

In each risk area, different types and grades of risk sources are overlaid to form compound risk sources and integrated risk probability:

$$P_k = \Sigma \beta_j P_{jl} \quad \text{...(7), in which}$$

P_k : integrated risk probability of area k–1

P_{jl} : probability of type j, grade I of risk source in area k–1

β_j : weight of type j risk source.

2. Integrated Ecological Damage Degree (D_k)

Although there is accordant status of risk source within a risk area, the receptor status is disaccord. It means that there may exist different habitats in a single risk area. The compound formula is:

$$D_k = \Sigma \left(S_{ki} / S_k\right) D_i \quad \text{...(8), in which}$$

D_k : integrated ecological damage degree of risk area k;
S_{ki} : area of habitat i in risk area k;
S_k : total area of risk area k;
D_i : ecological damage index of habitat 1.

3. Integrated ecological risk value (R_k)

Finally, we use the following formula to calculate risk value of every risk area according to the principles of risk measurement:

$$R_k = P_k D_k$$ ...(9), in which

R_k : risk value of risk area k;
P_k : integrated risk probability of risk area k;
D_k : integrated ecological damage degree of risk area k.

By using the above formulas, we calculate the risk value of every risk area. Then by using the method of quartile classification in Arcview, all risk values are classified into 5 grades and a map of integrated regional ecological risk assessment is generated automatically.

Management Countermeasures Against Regional Ecological Risk

The purpose of ecological risk assessment is to provide scientific basis for regional ecological environmental management.

The study shows that there are certain regularities of ecological risk intensity distribution in the River delta. As a whole, areas on the river mouth and along the coast are highest risk zones, the risk intensity gradually recedes from coast to inland, and from both side of river to beyond. Although the assessment is in virtue of bird-habitats, the result reflects the risk trend of all ecosystems in the region.

Risk management countermeasures can be put forward to aim at the 5 grades of ecological risk zones respectively.

In brief, ecological risk management of the Liaohe River delta must pay special attention to the following aspects: (1) ensure good management of the national nature reserve for protecting river mouth and sea beach wetland ecosystems; (2) develop a system of water conservation projects including irrigation, drainage and prevention; (3) improve ecological engineering infra-structures; (4) control and reduce pollution of oilfield and other industries; (5) set up a system to monitor ecological risks. For risk zones of different grades, corresponding risk management policies should be adopted to achieve harmony of regional ecological environment, protect biological diversity effectively, and stabilize ecosystems in both structure and function.

Conclusion

Through the wetland case study of the Liaohe River delta, we probed into the method of regional ecological risk assessment.

When making regional ecological risk assessment, the basic principles and methods are the same, and the process consists of the same 5 steps: regional definition and background analysis, risk receptor selection, risk sources analysis, exposure and hazard analysis, and integrated risk assessment. The results of the case study, which followed the method, can provide scientific basis for risk management of ecological environment and decision-making of regional sustainable development in the Liaohe River delta.

The case study has proved that the above method is feasible. The indices and formulas developed in this paper are successful because the assessment results reflect the situation of the research region very well. Of course, this method is not perfect yet and needs further improvement in some aspects. For example, the measurement of fragile index through ranking and normalization is somewhat too simple. Therefore, further improvement is necessary. Because risk receptor and risk sources in other regions may be different from our case study, and the compound formula may also be different, the validity of the indices and formulas needs to be tested in other regions for their wider applicability.

Acknowledgement

This research is a part of Project 49631040 supported by Natural Science Foundation of China (NSFC). The authors kindly acknowledge the NSFC committee for the support. Special thanks go to Prof. Diming Xiao for his advice, Mr. Xianli Wang for providing the basic information on the Liaohe River Delta.

References

Bartell, S.M., Gardner, R.H., O'Neill, R.V., (1992), *Ecological Risk Estimation,* Boca Raton, Lewis Publishers, 1–252.

Bolger, P.M., (1997), "Risk Assessment: What is the Question". Kamrin M.A., (ed.), *Environmental Risk Harmonization,* Toronto, John Wiley & Sons Ltd., 51–60.

Cao, H. and Shen, Y., (1991), "Summarization of Ecological Risk Assessment Research, *Environmental Chemistry,* 10(3), 26–30.

Gaudet, C.A., (1994), *Framework for Ecological Risk Assessment at Contaminated Sites in Canada: Review and Recommendations,* Ottawa, Minister of Supply and Services, Canada, 1–95.

Hunsaker, C.T., Graham, R.L., Suter, G.W., (1989), *Regional Ecological Risk Assessment: Theory and Demonstration,* Washington, Environmental Sciences Division Publication, 1–192.

Liu, Y. and Han, M., (1998), "Assessment of Trend and Impacts of Sea Level Rise in the Liaohe River Delta", *Acta Oceanologica Sinica,* 20(2), 73–82.

McDonald, I.A., Becker, R.A. and Ting, D.I. W., (1997), "A Review of the Consistency, Harmonization, and Scientific Basis of Risk Assessment Activities Within the California Environmental Protection Agency", Kamrin M.A., (ed.), *Environmental Risk Harmonization.* Toronto, John Wiley & Sons Ltd., 125–138.

Nath, B., (1996), *Environmental Management,* Beijing, China Environmental Sciences Press. p. 301.

Panjin Chorography Office, (1998), *Chorography of Panjin City*, Beijing, Chorography Publishing House, 1–468.

Yin, H., (1995), "Ecological Risk Assessment Program of Water Environment", *Environmental Sciences in Shanghai,* 14(11), 1–14.

4 Bioindustrial Watershed Management

A Case Study of Uttaranchal

J.S. Bali

India lives in the villages. Seventy per cent of the people are there. But are seventy per cent of the government resources being spent there, as it should be, if we call ourselves one national family? The actual figure is less than ten per cent! The need is for injecting massive doses of technology and investment. The basic panacea of: from rich to the poor has been forgotten. Revenue comes from the rich and is spent on them and their cities. The villages remain high and dry.

What does come to the villages, is through a plethora of scattered schemes. Villages even do not know of all of them. Rajiv Gandhi said: only 15 per cent of the funds of such schemes reach the people. Today the ground reality is worse. Delivery cost, inefficiency and corruption soak up the allotted funds.

Now the delivery is envisaged through the village Panchayat itself. Funds will flow from the Centre to the Panchayat directly. But the villagers are themselves not 'washed in milk'. They are capable of the worst leakages. A saying in rural Punjab goes: *Chor uchakka chaudhary te* lundi *runn pradhan* (who is the leader?—the thief among men and the immoral among women). Is our administrative set-up capable of the required vigilance?

From the state capital city to the village, each department has a vertical column. There is no horizontal integration at the field level. Agriculture itself is a cosmos. Even its various divisions have vertical structure from top to the bottom. How to integrate it with forestry? Land and the soil have a precarious perch in Uttaranchal. Steep slopes, fragile geology of the young Himalayas and the younger Shivaliks, and high rainfall amounts and intensities, all combine to create conditions of accelerated erosion, land slides, and severe land degradation in the hills and consequent sedimentation problems in the lower plains.

There is too much pressure on the natural resources: Human population of 59 lakh dependent on the meagre agricultural land of 6.7 lakh ha, or 8.9 persons per ha; 148 lakh animal units grazing over 34 lakh ha of forestland or 4.3 per ha. On pasture land of only 2.27 lakh ha, the grazing pressure comes to 65 animal units per ha. There is neither adequate control over human population nor on animal numbers and quality. In the words of S.L. Shah, the population becomes both a culprit and a victim.

Rainwater is not managed properly. The 8 million hectare-metres of the rain water which falls on the hilly areas of Uttaranchal, mostly runs off the land to cause erosion in the hills and floods and sedimentation in the plains. After the rains, there is water shortage for men, animals and for the parched land. Young men leave the hills for hope of greener pastures in the plains. The burden of the family falls on the women. They have to bring water, fodder and fuel wood, and even till the land.

Continuing conflict between forest managers and the people spells ecological danger and underutilization of the natural resource. During the last decade there has been jobless growth of the national economy affecting the rural poor adversely.

The Study Area

Uttaranchal Assets

Uttaranchal is rich in natural resources and biodiversity. Still the people are poor. Even for a peon's job men have to migrate to the cities to swell the shameful slums. Unemployment is the real name of poverty. Gone are the days when rural employment meant unskilled and low-paid manual labour. The youth of today, thanks to school education in the villages and the view of the world through TV, aspire for better paid skilled employment. Can agriculture alone provide it? The answer is No. Only a combination of agriculture and industry can do it.

Uttaranchal can count many blessings: The sacred Ganga and the Yamuna originate in the state and present a large potential for hydropower, both macro, meso and micro. It is the Dev-bhumi which means pilgrimage. It has mountains and the salubrious climate, attracting tourists. Rivers provide recreation and water sports. Herbs are an asset in the Himalayas. Herbs, when grown and processed scientifically and commercially, have the potential of attracting world level pharmaceutical and cosmetic industry and wealth to the state. The state is a climatic niche for fruits, flowers and off-season vegetables. Animal husbandry and fisheries have a large scope. Wool, silk, bio-fuel plants, bamboo, grass wealth and many other biomass types offer tremendous scope for development. People of the state are cooperative with one another and with development agencies. Development potential is thus great. Realization of this potential is a challenge for the academics, politicians, administrators and managers,

The Development Strategy

At the State Level

1. The foremost need is for tourism development.
2. Equally important would the hydropower development.

These state level macro programmes would enrich the state which could then invest the surplus funds for rural development.

At the Watershed and Village Level

The best way of synergic rural development at the village level, would be through taking up integrated bioindustrial watershed projects.

The bioindustrial sites would become growth centres and develop into future towns and cities. The difference between cities and villages would vanish. Schools, colleges, health facilities and other infrastructure would be centred around the project. Rural India would become fully developed

Bioindustrial Watershed Project Planning and Implementation

Product Choice

The first step in planning would be the choice of a bioproduct which can grow in that area or is already established there naturally. For example, suitable climate and facilities may exist for growing of vegetables or fruits as in most mountainous areas. It may be medicinal and aromatic herbs as in Uttaranchal. It may be milk and milk products. It may be paper and paper pulp based on the forests. It may be rice which can yield many food and industrial products through processing.

It may be simply packaging of wheat, maize, sorghum and other food grains. It might be a wheat area in which processed wheat products could be manufactured, packed and marketed.

Processing of medicinal and aromatic herbs is a profitable business for the rural areas. Uttaranchal has been declared by the government as a herbal state. Profitable industries based on herbs have already been set up in the 100 per cent export zone of Noida and many other places.

Milk and milk products are a promising line for bioindustrial watersheds. There is already a 'white revolution' in India. But many cities do not still have scientific milk schemes producing safe milk and milk products. AMUL has already set a good example. Given determination and zeal for social service many more AMULs can be set up profiting millions of our small and marginal farmers whose livelihood is dependent upon cows and buffaloes. Other products can be processed on the AMUL pattern.

Feasibility Reports

Planning for the bioindustrial watershed projects should begin with the preparation of a preliminary feasibility report and then a detailed feasibility report. The two feasibility reports should establish the viability of the venture. Once the basic feasibility of the selected product for processing and marketing is established, action on the project components should be started.

Project Components

Basic Infrastructure

- Health Facilities
- Education Facilities
- Power

- Roads and Transport
- Communications

Foundation Practices

- Forest Development and Conservation
- Water Development and Management (Irrigation & Drainage)
- Land Development and Management

Sustainable Production Systems

Based on Plants

- Agriculture
- Horticulture
- Vegetable Culture
- Floriculture
- Agroforestry
- Other Systems

Based on Animals

- Milk and Milk Products
- Poultry
- Meat
- Other Systems

Processing Industry

- Preliminary Feasibility Report
- Detailed Feasibility Report
- Land, Water, Power and Other Basic Needs of the Industry
- Plant Erection and Buildings
- Operational Components

Storage

- Grain Storage
- Cold Storage
- Low Cost Storage

Marketing

- Local Marketing
- National Marketing
- International Marketing

Finance

- Bank Loans
- Institutional Loans
- Government Liabilities
- Subsidies

Building Partnerships

- With National Companies
- With the Government

- Local, State and Central
- With the NGOs

Implementation Steps

- Building infrastructure and arranging finance
- Ecology: Apply soil and water conservation in the watershed
- Setting up agreement and contracts for sustainable production systems based on *plants*
- Animal shelters and facilities for sustainable production systems based on *animals*
- Farm buildings for agricultural production for *consumption* and for *processing*
- Processing industry installation
- Building storage structures
- Setting up marketing facilities
- Building management structures

Research Strategy for Uttaranchal Development

Researcher's Problems

When our research scientists go to Europe and USA, foreigners are surprised to see so much talent in a poor, underdeveloped (now renamed developing) country. They ask the question: With so much talent in India how do you manage to remain poor?

The answer seems to be relevance (or its absence). Most of our research is not relevant to the majority of us. Majority of India lives in the villages. Rural problems do not find priority place in the research programmes. Excellence and relevance do not go together in the Indian research establishments. Non-scientific and non-technical personnel in the Indian administration have a stranglehold on scientists and technologists. Our research equipment is not always up to date.

Some Problems for Research Studies

The departments of the government being vertically divided without effective horizontal links, it is becoming difficult to put integrated watershed models on the ground. It needs research studies to find out how best to integrate the implementing agencies at the watershed level. Methods so far adopted have proved ineffective.

Let our excellent personnel in the universities and institutes solve people's problems.

Seventy per cent of our people live in the rural areas. Barring a few artisans, they depend upon agriculture. Land holdings are small. Can they ever be lifted out of poverty by agriculture alone? Even where agriculture has prospered, people have not. Majority of the agriculturist families are too poor to afford the technology and the investments required by that technology. What is the carrying capacity of half hectare of land? How many are actually subsisting on that meagre asset?

If agriculture alone cannot lift people out of poverty, what can? Has it been a problem of research?

Should villages be ever entitled to village industries only? (There is a Village Industries Commission doing commendable work). How can the modern industry, with international standards and marketing reach them?

Young men do not stay in the Uttaranchal villages. What are their aspirations? How can they be persuaded to stay back. Villages at present have mostly old men and women. Any research?

What type of employment would make the young man stay in the village? Is it desirable at all to keep the youth tied down to the subsistence level rural profession of agriculture? How many want to leave and where?

What is the carrying capacity of present and recommended land use systems? At what levels of technology and investment? Is there any guidance available from research on these problems?

Women carry a back-breaking burden in rural areas of Uttaranchal. Any research results on reducing the burden?

What does a village young man want today? Can he ever get it in the village?

Socio-economic problems are more important for Uttaranchal today than physical technology problems. Are proportionate research personnel and funds devoted to them?

Can the researchers take up preparation of feasibility reports for watershed development in Uttaranchal and assure lifting of people from poverty to prosperity? Can they accept this challenge or would they say, it is not their job?

Uttaranchal has 15 or so major scientific universities, institutes and departments. Can they divide the state into 15 major watersheds and take charge of one each to take science and technology there within a certain period, say 10 years? Can they take people's income to the level of at least Rupees one lakh per family annually? What funds would be required from the state?

Conclusion

Agriculture under Indian conditions, and more so in Uttaranchal with small land holdings, would keep the poor, poor. Only when industry is mixed with agriculture, shall we see the rural youth getting skilled employment with urban wages near their homes. There would be no need to migrate, leaving only the women and old men in the villages. But the foundation of agriculture and processing industry is protection of ecology. It means forest, biodiversity, and soil and water conservation, and management. Only on this foundation, sustainable and balanced systems of production, based upon plants and animals, can be built. Once a balance has been restored to ecology and production systems, a stage is set for bioindustries, for value addition and profitable marketing. Infrastructure of health, education, power, roads, transport and communications, is the first necessity in the rural areas. Their neglect amounts to inflicting a disaster upon a majority of our population. The integrated implementation of all these components constitutes bioindustrial watershed management. It would require massive infusion of technology and investment. Research has to combine excellence with relevance. Seventy per cent of our population lives in the villages. Only when rural people's problems are solved and they are lifted from poverty to prosperity, shall we see India becoming a great economic power.

Researchers' problems, as indicated above, should be solved by the government. Let there be a Commission appointed to do it, which should be asked to submit a Report within a year. Let greater priority be given to relevance in research. Scientists and technologists should be paid much more than others. Only then can we continuously tap the best talent. At present, best talent is not going to science and technology. It is going to commerce. National honour and prestige should be liberally bestowed upon those who solve the real problems of the people.

Bioindustrial watershed management may be adopted at the village level, integrating all the development programmes. The nation may heed what M.S. Swaminathan, the 'Man of the Millennium', says: "Various International Conferences have endorsed Prof. Bali's concept. I hope soon every watershed in the country will become a bioindustrial watershed, in order to ensure work and income security to rural families". (Foreword to the book: *Bioindustrial Watershed Management: Concepts and Strategies*; Bali, 2005)

References

Bali, J.S. (1975) Lecture at the Indian Institute of Public Administration, New Delhi, containing the first mention of the concept of Bioindustrial Watershed Management under the Chairmanship of Prof. Anil Gupta (now Professor at the IIMA, Ahmedabad).

— (1984/87) "Agroindustrial Watersheds, A Model for Removal of Poverty", *J. of Soil and Water Cons.*, NAAS Campus, IARI, New Delhi.

— (2000) "Rural Employment and Land Resource Management through Bioindustrial Watersheds" *Advances in Land Resource Management*, pp. 378–387, Soil Conservation Society of India.

— (2003) "Bioindustrial Development and Management for A Better India", *Advances in Land Care Movement*, pp 285–293, Soil Conservative Society of India, Tamil Nadu Chapter, Deptt. of Agri. Engg., Chennai.

— (2003) "Bioindustrial Watersheds for Poverty Alleviation in Uttaranchal", Proceedings of the Seminar of HIMCON, SCSI, Mountain India People's Forum and G.B. Pant University of Agriculture and Technology, Ranichauri Campus, Tehri, Uttaranchal.

— (2003) "Save Ganga from the People for the People", in *Save Ganga, Save Himalaya, Save India*, pp. 34–37, Delhi Environment Education Project (DEEP).

— (2005) "Bioindustrial Watershed Management", *Concept and Strategies,* Book, 97 p, Soil Cons. Soc. of India, NAAS Campus, New Delhi.

5 Communicating Education for Sustainable Development with Less Articulate and Underprivileged Communities

Innovative Approaches to Socially Critical Environmental Education

Shailesh R. Shukla

This paper is divided into four sections. The first section provides a cursory look at emerging trends in Environmental Education (EE) and discusses the dynamics of communication among local communities in one of the most relevant forms of EE i.e., socially critical EE. The second section reviews the examples of the different modes of communication in socially critical EE across the globe and in India. The third section briefly describes three case studies that show innovative methods from India. The final section concludes with a cursory evaluation of these methods by suggesting the ways for improved communication targeted at less articulate and underprivileged rural children and local communities and addresses some of the limitations of socially critical EE.

Communicating for the Environment: The Relevance of Socially Critical EE

The recent trends in programmes and policies of EE reveal that EE can be more effective if it can make use of prior experiences that children bring to schools, often acquired within their communities through informal learning modes (Palmer, 1998; Chawla, 2001). Palmer also concluded that the world's most successful programmes in the 21st century would be those in which formal and informal elements of education are supported alongside each other, with the involvement of major stakeholders. Sauvé (1996) notices, "many aspects of socially-critical environmental education, as defined by Robbotom and Hart (1994) correspond to this vision". Walker (1997), based on these works, lists four criteria for socially critical EE viz., (a) Involvement of students, teachers and communities in collaborative investigations of local environmental issues; (b) Schools working together with communities to develop a new critical awareness of the roles that communities play in influencing the course of such issues; (c) Making explicit the values and interests of various groups; and (d) Unfolding of the local environmental issue by prior commitment to teaching a body of knowledge/skills.

The socially critical approach draws its conceptual underpinnings from critical pedagogy (Freire, 1968, 1985). The desired result is the empowerment of students and teachers to bring about social transformation and change on the basis of democratic and socially just ways. In the Indian context, Gandhiji's philosophical and political activism has been greatly in line with the Freirian critique of formal education as a tool of oppression and political power. This vision gave birth to an alternative education model to that of the basic education model *nai talim,* which revolves around local knowledge, practical experience and emphasizes the importance of regional languages (Sherry Chand, 1996). Critical pedagogy's new philosophical form is socially critical EE which, according to some scholars (Palmer, 1998) constitutes a 'radical reform' approach to environment improvement, one that views the environmental crisis as a larger problem of society.

In her most recent critique on socially critical EE, Walker challenged the application of socially critical theory in EE due to its impracticality in bringing about change in existing school practices and its inability to explain educational change fully. She concludes, "for many practitioners, socially critical theory not only fails to give them an implementation theory, it de facto denies their own practical knowledge" (Walker, 1997: 5). This gap in recognizing the potential of local knowledge is also seen in Robbotom and Hart's proposed community-based model of socially critical EE, wherein the role of the community's own knowledge has not been highlighted clearly. Both of these limitations also add up to the major challenge of communicating with and by 'underprivileged' and 'less articulate' group of stakeholders, including tribal rural communities and most importantly, children and women.

To counteract these two main theoretical criticisms viz., impracticality and making explicit use of local knowledge systems, it may be useful and interesting to review and discuss the examples, which have demonstrated at least one of the methods of socially critical EE.

Communicating by Putting Socially Critical EE in Practice

Review and Examples

The issue of combining different knowledge systems of stakeholders (and more specifically bringing in the local knowledge of the community to partner the formal systems of knowledge) may be suggested as a stepping-stone, which can also make socially critical EE practical. The literature on the ways and methods of facilitating such fusions of knowledge however, is scanty (CEE, 1999; Semali and Kincheloe, 1999; Gardner and Shukla, 2002) and focused on non-formal EE programmes for the high school level or for post-secondary education. Non-formal EE programmes seem to have evolved from the camping programmes in the 1980s in North America. The North American Association for EE has designed instructional programmes at community or school sites, which have become popular methods of incorporating EE in formal schooling through outdoor EE programmes (Lori et al., 1988). Some of these programmes have intentionally or unintentionally tried to separate informal learning from local knowledge systems and formal schooling as documented in the international case studies on non-formal EE. Some of these examples are provided in the following box (Box 1), to give an idea of how students, teachers and communities have joined hands, established common grounds of communications and have engaged in reciprocal learning in innovative ways.

In line with these efforts in North America and Europe, the response from a developing country like India was gradual but noteworthy. To fulfil its obligation as a signatory of the Convention of Biological Diversity, the Government of India has drafted the National Biodiversity Strategy and Action Plans (NBSAP) for the states. NBSAP[1] notes, that the formal education system in India provides great scope for creating awareness about biodiversity and its conservation among children and youth. However, there are gaps in terms of the coverage of biodiversity, and the methodology of teaching. In particular, the formal education system needs to create space and scope for the community's traditional knowledge systems and practices' (www.ceeindia.org).

The incorporation of local knowledge systems into the school curriculum in India has largely been attempted by development NGOs and a few isolated but motivated educators. Notable in this regard are the efforts of SRISTI (Society for Research and Initiatives for Sustainable Technologies and Institutions), based at the Indian Institute of Management, Ahmedabad. They have valorized the local ecological knowledge of tribal people and the poor children and communities across India through innovative scouting methods like biodiversity and recipe contests (For details see SRISTI, 1994). The Peoples Biodiversity Registers Programme (Gadgil et al., 2000) was launched in southern India with the aim of providing intellectual property rights protection to the local communities through protection of local biodiversity and associated knowledge systems. This grassroots programme has demonstrated that the local ecological knowledge of village communities can communicate and collaborate with village level formal institutions such as the panchayat or schools, and thereby create socially critical EE for conservation. Centre for Environment Education (CEE) (an autonomous organization supported by the Ministry of Environment Education, Government of India) reviewed the primary school level curriculum of EE in order to recognize and support innovative EE programmes in 1987 and initiated efforts to correct certain inadequacies through programmes like BAIDIK[2] (CEE, 1999) at the primary school level. At school level, an innovative experiment in Hoshangabad, Madhya Pradesh was initiated by a local NGO called Ekalavya with support from university scientists. This experiment emphasized the use of educational aids made from local and low-cost materials, as well as learning from local communities' daily rural life experiences. The experiment was expanded to more than 500 schools in Madhya Pradesh by mobilizing political support from the state government, and is being regarded as a successful

Box 1

Some Examples Linking Local Knowledge Systems with Formal EE

- In a primary school in Bairnt Antholing, Bavaria (FRG), a class teacher together with his pupils (aged 8–10) developed a cultivated plot, a natural meadow, village pond, field with historic types of grains and wild herbs and protective hedge were created in cooperation with parents and local farmers.
- At Loreto Apruntino, Italy, school students and teachers observed relationship between aspects of the rural environment and socio-economic and cultural living conditions of local people (professions, dialects, customs, everyday family life etc.). Excursion and interviews etc., have been used as date collection tools. Results were then displayed through an exhibition on rural civilization in the local museum.
- In Czechoslovakia 'golden leaf' competition for young pioneer group (autumn, 1972) were organized comprising of practical activities little-cleaning, tree/shrub planting, promotion and practising biological control of pests etc., undertook by young people in conservation management and environmental studies. The object of the campaign is to involve youth in environmental issues. Each competition was judged on this practical activity and tested on actual nature trail where competitor's knowledge and experience was tested. After the district and regional level, competitions were organized at national level for up to less than 12 years group and up to 15 years group. The final contest was held in the form of a camp where the best groups contested. During camp, discussions, film-shows and excursions were also organized.
- Similarly, for the school children of 12–15 years workshop was organized in Prague in 1973–74. The 60 minutes workshop had different themes of environmental education consisting of film presentations and puzzle-solving sessions followed by nature trail competition among teams of participants. For instance, workshop on 'Plants and animals as natural resources' started with film on conservation of protected species then children were asked to complete the scheme of food-chain by connection lines through several living things and then natural trail competition.
- West Virginia case study provides a pedagogical method which imparts the students to combine out-door nature studies and Native American studies. Winter Lodge is a 2-day and one night programme developed for the grade 5 students. The uniqueness of the programme lies in its approach, which combines educational techniques and principles with the use of dramatic techniques.
- A teacher in Junagadh forest district of Gujarat India, inspired students to collect different shapes and types of plant thorns with support of

local shepherds and village communities. These thorns were later organized in educational charts, and small less-known village schools received honour and award in state science competitions.

- The organizations learning through landscapes in UK (www.ltl.org.uk) and natural learning in the United States (www.naturallearning.org) have encouraged local communities and students to explore local natural resources by turning their school yards into experiential open access natural areas.
- The foxfire approach in Appalachian Georgia State of USA have inspired high school students and teachers to learn Basic English through oral histories and folkways of native Appalachian everyday life in nature. Based on experiential learning approach, this experiment has produced magazines and series of books, which are valuable references for schools to learn about native knowledge. The foxfire approach, which was originated 30 years ago, is still recognized as model for school reform demonstration programmes in the United States.
- Rediscovery is an outdoor environmental education programme initiated in 1978 in British Columbia, Canada. It has expanded into an international activity with 40 or more such regular annual programmes in North America. The native elders and local experts provide ecological teaching in a camp through folksongs, nature trails and local cultural forms for native and non-native young students.

Source: Centre for Educational Research and Innovations (1991), Chand and Shukla (1998), Starnes (1991), Henley (1996), Chawla (2001).

educational reform with regard to teaching science in schools (Rampal, 1994). In the field of school-based EE, a practical course titled 'Our land our life' was designed and implemented by a local NGO, the Uttaranchal EE Centre, in 1987 with the help of state and national departments of education in the Himalayan region. This course curriculum encourages high school students to engage in solving local environmental issues such as land degradation using both the local knowledge of the elders in the villages and the concepts learned through the formal EE curriculum in schools. It has now expanded to over 530 schools in Uttaranchal, covering more than 68,000 high school students (Pande, 2001). Many of these NGO-led initiatives targeted at the school level correspond to the Gandhian philosophy of alternative education models for economically and socially disadvantaged groups like women and tribal communities.

Unfortunately, these efforts seldom attracted the attention of policy makers, development agencies or mainstream educators (and more specifically environmental educators) and remained neglected. Barring a few exceptions, most of these efforts have been limited to short-term projects and somehow could not become a regular part of mainstream or formal education due to variety of reasons, including the lack of effective communication among various stakeholders and most importantly between knowledge creators, supporting institutions and policy makers. Gupta (1999) highlighted ten barriers for effective articulation, dialogue and utilization of formal and informal science in public policies. Some of these communication barriers, if addressed, could help overcome the impracticality and limitations of socially critical EE in making use of local knowledge systems. The three case studies presented in the next section, are an attempt to demonstrate the ways in which such barriers were handled.

Community-Based Socially Critically EE: Case Studies

Three innovative attempts experimented among the rural, poor and tribal communities in Gujarat and Maharashtra viz., biodiversity contests, recipe contests and community plant diversity register are highlighted here to demonstrate the potential of these methods in successfully establishing communication among various stakeholders building on local knowledge systems thereby, creating a case for applied socially critical EE for sustainable conservation of plant diversity. Both the biodiversity and the recipe contests have been innovated and experimented by SRISTI (www.sristi.org) in various parts of India to document and recognize the local knowledge of rural school children and women. The community plant diversity idea has been inspired from the People's Biodiversity Register Programme (Gadgil et al., 2000) and is being self-experimented by tribal village communities of Baripada village, in the Dhule district of Maharashtra.

Biodiversity Contests

Biodiversity contests focus on children and their traditional ecological knowledge. The philosophy that guides them is that children are the future guardians of biodiversity conservation and resource management at the local level. The idea of biodiversity contests, as articulated here, originated at SRISTI, and was implemented successfully in various parts of India, Bhutan and Vietnam through members of the Honeybee Network (www.sristi.org). The biodiversity contests were aimed at school-going children and their teachers.

The process undertaken is initiated at the local level through public announcements, pamphlets and a group meeting with teachers, children and community members several days prior to the contest. On the day of the contest, the children bring plant specimens they have collected to the local school. A jury consisting of teachers, knowledgeable and interested community members, and forest staff interviews each child and scores them on their knowledge according to the following criteria: the number of specimens presented and their novelty, the knowledge about plants (such as habitat, growing habits etc), the uses of the plants and presentation style. Winners are awarded prizes such as teaching/learning materials. Follow-up activities include: the creation of a school herbal garden, the creation of a school herbarium, and various displays. The results showed that children possess a remarkable body of informal knowledge about plants, particularly their medicinal uses for humans and cattle. Biodiversity contests of this type have taken place for a number of years in a variety of rural settings but the insights for this paper were mainly gained from two recently concluded contests in rural schools in Maharashtra.[3]

Recipe Contests: Triad of Culinary Creativity, Health and Conservation

The idea of a recipe contest was conceptualized and implemented by SRISTI. Recipe contests designed *mainly by and for women* to document their local knowledge about uncultivated flora used in cuisine, with some medicinal or health value. The rural and tribal women, engaging in such interactive knowledge creation and exchange, have opportunities to demonstrate their culinary creativity and local knowledge with regard to wild plant uses. A potential direct benefit is the improvement in health (nutritional and medicinal values of recipes or ingredients used therein) of rural families where quality health facilities set up by the governments are either not accessible or if available, are in a sad state. It also maintains the health of economically poor girls and women, who often have to travel through the woods often in harsh weather conditions as part of their domestic chores and can collect medicinal plants at the same time. In forests, deserts and mountains, women from poor tribal areas often possess an outstanding knowledge of local plants and their uses in sustainably managing their domestic health, agriculture, livestock and natural resources. The knowledge of rural women, does not just contribute to the improvement of livelihood, health and nutrition, but holds potential for the sustainable management of natural resources.

These contests are often organized among village women,[4] who are asked in advance to make new recipes, using unconventional and uncultivated/wild flora. Each dish will be judged using criteria such as: (i) number of different plants/plant parts used, (ii) conservation value, (iii) taste, and (iv) the participant's knowledge of the different plants/plant parts used in the dishes. The idea of the recipe contest is publicized through pamphlets or appropriate available local media such as puppets, village gatherings etc. On the day of the contest, the participating women bring the dishes and/or the methods of preparing them (if these plants are not available in a given season/year/location). A local jury consisting of local knowledgeable women experts, community leaders, local healers, and NGO representatives is set up to evaluate the participants. The winners are awarded prizes in the form of a certificate and/or kitchen utensils etc., in a public gathering in the village. The winning recipes are displayed and shared with local communities and schools. A successful recipe contest would involve about 12–15 women from a village or 4–5 mother-daughter teams. Such dishes often represent the wise use of wild vegetables such as dishes with special nutritional values; dishes that help to cope with extreme seasons and calamities, dishes with medicinal values etc.

The first contest was organized in January 2004, at Amboli village, Sidhudurg district, and the second was organized in August 2004, in the tribal village of Baripada in Dhule district.

The Community-Based Plant Diversity Monitoring Register of Baripada

Baripada is a small remote, tribal village of 101 houses (population less than 700), is located at the border between Maharashtra and Gujarat. In response to the emerging crisis of shortage of firewood, food and water in the nearby forests of 445 hectares, caused by indiscriminate cutting of teakwood trees by outsiders, a village youth called Chaitarm Pawar had mobilized the community and formed an informal village forest protection committee in the year 1993. They crafted and enforced an interesting set of access, resource-use, and conflict-resolution rules to prevent pressure on this reserved forest.[5] The forest department after realizing the community spirit legitimized the informal village forest protection committee into a joint forest management protection committee in 1999. The results were revealing and brought international recognition when the Rome-based International Fund for Agricultural Development (IFAD) honoured them for their outstanding community-based conservation and development work in 2003. After this success the villagers realized the need for continuous monitoring of the plant diversity within the

Table 1

The Structure of Plant Diversity Register in Baripada

Site 1			*Plot 1*	
Local name of plant	*Number of particular plants*	*Botanic name*	*Local use (most frequent) as reported by healers, local experts and members of forest protection committee*	*Remarks (status such as rare, significantly reducing, new growth etc.)*
Plant 1	1 … n			
Plant 2	do			
Plant 3: Plant N	do			
Site 1			*Plot 2*	
Plant 1	1 … n			
Plant 2	do			
Plant 3: Plant N	do			
Site 1			*Plot 3*	
Plant 1	1 … n			
Plant 2	do			
Plant 3: Plant N	do			

Note: Same format applied to all fourteen sites.

forest and therefore engaged in the collection of baseline data on plant diversity in the forest. The next stage was that of community-based management of plant diversity with the forest department. This led to a joint exploration by local youths, healers and university teachers from the botany and zoology departments from Pimpalner and Sakri towns for collaborative learning about the mapping of plant diversity. In September 2004, a focus group discussion among all partners to decide on a site for demonstration was held. An area of 0.50 acres consisting of dense mixed vegetation was thus identified by the local healers, Baripada villagers and the team of scientists. After careful consideration of potential sites, it was finalized that a plot of 10 sq m would be ideal for laying down the quadrant. The skill of marking, plotting and recording of plants in a zigzag fashion was demonstrated on the selected site. It was decided that individual healers and village youths who have knowledge of local plants would carry out the recording of the observations. The recording format was collectively decided, as described in Table 1.

A simple code of conduct was deliberated and followed. For example, each local healer would take turns in counting the number of the plants that she/he first reports. Once she/he completes the counting of a particular plant, she/he would then move on to the next plant that has not been reported so far. The botanists and zoologists guiding the exercise, decided to record the information in Marathi. Each plant should be collected with flowers and fruits, for easy identification. The order of counting should be from smaller size to bigger ones (first creepers and lastly trees). After the counting is done, all the members of the team would move randomly within the plots, so that any unlisted plants would not be left out.

The forest protection committee after a couple of meetings with local healers and shepherds finalized the 14 different sites capturing the geographical variations, as samples. From each such site, three plots (10 sq m each) would be laid. Three teams of fifteen people, each team having five members consisting of local healers, village youths and members from the forest protection committee were formed. Within two weeks, a vegetation mapping of all the sites was carried out. The register was formally inaugurated in a village level workshop in October 2004, which was attended by local healers, primary school teachers and children, forest staff, Ayurvedic practitioners, government officials from the district agriculture department and local NGOs. It helped generate useful policy recommendations for community-based conservation. Some simple rules for the use of the information provided in the community register were also formed. The villagers have decided to monitor the status of plant diversity every two years from the same sites.

The features of communication strategies, the target audience (in this case the underprivileged groups) and the key barriers addressed have been highlighted in the following Table 2.

Table 2

Evaluating Methods: Have they Served the Purpose?

Features	*Biodiversity contests*	*Recipe contests*	*Community Plant Diversity Register*
Communication strategies	• Printed pamphlets and verbal announcements followed by informal discussion with school children and walks. • Orientation to teachers • Orientation to local community leaders including local experts.	• Printed pamphlets and verbal announcements followed by informal discussion with women • Orientation to local village women's groups such as self-help • Orientation to local women experts and community leaders	• Orientation to village community leaders and local healers • Discussions with formal scientists—botanists and ecologists • Local demonstration/hands-on training on 'how-to' and 'where'
Targeted underprivileged group	• Primary school children and girls in rural and tribal areas • Local ecological experts such as healers • Primary school teachers	• Rural and tribal women, particularly illiterate girls and women • Village women's groups	• Economically-disadvantaged rural community • Tribal communities • Local healers and unemployed village youths
Barriers addressed	• Language (regional) • Engaging local participation: the less-articulate (healers and educationally- challenged children) • Bringing together multiple stakeholders (e.g., teachers, local healers and school children) with varied interests and knowledge • Recognition to local knowledge systems • Platform for exchange and positive action between local knowledge and school-based knowledge for ensuring local conservation • Updating school-based environmental education by chronicling local uses of plants (such as eco-indicators) and collaborative action with communities for conservation • Intergenerational transmission of local knowledge maintains the stake of the younger generation	• Language (regional) • Recognition of women's local knowledge • Engaging local participation of the less-articulate (women, particularly tribal, and from weaker social groups) • Bringing together multiple stakeholders (e.g., knowledgeable elderly women, community leaders and village women groups) having varied interest and knowledge • Intergenerational transmission of local knowledge maintains the stake of the younger girls in local knowledge and conservation	• Language (regional) • Intergenerational transmission of local knowledge maintains the stake of village youths • Platform for exchange between local knowledge and formal knowledge for ensuring local conservation • Intergenerational transmission of local knowledge maintain stake of village youths in local conservation and knowledge • Educating the forest department by chronicling local uses and monitoring of plants leads to collaborative action for sustainable co-management • Recognition to local knowledge systems

There is a great deal of similarity in the communication strategies adopted by all three approaches. Keeping the target groups informed right from the inception of the programme, through a mix of locally appropriate communication media is the most common communication strategy across all three methods. Since all three were targeted at different village level stakeholders, their priorities were naturally different. For example, the biodiversity contest was designed for school children and targeted at informing, interacting and transforming school-based EE, while the recipe contest was targeted towards facilitating intergeneration transmission and recognition of commonly-discounted women's knowledge systems in the local conservation agenda. The use of the local language and resources including knowledge, creating a platform for dialogue among multiple stakeholders, encouraging intergenerational interactions by building upon local knowledge systems are a few of the enabling conditions for managing the barriers for effective communication and thereby creating a niche for socially critical EE.

Conclusion

Socially critical EE will be more useful and practical, if it relies on approaches that build on local knowledge systems. Approaches like biodiversity contests, recipe contests and community-based plant diversity registers help to create a platform on which local knowledge systems can inform, interact and transform with formal knowledge and connect the holders of these multiple knowledge systems in a collaborative and cooperative learning and action. To do so, these innovative approaches will have to be tailored to remove constraints in establishing effective communication with the underprivileged and less-articulate group of children, women and socio-economically disadvantaged tribal communities. The enabling conditions under which these methods function are their simplistic local ways and down-to-earth approaches of communication, promoting links between and interests in more than one generation and their knowledge systems and creating a congeniality for local knowledge systems, to nurture and interact with formal knowledge systems. Biodiversity and recipe contests are however, more process-oriented methods, which helped create platforms for future actions, while the community-plant diversity register is more action-oriented method, which not only helped in recognition of local knowledge but also combined it with formal knowledge and created a model for sustainable co-management of local biodiversity. These methods are however, means and not ends and therefore, require constant experimentation and innovations considering the different ecological and socio-cultural contexts.

Notes

1. The State Action Plan (SAP) of Gujarat is a part of the National Biodiversity Strategy and Action Plan (NBSAP) being prepared by the Ministry of Environment

and Forests (MoEF) with support from the Global Environment Facility (GEF). The NBSAP is to be a detailed micro-level plan for the conservation of the country's biodiversity. Undergraduate and postgraduate students in different regions can take up a variety of biodiversity studies, close to their own place, along with their teachers. There are examples of colleges having taken up such studies. In Pune RANWA has prepared an assessment of species diversity in the city, through college students. In Karnataka, CEE South has carried out an Environmental Quality Monitoring programme in nine districts with 20 undergraduate colleges.

2. The BAIDIK (Biodiversity Awareness Integration through Documentation of Indigenous Knowledge) initiative implemented by CEE has noted that knowledge regarding ethnobotany; plant-based home remedies and eco-indicators were recurrently reported by children of five different agro-ecological zones in India.
3. Funding support from IDRC (International Development Research Centre) and University of Manitoba, Canada for organizing these events at various stages of my research is gratefully acknowledged.
4. In a recent modification, a recipe contest broadened its scope in the form of a traditional food festival organized by SRISTI that invited urban women with the intention of revitalizing the traditional recipes, the knowledge of which is eroding at a rapid rate due to various reasons.
5. For description on these rules, please see Honeybee, Vol. 15 (2), 7–9.

References

Centre for Environment Education (CEE). 1999. *Environmental Orientation for School Education: Some Experiences and Learning.* A Programme of Ministry of HRD, Ahmedabad, India: CEE India.

Centre for Educational Research and Innovations. 1991. *Environment, Schools and Active Learning*, Paris: Organization for Economic Cooperation and Development (OECD).

Chawla, L. (2001). "Significant Life Experiences Revisited Once Again: Response to Vol. 5 (4) Five Critical Commentaries on Significant Life Experience Research in Environmental Education", *Environmental Education Research (7)* 4: 451–61.

Freire, P. (1985). *The Politics of Education: Culture, Power and Liberation*, South Hadley, MA: Bergin and Gavery.

Freire, P. (1968). *Pedagogy of the Oppressed*, New York: Seabury Press.

Gadgil, M., P.R. Seshagiri Rao, Utkarsh Ghate, P. Pramod., A. Chhatre and members of the peoples' Biodiversity Register Programme (2000). "New Meanings for Old Knowledge: The Peoples' Biodiversity Register Programme". *Ecological Applications, 2000.*

Gardner, James and S. Shukla (2002). The Use of Indigenous Knowledge in Post-Secondary Environmental Education: Strengthened Global Environmental Management Through Local Understanding. Paper presented at Canadian Congress, June 2002 at Toronto: University of Toronto.

Gupta, Anil, K. (1999). "Science, Sustainability and Social Purpose: Barriers to Effective Articulation, Dialogue and Utilization of Formal and Informal Science in Public Policy", *International Journal of Sustainable Development*, 2(3), 368–71.

Hanley, Thomas (1996). *Re-discovery: Ancient Pathways, New Directions,* Calgary, Alberta: Lone Pine Publishing Inc.

Mann, Lori D., Edward J. McCrea and Augusto Q. Medina (eds.) (1988). Excellence in Environmental Education: Gaining Momentum for the Challenges Ahead, Selected Papers from the Sixteenth Annual Conference of The North American Association for Environmental Education (NAAEE), Troy, Ohio: NAAEE.

Palmer, Joy, A. (1998). *Environmental Education in the 21st Century: Theory, Practice, Progress and Promise*, New York: Routledge.

Pande, Anuradha (2001). "Environmental Education in Rural Central Himalayan Schools", *The Journal of Environmental Education,* 32(3), 47–53.

Rampal, Anita (1994). "Innovative Science Teaching in the Rural Schools of India: Questioning Social Beliefs and Superstitions" (pp. 131–38) In J. Soloman and G. Aikenhead (eds.), *STS Education: International Perspectives on Reform.* Teachers College, USA: Columbia University.

Robottom, I. and P. Hart (1994). *Research in Environmental Education: Engaging the Debate*, Victoria: Deakin University Press.

Sauvé, L. (1996). "Environment Education and Sustainable Development: Further Appraisal", *Canadian Journal of Environment Education,* I: 7–35.

Saveland, Robert (1976). *Handbook of Environmental Education (with international case studies),* London, Toronto: John Wiley and Sons.

Semali, L.M., and J.L. Kincheloe (eds.) (1999). *What is Indigenous Knowledge? Voices from The Academy,* New York: Falmer Press.

Sherry Chand, Vijaya (1996). Innovations in Rural Higher Education: Gramvidyapiths of Gujarat, Unpublished doctoral dissertation, Ahmedabad: RJMCEI, Indian Institute of Management.

Sherry Chand, Vijaya and Shailesh Shukla (1998). *Teachers as Transformers: Innovations in Primary Education*, United Nations Childrens' Emergency Funds (UNICEF).

Walker, Kim (1997). "Challenging Critical Theory in Environmental Education", *Environmental Education Research*, 1997, 3(2), 155–63.

Websites

Centre for Environment Education (CEE). 2002. National Biodiversity Strategies and Action Plan <http://www.ceeindia.org/SAPinformat.htm>. (12 March, 2002).

Society for Research and Initiatives for Sustainable Technologies and Institutions (SRISTI). 1994. *Scouting Indicators for Sustainability: Conservation Through Competition.* SRISTI. 2003. Scouting methods. <www.sristi.org> (12 November, 2003).

Starners, Bobby, A. 1999. The Foxfire approach to Teaching and Learning: John Dewey, Experiential learning and Core practices. ERIC Digest, January 1999 <http://www.ael.org/eric/digest> (January, 2002).

6 Genetic Resources from Micro-organisms Living in Extreme Marine Environment

Anu

The coastline of world ocean is 3,12,000 km (1,93,000 miles) and a volume of 137×106 km^3, is the largest ecosystem on earth, and has been used for a variety of purposes by human beings for millions of years.

Micro-organisms occur nearly everywhere in nature and occupy an important place in human view of life. In the field of microbiology, scientists have discovered novel 'extremophiles', micro-organisms that are capable of living in extreme environments such as highly acidic or alkaline conditions, at high salt concentration, with no oxygen, extreme temperatures (as low as –20°C and as high as 300°C), at high concentrations of heavy metals and in high pressure environments such as the deep-sea. The deep-sea environment is characterized by high pressure and low temperature but in the vicinity of hydrothermal vents regions of extremely high temperature exist. Deep-sea micro-organisms have specially adapted features that enable them to live and grow in this extreme environment (Photo 1).

Photo 1

Oceanic organisms are of enormous scientific interest, for two major reasons:

1 They constitute a major share of the earth's biological resources, and
2. Marine organisms often possess unique structures, metabolic pathways, reproductive systems, and sensory and defence mechanisms because they have adapted to extreme environments ranging from the cold polar seas at –2°C to the great pressures of the ocean floor, where hydrothermal fluids spew forth. Most major classes of the earth's organisms are primarily or exclusively marine, so the oceans represent a source of unique genetic information.

Exploration of the marine environment began as an exciting resource during the 1960s to isolate new compounds. Much attention has been paid to identify, isolate, and purify them for useful biochemicals and utilize them in human welfare and development. The oceans are a rich source of biological diversity and potential for discovery of novel drugs.

Marine microbes represent a potential source for commercially important compounds and their bioremediation capabilities are also remarkable. They also play a crucial role in decomposition of organic matter and cycling of nutrients. Microbes also serve as food for some bottom-living organisms. Our knowledge of marine microbial diversity has, however, been severely limited by relying on micro-organisms that have been cultured.

Extreme Environments

Any environmental condition that can be perceived as beyond the normal acceptable range is an extreme condition. Extreme environments can be found in many parts of the world and all these environments are colonized by micro-organisms adapted to these conditions.

There are four general phenotypic groups of archaea: the methanogens, the extreme halophiles, the sulfate-reducing archaea, and the extreme thermophiles. In the marine environment, archaeal habitats are generally limited to shallow or deep-sea anaerobic sediments (free-living and endosymbiotic methanogens), hot springs or deep-sea hydrothermal vents (methanogens, sulfate reducers, and extreme thermophiles), and highly saline land-locked seas (halophiles).

Little is known about the fungi living in the extreme environments. The hyper saline Dead Sea environment, which is one of the most extreme environments for micro-organisms on the earth (Table 1). Highly saline conditions in waters and soils exert a strong selective pressure on the biota, favouring the development of and halophilic forms. Biodiversity of fungal life drastically decreases with increasing salt concentration.

Table 1
Different Physiological Groups of Marine Bacteria

Group	*Physiology*	*Role in marine environment*	*Example*
Archaebacteria Sulphate-reducing bacteria	Chemoautotrophs anaerobic, themophilic and mesophilic.	Contribute over 50% of the carbon turnover of coastal marine sediments: take part in the cycling of sulphur compounds in sea water.	*Desufomonas, Desulfovibrio, Desulfobulbus, Desulfotamaculum and Desulfococcus*
Methanogenic bacteria	Chemoautotrophs, strictest anaerobes, utilize a limited number of simple carbon compounds (hydrogen, carbon dioxide, formate, acetate and methanol) as their carbon and energy sources for methanogenesis.	Utilize trimethylamine in the marine environment as substrate and produce mathane as an end-product of their energy-generation metabolism.	*Methanococcus, Methanosarcina, Methanomicrobium, Methanogenium, Methanoplanus, Methanocaccoides and Methanobolus*
Halophilic bacteria	Require at least 12–15% NaCl to survive and grow well even at concentrations up to saturation.	Red colonies formed due to high carotenoid content and dominate in high salt environments, such as salterns and salt lakes; regulate the osmotic pressure thereby resisting the denaturing effect of salt in their environment.	*Haloarcula, Halobacterium, Haloferax and Halococcus*
Eubacteria Luminous bacteria	Produce light by a simple protein-like substance called luciferin in contact with the oxygen molecule; Gram-negative and motile heterotrophic rods.	Bioluminescence in the deep ocean helps the organisms defensively to startle and divert predators (defence), to attract prey (offence) and to camouffage Luminous bacteria help in cycling of nutrients in the sea and contribute in the nutrition of marine organizma as gut microflora.	Photobacterium Leiognathi, Photobacterium Phosphoreum, Vibrio fusheri and Vibrio harveyi
Nitrifying bacteria	Oxidize either ammonia to nitrite (*Nitrosococcus*) or nitrite to nitrate (*Nitrocaccus*) and convert nitrogen to a form readily available for other biological processes.	Extremely important process, since positively charged ammonium ions bind to acidic sediment particles, where they become available for biological processes; more abundant in nearshore waters than in offshore regions.	*Netroococcus, Nitrococcus*, etc.

Recent research on the physiology and molecular biology of deep-sea bacteria has identified pressure regulated operons and shown that microbial growth is influenced by the relationship between temperature and pressure in the deep-sea environment. Barophilic bacteria, which are able to grow at high hydrostatic pressure, but that are unable to grow at atmospheric pressure.

Piezophiles are typical micro-organisms adapted to high-pressure and cold temperature environments, and located in deep-sea bottom. Otherwise, hyperthermophiles are living in high temperature environment, and located at around the hydrothermal vent systems in deep-sea. They are not typical deep-sea micro-organisms, but they can grow well at high-pressure condition, just like piezophiles.

Hydrothermal-vents are geysers on the ocean floor and associated with mid-ocean ridges in which extreme environmental conditions, in terms of temperature, pressure and toxicity prevail.

Because of the peculiar characteristics in which life develops in these ecosystems, hydrothermal vent organisms represent a subject of interest from both a and a commercial point of view. The term 'black smokers', commonly used to designate hydrothermal vents, indicates intense and dense fluid emissions from the ocean floor. These emissions are the result of magmatic activity and are characterized by very high temperatures (300°C and more). They support a dense microbial community. All vents are characterized by extremely high pressure due to the depth at which they are located, by extreme temperatures and pH values, and high by extreme and due to the minerals that salinity toxicity escape from the earth's crust.

Marine and Extreme Genomic Research

Marine biotechnology can enhance a nation's competitive power and provide many solutions for food and resource shortages predicted in the 21st century. Korea possesses rich marine living resources and it has recently developed capabilities to access living resources in extreme environments, such as the Antarctic and the Arctic regions, hydrothermal vents, tropical oceans, and the deep sea (Photo 2).

Photo 2

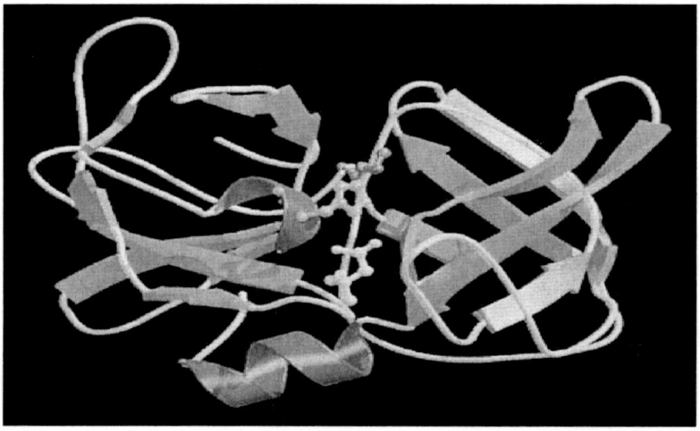

The goal of the Marine and Extreme Genome Research project is to analyze the genomes of diverse living resources collected from different ecosystems of many marine and extreme environments, understand their biological processes, and ultimately develop new biomaterials.

The centre instituted the Marine and Extreme Bioresources Bank (MEBB) to collect, store and distribute resources. The bank has now collected about 3,500 strains of micro-organisms. In 2005, it added species of brown algae that are commonly found along the Korean coast. Micro-organisms were also sampled from sediment, water columns and surfaces of animals and plants in environments such as the Arctic, the Antarctic, hydrothermal vents, cold seep areas, tropical areas, and the Korea near shore area. The Centre also established a genetic information bank, Marine Bio Base, to maintain gene sequence data produced by the project (Figure 1).

Figure 1
The Process of Separation, Incubation, and Functional Analysis of Marine Microorganisms.

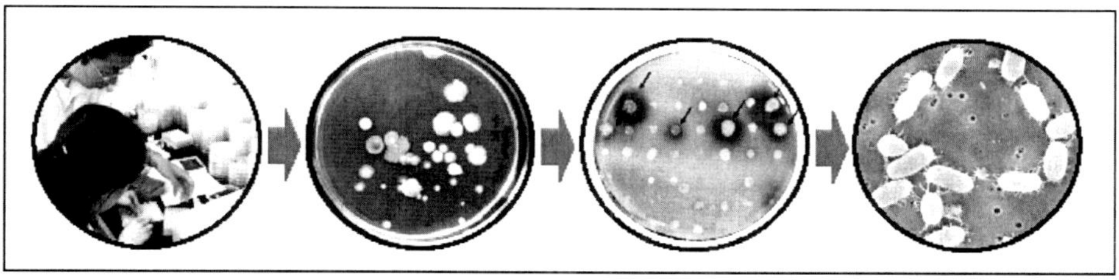

Studies on marine and extreme micro-organisms led to the discovery of many new strains or species. Approximately 10 per cent of the 3,500 isolates collected could be a novel genus or species based on 16S rRNA sequences. The collected strains include gamma-proteobacteria (42%), gram-positive bacteria (37%), alpha-proteobacteria (14%), and CFB-group (7%). A bacterium *Kordiimonas gwangyangensis* capable of degrading (compounds composed of erase) polycyclic aromatic hydrocarbons (PAHs) was identified as a novel order of alpha-proteobacteria, named Kordiimonadalesord. Additional novel micro-organisms are currently being identified: HJ114 from the seashore, DOKDO007 from Dokdo Island, and AKS600 from the Antarctic. The collected resources will be distributed to (research programme: erase) collaborators and participants of the research programme to search for novel pharmaceuticals and useful biocatalysts.

Biotechnology in Use of Extreme Marine Genetic Resources

Potential applications for marine organisms include: pharmaceuticals; enzymes; cryoprotectants; cosmaceuticals; agrichemicals; bioremediators; nutraceuticals; and five chemicals. All the major marine drugs can be used as antioxidant, anti-fungal, anti-HIV, antibiotic, anti-cancer, anti-tuberculosis and anti-malarial. Applications for the treatment of Alzheimer's disease, cystic fibrosis and impotence are also considered.

Several marine compounds are currently at various phases of development, including anti-cancer agents and immuno-suppressants. Considerable hope is placed in drugs based on marine organisms in light of the shortcomings of current anti-cancer drugs, which have either been limited to the treatment of specific cancers or to which patients have often become resistant (Figure 2).

Figure 2

Novel Bioactive Compounds from Ocean

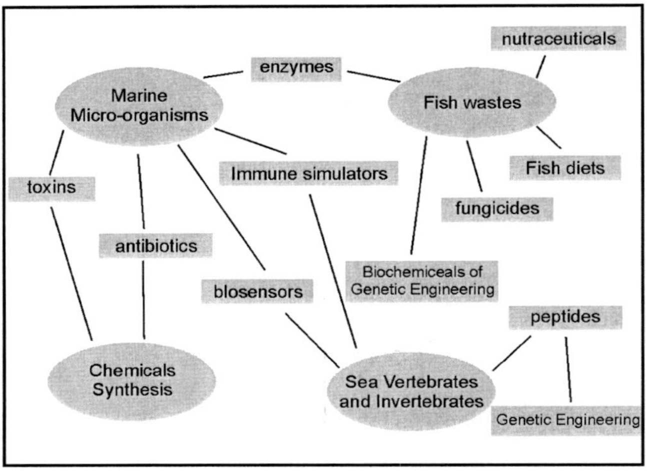

Microbial populations, their distributions, and their aquatic environments were studied over a year (1997) at an acid mine drainage (AMD) site at Iron Mountain, Calif. The two species most widely studied and implicated for their role in AMD production, are *Thiobacillus ferrooxidans* and *Leptospirillum ferrooxidans*.

Istamycin, a new antibiotic, is formed by *Streptomyces tenjimariensis* isolated from the mud of a shallow bay in Japan. It appears to be an aminoglycoside that can inhibit a fairly large number of gram-positive and gram-negative bacteria.

Aplasmomycins produced by *Streptomyces griseus* inhibits gram-positive bacteria and shows an effective antimalarial activity.

Flavobacterium ulginosum, isolated from the seaweed of Sugami bay in Japan, produces a substance named mariactin that inhibits the development of tumors in various animals by up to 95 per cent.

Rubrenoic acid is extracted from the marine Alteromonas rubre, which shows bronchodilator activity, neuromuscular blocker, antiserotonin, anti-amphetamine, and hypotensives.

Few marine micro-organisms are known to produce antiviral compounds. *Vibrio* marinus has antiviral activity and inactivated enteroviruses, and Flavobacterium. Pseudemonas and Vibrio has anti-poliovirus activity. The antiviral effect of algae is most important as very few antiviral drugs presently exist.

Porphyosin is a substance that isolated from Porphyra, exhibited anti-shay ulcer activity. Dinoflagellates and cyanobacteria are major sources of neurotoxins and neuropeptides and play a vital role in understanding the development and function of the nervous system.

Research has also shown that cryptophycins produced by cyanobacteria also have anti-cancer potential, as well as being effective against viral diseases such as HIV. Other useful compounds include anti-inflammatory compounds such as manoalide and topsentin, and the cosmaceutical anti-irritant pseudopterosin.

Chloroflexus aurantiacus is a photosynthetic bacterium isolated from hot springs, belonging to the green non-sulphur bacteria. This organism is thermophilic and can grow at temperatures from 35°C to 70°C.

In the acidic continental hot springs and deep-sea hydrothermal vents, where the pH optimum for growth at 5.5–6.0 and the pH minimum at 3.0, temperature is around 85 degrees C, salinity optimum at 0–2.5 per cent NaCl and the maximum at 7 per cent, favours the growth of genus Thermococcus, Thermoanaerobacter siderophilus and Thermoanaerobacterium (Photo 3).

Photo 3

Thermophilic Organisms

Extremely high alkaline protease form a deep-subsurface bacterium, *Alkaliphilus transvaalensis.*

Deinococcus radiodurans ('strange berry that withstand radiation', formerly called Micrococcus radiodurans) is an extremophilic bacterium, and is the most radioresistant organism known. It has been used for bioremediation to consume and digest solvents and heavy metals, even in a highly radioactive site. The bacterial mercuric reductase gene has been cloned from Escherichia coli into Deinococcus to detoxify the ionnic mercury frequently found in radioactive waste generated from nuclear weapons manufacture.

Snottites or snotties are colonies of single-celledextremophilic bacteria that hang from the walls and ceilings of caves. These organisms derive their energy through chemosynthesis of volcanic sulphur compounds and warm water solution dripping down from above.

Pyrococcus furiosus was originally isolated anaerobically from geothermally heated marine sediments with temperatures between 90°C and 100°C collected at the beach of Porto Levante, Vulcano Island, Italy. The enzymes of *Pyrococcus furiosus* are extremely thermostable and used in the Polymerase Chain Reaction, with the relevant enzyme is known as Pfu (Photo 4).

Photo 4

P. Furiosus

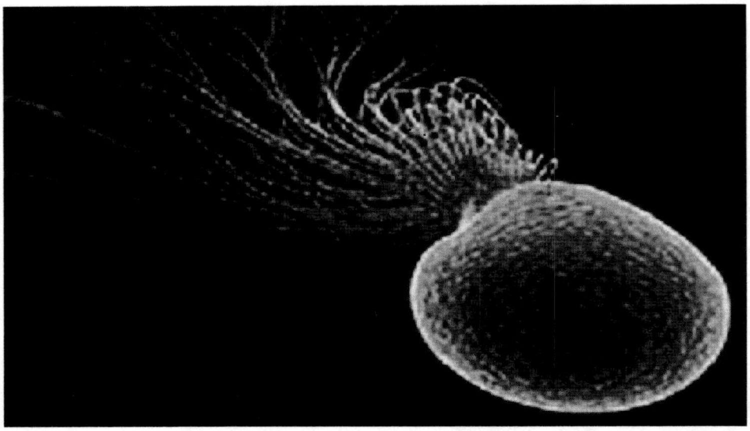

The scientists say P. furiosus is a microbe that can survive in extreme temperatures. It grows and dwells in underwater sea volcanoes where temperatures reach that of boiling water. It also can survive in near freezing deep-sea water.

The NASA-funded research entails extracting a gene from P. furiosus and expressing it in plants.

"This is very fundamental research" Boss said. "If we could add new genes to plants, we could potentially make the plants more resistant to extreme conditions such as drought and extreme temperatures that we have on earth, but also to the extreme conditions that one might find on Mars."

Enzymes

Marine organisms are richly endowed with many enzymes and have unique functions. This has boosted marine microbial enzyme technology in recent years, offering valuable products. These are used as pharmaceuticals, food additives, and fine chemicals. they have a diverse range of enzymatic activity and are capable of analyzing various biochemical reactions with novel enzymes.

Enzymes must be stable to extremes of temperature, pH, and salt concentration. For this reason, the isolation and characterization of enzymes form a special class of micro-organisms known as 'extremophiles' may yield useful new products for this industry. The deep-sea bed provides many example of these extremophiles and consequently may be of interest to companies involved in developing enzymes for this sector:

1. Proteases enzymes are able to degrade proteins, so this is particularly useful in commercial laundry detergents for removing stains containing protein.
2. The textile industry uses enzymes called cellulases to break down cotton fibres for such processes as 'stonewashing', surface polishing, and softening.

Among them, aerobic bacteria appear to be capable of producing several enzymes, such as amylases, which are well recognized to have applications in various industries.

A mutant of amoeba *Trichospaherium sp.* is found to degrade and destroy plastic in the field as well as in the laboratory. Viral DNA polymerase produced by *Thermococcus litoralis*, a type of archebacterium, is active for more than two hours at 100°C.

The Orissa Coast of India reports that a few potent strains of marine organisms could produce novel enzymes like amalyse, protease, cellulase, L-asperginase, acetylcholinesterase, and urethanase. The results indicate that these micro-organisms show potential for the production of salt-tolerant enzymes that could be used in the biotech-based food, beverage, and pharmaceutical industries.

Beta carotene is used extensively as a food colourant by the margarine, cheese, soft drink, and baking industries. Studies undertaken currently indicate that beta-carotene may also be a cancer-preventing agent. Some marine bacteria belong to species of *Pseudomonas, Vibrio, Aeromonas, Bacillus,* and *Micrococcus* are capable of synthesizing vitamin B_{12}. They are reported to be about 50 to 95 per cent of heterotrophic bacteria and are capable of secreting active thiamine. This source can be exploited for therapeutic use.

The goal of biotechnology enzymes research and development is to identify enzymes that are capable of carrying out very specific molecular tasks, usually related to the modification of DNA or RNA, for the creation of genetically modified organisms or for diagnostic procedures. These enzymes go by a variety of names, including restriction endonucleases, RNA and DNA polymerases, alkaline phosphatases, kinases, reverse transcriptases, ligases, and so on.

A ubiquitous and powerful biotechnology process known as the 'polymerase chain reaction' (PCR) depends upon an unusually thermostable enzyme known as Taq DNA polymerase to replicate DNA in a test tube. The enzyme must be able to withstand the alternating cycles of heating and cooling inherent to the PCR process. Taq DNA polymerase is derived from a species of thermophilic bacteria, *Thermus aquaticus,* originally isolated from a hot spring in the western United States (Photo 5).

Photo 5
Thermus Aquaticus

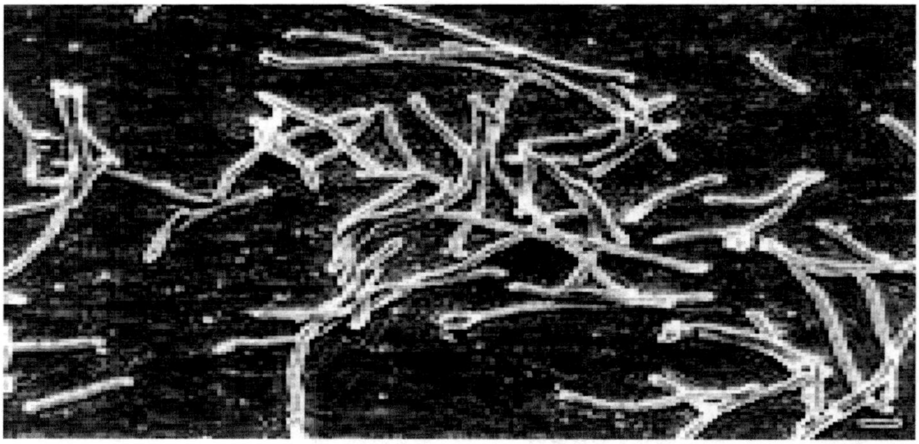

The extracellular proteases are of particular importance and can be used in detergents and industrial cleaning applications, such as in cleaning reverse-osmosis membranes. *Vibrio* species have been found to produce a variety of extracellular proteases. *Vibrio alginolyticus* produces six proteases, including an unusual detergent-resistant, alkaline serine exoprotease. This marine bacterium also produces collagenase, an enzyme with a variety of industrial and commercial applications, including the dispersion of cells in tissue culture studies.

Biomonitors

Marine organisms can provide the basis for development of biosensors, bio-indicators, and diagnostic devices for medicine, aquaculture, and environmental monitoring. One type of biosensor employs the enzymes responsible for bioluminescence. The *lux* genes, which encode these enzymes, have been cloned from marine bacteria such as *Vibrio fischeri* and transferred successfully to a variety of plants and other bacteria. The *lux* genes typically are inserted into a gene sequence, or operon, that is functional only when stimulated by a defined environmental feature. The enzymes responsible for toluene degradation, for example, are synthesized only in the presence of toluene. When *lux* genes are inserted into a toluene operon, the engineered bacterium glows yellow-green in the presence of toluene. This genetically engineered system 'reports' that biodegradation of a specific chemical, in this case toluene, is proceeding.

Another type of biomonitor that holds great promise is the gene probe, which can be used to identify organisms that pose health hazards or may be useful in research. Specific gene probes can be employed, for example, to detect human pathogens in seafood and recreational waters; fish pathogens in aquaculture systems; micro-organisms capable of mediating desired chemical transformations (e.g., toxic chemical degradation, CO_2 assimilation, metal reduction); and specific fish stocks in fish migration and recruitment studies.

Piezophiles micro-organisms are used as industrial wastewater treatment, municipal wastewater treatment, bioremediation of contaminated soils, bioleaching of mineral-rich ores, food processing.

Biomaterials

Extreme marine natural products may also offer clues to the 'bioengineering' of surfaces to reduce or prevent biofouling of ship hulls and other submerged structures by sessile marine organisms.

Patent Based Products

Sederma Company, located in France, has used enzymes isolated from deep sea bacteria to develop commercial skin protection products providing higher resistance to UV and heat exposure. The enzymes used for these products, isolated from the extremophile *Thermus thermophilus*, are effective in counter-acting free radicals that form as a result of UV action, especially at high temperatures, thus preventing the skin from damage.

T. thermophilus and other species of thermophiles are also the subject of research by the Company, Roche. Roche's patents relate to improvements in the amplification of nucleic acids, and include a number of inventions based on thermophiles such as a DNA sequence using a Thermus aquaticus DNA polymerase (patent US5075216), a Thermus thermophilus polymerase (patent US5407800), a mutated thermostable polymerase from Thermotoga maritima (patent US5420029), a mutated thermostable polymerase from Thermus sp. (patent US5455170), a thermostable polymerase from Pyrodictium sp. (patent US5491086), and a thermostable polymerase from *Thermosipho africanus* (patent US 5968799). These products are used in conventional molecular biology such as sequencing.

Diversa is a California-based Company/Corporation has been granted a significant number of patents. Diversa has commercialized several products developed on the basis of deep sea organisms, including: the PyrolaseTM 160 enzyme, derived from a hydrothermal organism and used in industry to reduce viscosity; and the thermostable Thermal AceTM DNA Polymerase, derived from a non-declared Archaea species, and used in DNA sequencing.

The Company, New England BioLabs Inc., headquartered in Beverly, Massachusetts, US, has an extensive list of commercialized products derived from deep seabed organisms. Examples include: the Deep VentR® DNA Polymerase, obtained from a bacterium carrying polymeras genes, Pyrococcus sp – a hydrothermal vent species found at 2,010 m of depth and growing at temperatures of up to 104°C; and the Therminatora DNA Polymerase, obtained from a genetically- engineered form of the DNA polymerase of Thermococcus sp.

Candida Antarctica

The preparation of esters in the presence of *Candida antarctica* lipase A, the ester are useful as ingredients in fat blends such as margarine (WO0153511).

The enzymatic synthesis of polyesters in the presence of a lipase derived from, amongst others, Candida antarctica. The polyesters are useful in formulating products such as skin cream and cosmetics as they normally function as thickeners or softeners in such formulations (US5962624) (Photo 6).

Photo 6

Candida Antarctica

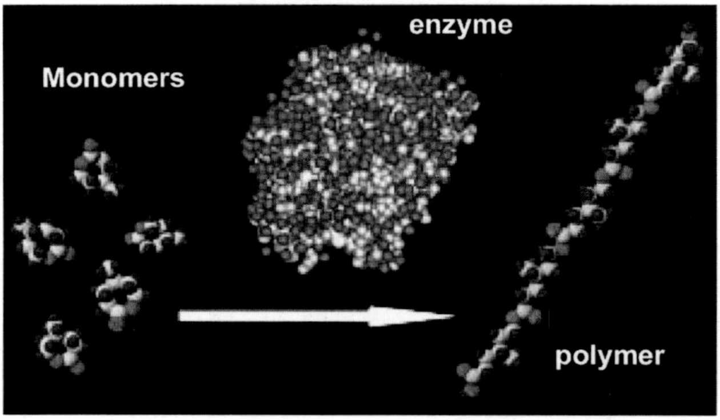

The preparation of an optically-active ester using an enzyme originating from Candida antarctica. The ester can be used for preparing pharmaceuticals such as benzothiazepines and benzazepines (US5407828).

The hydrolysis of water-insoluble ester in the presence of a lipase derived from a strain of Candida antarctica. This is useful as some types of pulp made from wood have high resin content, and the resin can create disturbances in the process of pulp manufacture and may have negative effects on the properties of the final pulp product (WO9218638). *Pseudoalteromonas antarctica* produces glycoprotein which is used as in the preparation of pharmaceutical, veterinary and cosmetic compositions for topical or mucosal application aimed at the treatment and re-epithelialization of wounds (WO02102406).

A glycoprotein obtained by culture of the bacterial species Pseudoalteromonas antarctica CECT4664, which is useful for coating liposomes in order to improve their stability in relation to external factors such as surfactants (WO9842731).

The production of a particular stimulating agent containing the extract of an alga belonging to the genus Durvillea, including Durvillea antarctica. The stimulating agent prevents the development of skin wrinkles (JP9176036) in Turku, Finland. HyTest manufactures and markets cardiac markers, hormones, toxins, human proteins, and infectious and autoimmune disease reagents. Taq Red is used as an enzyme for molecular biology.

The thermostable Tth DNA Polymerasea is commercialized by the Company, Promega, headquartered in Madison, Wisconsin, US. This product is also used as an enzyme for molecular biology.

Diversa is among the applicants of patent WO03093434, concerning the genome of the hyperthermophilic *Nanoarchaeum equitans*, its proteins (including enzymes), identified genes encoding these proteins, and also the isolated Nanoarchaeum equitans.

Patent US2003235902125 concerns the production of thioredoxin from the hyperthermophilic *Aeropyrum pernix* and *Pyrococcus horikoshii*, growing at temperatures between 90 and 100°C. Thioredoxin has an interest for the pharmaceutical industry, including as an agent detoxifying snake venom protein and preventing skin inflammation caused by UV radiation. This protein is relevant to the food industry as a compound that eliminates food allergens, and to the cosmetic industry for development of products used as protection against adverse effects of UV radiation.

Patent US2003129734127 relates to the copper tolerant yeast *Cryptococcus sp*, isolated from deep sea sediments.

Patent US5989587128 concerns the production of novel ether lipids from the Archaea *Methanospirillum hungatei*, *Methanococcus jannaschii*, *Methanococcus voltae*, *Methanosarcina mazei*, *Methanobrevibacter smithii* and *Halobacterium cutirubrum*. These liposomes (lipid vesicles) prove to be very stable, making them good candidates for many liposome applications, including as components of biological membrane systems for the study of processes such as transmembrane transport, immunological adjuvant, carriers of drugs, skin care compounds, and insecticides (Photo 7).

Photo 7

Methanococcus Jannaschii was Isolated near a "Smoker", a Hydrothermal Vent 2,600 m Deep in the Pacific Ocean.

Patent application US20020106660129 relates to the structure of the 30S ribosome unit of *Thermus thermophilus*, on the basis of which 30S inhibitors can be developed, and potentially used as antibiotics.

At least 21 patents meet the search criteria 'Archaea' under the metadatabase of the European Patent Office 130. These span form methods for detecting and identifying DNA in a sample (patent US2004176584) to novel extreme halotolerant an halophilic (patents TW579390 and WO0130934) as well as thermostable enzymes (patents US2004002075, US6391604,

WO9833895),the formation of stable lipids from Archaea's lipid extracts (WO9308202), and the genome of particular species, such as patents related to the genome of the hyperthermophile *Methanopyrus kandleri* (Photo 8).

Photo 8
Methanopyrus Kandleri

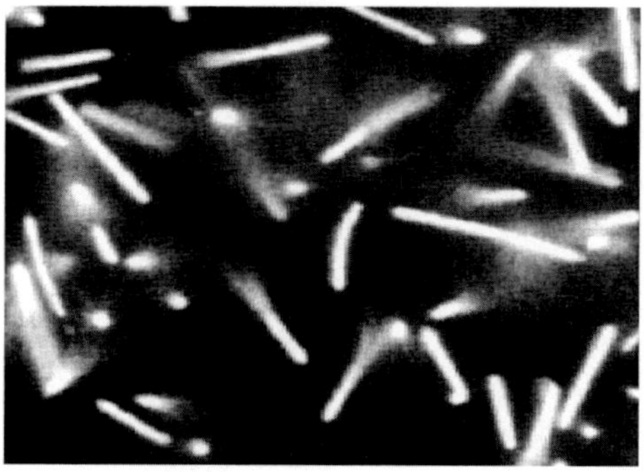

Conventions for the Protection of Marine Resources

United Nations Convention on the Law of the Sea (UNCLOS), which was adopted in 1982 and entered into force on 16 November 1994, aims to establish "a legal order for the seas and oceans which will facilitate international communication, and will promote the peaceful uses of the seas and oceans, the equitable and efficient utilization of their resources, the conservation of their living resources, and the study, protection and preservation of the marine environment".

The Convention on Biological Diversity(CBD) was adopted in June 1992, and entered into force on 29 December 1993. The CBD aims at the conservation of biological diversity, the sustainable use of its components and the fair and equitable sharing of benefits arising from the utilization of genetic resources, including by appropriate access to genetic resources and transfer of relevant technologies.

Convention for the Protection of the Marine Environment of the North-East Atlantic (OSPAR), which entered into force on 25 March, 1998, is the instrument providing for international cooperation for the protection of the marine environment of the North-East Atlantic. The Convention applies to the area situated not only within the internal waters, territorial seas and Exclusive Economic Zone (EEZ) of its contracting parties, but also to a significant proportion of the High Seas and the underlying seabed and subsoil in the North-East Atlantic and Arctic Oceans as delineated by the Convention. In fact, more than 50 per cent of the OSPAR area is beyond national jurisdiction. There are at least four known hydrothermal vent fields in the OSPAR maritime area, including the Menez Gwen, Lucky Strike, Saldanha and Rainbow vent fields.

Mediterranean Action Plan (MAP) includes maritime areas in the High Seas, beyond the national jurisdiction of the 22 Parties to the Barcelona Convention for the Protection of the Marine Environment and the Coastal Region of the Mediterranean (Barcelona Convention). The MAP draws on the Barcelona Convention and its six protocols, which address specific environmental issues. Under the Convention, parties are to apply the precautionary principle and the polluter-pays principle, as well as undertake environmental impact assessment for activities that are likely to have a significant adverse effect on the marine environment.

References

Biocatalysis: A New Pathway to Polymeric Materials, Interview with Dr. Andreas Heise, DSM.

Convention on Biological Diversity. (1996)., Bioprospective of Genetic Resources of the Deep Sea-Bed.

Das, S., Lyla, P.S., and Khan A.S., General Article, "Marine Microbial Diversity and Ecology: Importance and Future Perspectives".

Delong, E.F., (1992) *Ecology*. Vol. 89, pp. 5685–89.

Edwards, K.J., Gihring, T.M., Banfield, J.F., (1999), Applied and Environmental Microbiology, Seasonal Variations in Microbial Populations and Environmental Conditions in an Extreme Acid Mine Drainage Environment, pp. 3627–3632, Vol. 65.

Horikoshi, K., (1998), Curr.Opini. Microbiol., Barophiles: Deep-sea Micro-organisms Adapted to an Extreme Environment, June, 1(3), pp. 291–5.

Horikoshi, K., (1998) Barophiles: Deep-sea Micro-organisms Adapted to an Extreme Environment. Curr. Opin. Microbiol. pp. 291–5.

Horikoshi, K., Kobayashi, T., Hung, V.S., Kurata, A., Hatada, Y., Takai, K. (2007) Appl. Microbiol Biotechnology, Extremely High Alkaline Protease from a Deep-subsurface Bacterium, Alkaliphilus transvaalensis.

Kato, C., Takai, K., Microbial Diversity of Deep-sea Extremophiles-Piezophiles, Hyperthermophiles, and Subsurface Micro-organisms.

Korea Ocean Research & Development Institute. (2005). Marine and Extreme Genomic Research., Annual Report. pp. 30–32.,

Kublanov, I.V., Prokofeva, M.I., Tourova, T.P., Kolganova, T.V., Lebedinsky, A.V., Bonch-Osmolovskaya, E.A., Spring, S., Jeanthon, C., Cultivated Anaerobic Acidophilic/acidotolerant Thermophiles from Terrestrial and Deep-sea Hydrothermal Habitats.

NASA Discovers Life's Building Blocks are Common in Space, Moffett Field CA (SPX) Oct. 12, 2005

Prokofeva, M.I., Kublanov, I.V., Tourova T., P., Kolganova T.V., Lebedinsky A.V., Bonch-Osmolovskaya E.A., Spring, S., Jeanthon, C., (2005), Extremophiles, Cultivated Anaerobic Acidophilic/acidotolerant Thermophiles from Terrestrial and Deep-sea Hydrothermal Habitats.

Reysenbach, A., L., Liu, Y., Banta, A.B., Beveridge, T.J., Kirshtein, J.D., Schouten, S., Tivey, M.K., Von Damm, K.L., Voytek, M.A., A Ubiquitous Thermoacidophilic Archaeon from Deep-sea Hydrothermal Vents.

UNU-IAS Report. Bioprospecting of Genetic Resources in the Deep Seabed: Scientific, Legal and Policy Aspects.

7 Human Impact on High Altitude Forests in Northern Pakistan

Degradation Processes and Root Causes

Udo Schickhoff

It is widely recognized that forests, including wooded lands, simultaneously provide a wide range of economic, social, environmental and cultural benefits at the local, regional and global levels, and that maintaining all types of the world's forests in a healthy state is in the collective interest of the present and future generations. However, tropical deforestation and degradation of forests in many parts of the world are negatively affecting the availability of forest goods and services. While forest area in developed countries has stabilized and is slightly increasing overall, deforestation has continued in developing countries (FAO, 1999). Deep concern is expressed about the ability of forests to meet the multiple needs of societies if the current trends continue and about the resource gap for meeting the needs of sustainable forest management.

In the world's high mountains, environmental effects of overexploitation of forest resources are often disastrous. In order to maintain the manifold protective functions of high altitude forests, sustainability of forest utilization and stability of forest ecosystems have to be secured (Hamilton et al., 1997; Mountain Agenda 2000; Price and Butt, 2000). In the high mountains of South Asia—Hindukush, Karakorum, Himalaya—it is still a matter of much debate how to solve the problem of sustainable forest management (Rao, 1997; Agrawal 2001; Wakeel et al., 2005). In this region, overuse of forest resources and forest degradation is causing great concern and is often directly connected with increased runoff and soil erosion. Some authors even relate it to inundations in the lowland plains (Zurick and Karan 1999).

Despite a growing number of research initiatives and publications within the so-called 'Himalayan Dilemma' context—the term refers to the alleged incompatibility of the recent socio-economic development and sustainable resource use (cf. Ives and Messerli, 1989; Messerli et al., 1993; Ives, 2004)—the present state of knowledge of degradation processes in mountain forests and its management implications still reveals some basic deficits:

1. Forest changes are almost exclusively descriptively assessed and are termed as forest depletion or deforestation without defining what is

explicitly expressed with these terms. The problem is that large-scale clearcuts are only exceptions. Instead, structural alterations in forest stands are widespread that are more difficult to quantify and are only gradually taking place (cf. Schmidt-Vogt, 1990; Schickhoff, 2002). Not only the extent of forest degradation, also the consequences for the site-ecological potential, for the subsystems of forest ecosystems, are often derived without a solid data basis. The complex reality of degradation processes cannot be assessed in this way.

2. Ecological studies are rarely connected with socio-economic research inquiries into the driving forces of alteration processes in the forests. Moreover, the complex of closely interrelated socio-cultural and economic factors causing degradation is not sufficiently analyzed (cf. Blaikie and Brookfield, 1987). Single variables, e.g., high population growth, are often picked out of this complex to explain forest degradation (cf. Geist, 1992). It is evident that sustainable forest management cannot be successful when the scientific basis, from which to derive respective management strategies, has not been elaborated in detail.

In general, there is a considerable lack of well-founded, regional, interdisciplinary, quantifiable and applied studies on degradation processes and their societal causes in high altitude forests. On the other hand, applied studies that integrate approaches from natural and social sciences are indispensable for a better understanding of evolution, structure and function of cultural landscape ecosystems and for contributing relevant findings with regard to the process of sustainable development (Forsyth, 1998; Kreutzmann, 1998).

In the Hindu Kush-Himalayas, research on forest degradation and forest management is concentrated on the Nepal Himalayas so far. Much less is known about the situation in the Karakorum and adjacent high mountain ranges in northern Pakistan. However, recent studies in the framework of the Pak-German 'Culture Area Karakorum' (CAK)—Project indicate widespread misuse and degradation of forest resources (Schickhoff, 1995, 1996, 1997, 1998, 2000a, b, 2001, 2002; Winiger 1996; Clemens and Nüsser 1997). This paper focuses on causes and consequences of forest degradation processes and on the interrelationships between the ecological and the socio-economic system within the Karakorum region. It advocates integrating natural and social sciences into an interdisciplinary approach in order to achieve deeper insights into environmental degradation. On the basis of a landscape ecosystem model the paper examines whether the goal of economically efficient, socially balanced and environmentally sound forest management in the Northern Areas of Pakistan can be achieved.

Study Area and Methodical Approach

The study area is situated in the arid-semiarid high mountain region north of the main Himalayan range in northern Pakistan. It extends throughout the administrative region of the Northern Areas (73–77° E; 35–37° N), including the areas of Darel-Tangir, Chilas, Astor, Baltistan, Hunza, Nager, Gilgit, and Ghizer. Within this high mountain region, closed forests reach their regional dry limit towards the north and east. The major part of the region is occupied by the NW-SE—striking Karakorum Mountains, the highest mass elevation worldwide situated in the hinge area of Himalaya, Hindukush and Pamir. These high mountain ranges between Central and South Asia rise out of arid-semiarid valley bottoms with tremendous relief intensity (K2: 8,611 m; Nanga Parbat: 8,126 m a.s.l.), resulting in pronounced ecological vertical gradients and in sophisticated land use systems. The study area can be broadly differentiated into a drier northern part and a more humid southern part (Indus-Himalaya sensu Troll, 1967), where monsoonal influence is much more effective.

In Pakistan's Northern Areas it suggests itself to conduct an integrated analysis of (i) forest-ecological alteration processes, and (ii) of the socio-economic matrix of forest degradation. On the one hand this is due to the natural setting. The ecological conditions in northern Pakistan are primarily controlled by a SW-NE—directed gradient of decreasing humidity, caused by decreasing monsoonal influence towards the interior mountain region (cf. Weiers, 1998). On the Himalayan south slope (Kaghan and neighbouring valleys), potentially dense forest vegetation is developed in all aspects from the valley floors up to the upper treeline in areas fully exposed to the monsoonal precipitation regime (Schickhoff, 1994). By contrast, in the Karakorum only high above the arid valley floors humidity is sufficient to support forests (Miehe et al., 1996; Schickhoff, 2000c). A pronounced vertical precipitation gradient is well reflected in a drought-caused lower forest limit (Cramer, 1997). So the forest stands, mainly composed of *Pinus wallichiana*, *Picea smithiana*, *Abies pindrow*, and *Cedrus deodara*, are restricted to a narrow altitudinal belt of 1,000–1,200 m. Moreover, forests are mainly confined to north-facing slopes, whereas southern aspects, due to the summation of unfavourable climatic conditions, are covered with xerophilous *Artemisia*—dwarf scrublands, interspersed with juniper trees. Thus, the state of the small forest stands can be accurately assessed on representative plots.

On the other hand, the suitability of the study area for an integrated approach is to be attributed to the current socio-economic development (cf. Kreutzmann, 1989, 1995; Ehlers, 1995; Stellrecht, 1998; Stellrecht and Bohle, 1998; Dittmann, 2000; Ehlers and Kreutzmann, 2000), accelerated by the recent opening up to the outside world. The formerly secluded high mountain region became only accessible in the 1970s, when the Karakorum Highway (KKH), that runs from the Pakistani lowland across the mountain ranges to the autonomous Chinese province of Sinkiang, was completed. This road building caused dramatic changes in the region in terms of socio-economic development and in terms of natural resource utilization.

The KKH enabled easy transport of timber to lowland markets (Schickhoff, 1997). The commercial forestry in private forests in the SW of the study area (Chilas and Darel-Tangir Forest Divisions) has benefited greatly from the improved transport facilities. Exploitation of forest stands for lowland markets, triggered by short-term profit interests, has been reinforced considerably. But also pressure on the remaining forest stands in the Northern Areas, classified as 'Government Protected Forests' (Sheikh and Aleem, 1975), has steadily increased. In the course of socio-economic transformations, the interdependent system of agriculture and animal husbandry called mixed mountain agriculture is loosing importance, whereas at the same time non-agrarian sources of income play a steadily increasing role in household economy. Due to the monetarization of living conditions many small households are now forced to use timber as a source of cash and to participate in illicit cutting and smuggling of wood (Dittrich, 1998; Schickhoff, 2002).

In order to achieve a comprehensive understanding of the interrelationships between ecological and socio-economic systems with respect to forest utilization, and to create a basis for the development of strategies towards sustainable resource management, the approach of the study was twofold:

1. A *forest-ecological* approach aimed at (i) assessing alterations of stand structures and dynamics as indicators for the extent of forest degradation, (ii) analyzing consequences for structure and function of forest ecosystems, and (iii) revealing spatial and temporal differentiations of alteration processes.
2. A *political ecology* approach aimed at (i) analyzing the impact of politico-social and economic structures and developments on the intensity of forest resource use, (ii) recognizing connections and interactions of causal factors, and (iii) identifying main protagonists and interest groups of forest exploitation.

To address these problems, an analytical framework was developed comprising a forest-ecological and a socio-economic level of analysis. The first level was based on the linkage of forest ecological, vegetation ecological, and site ecological approaches. Field sampling of vegetation and forest ecological data (phytosociological relevés according to Mueller-Dombois and Ellenberg, 1974; Dierschke, 1994 and stand structural analyses according to Leibundgut, 1959, 1993; Lamprecht, 1980) was carried out on 62 sample plots in 36 separate valleys in the region. Changes of stand structural complexity initially served as indicators of the extent of degradation. The investigated forest stands were differentiated into five disturbance classes (DC I-DC V) according to the degree of their exploitation. The second level of analysis comprised comprehensive data collection on socio-economic aspects of forestry and forest utilization following the aforementioned 'political ecology' approach (cf. Blaikie and Brookfield, 1987; Bryant and Bailey, 1997). The outcomes of both lines of investigations were directly related to each other and linked in an integrative, spatial-functional perspective. As a result of the synthesis, a holistic regional landscape ecosystem model showing the complex interrelationships between the socio-economic and the ecological system could be established. On this basis, development strategies towards sustainable resource management were derived.

Forest Degradation and its Socio-Economic Background

Stand structural analyses of forests in Pakistan's Northern Areas revealed widespread and advanced degradation. More than one-third of the investigated stands were heavily degraded and had to be assigned to DC IV and V. The decrease of stem number, basal area, stand volume and volume increment due to anthropogenic impacts (Figure 1) is a vivid illustration of the degree of exploitation. These parameters are nearly unchanged in the near-natural stands of DC I. Along the disturbance gradient the decimation continuously increases and reaches mean values of above 50 per cent in DC IV and V, i.e., more than half of the stems was cut in these stands, more than half of the stand volume was destroyed. If we transfer these results into forestry categories, respective figures of very intense thinnings are reached and partially surpassed. The degree of degradation has reached higher levels compared to other exploited and degraded Himalayan forests (Schmidt-Vogt, 1990; Sundriyal and Sharma, 1996). The considerable loss of stand volume increment, which is concomitant to the decrease of stand volume, aggravates the situation since it results in severely scarce future amount of wood resources.

Figure 1

Mean Decrease of Stem Number N (N/ha), Basal Area G (m^2/ha), Volume V (m^3/ha) and Total Volume Increment I_v (m^3/ha) According to Disturbance Classes in per cent

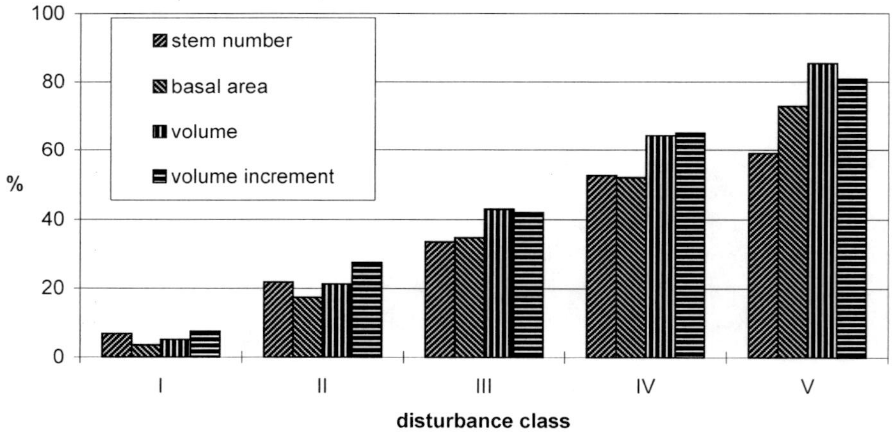

Structural alteration processes and forest dynamics along the disturbance gradient can be graphically summarized with help of profile diagrams that represent specific anthropogenic structure types. The profile diagram of plot Manu Gah II (Figure 2) represents a structure type of extremely low anthropogenic interference. Due to its structural characteristics this type can be termed near-natural. The stands of this type show high stand stability and indicate a stable steady state. Such stand structures might have characterized the high altitude forests of the study area over long time periods under weak human impact. It has become very rare nowadays.

Figure 2

Profile Diagram and Crown Map of Plot Manu Gah II

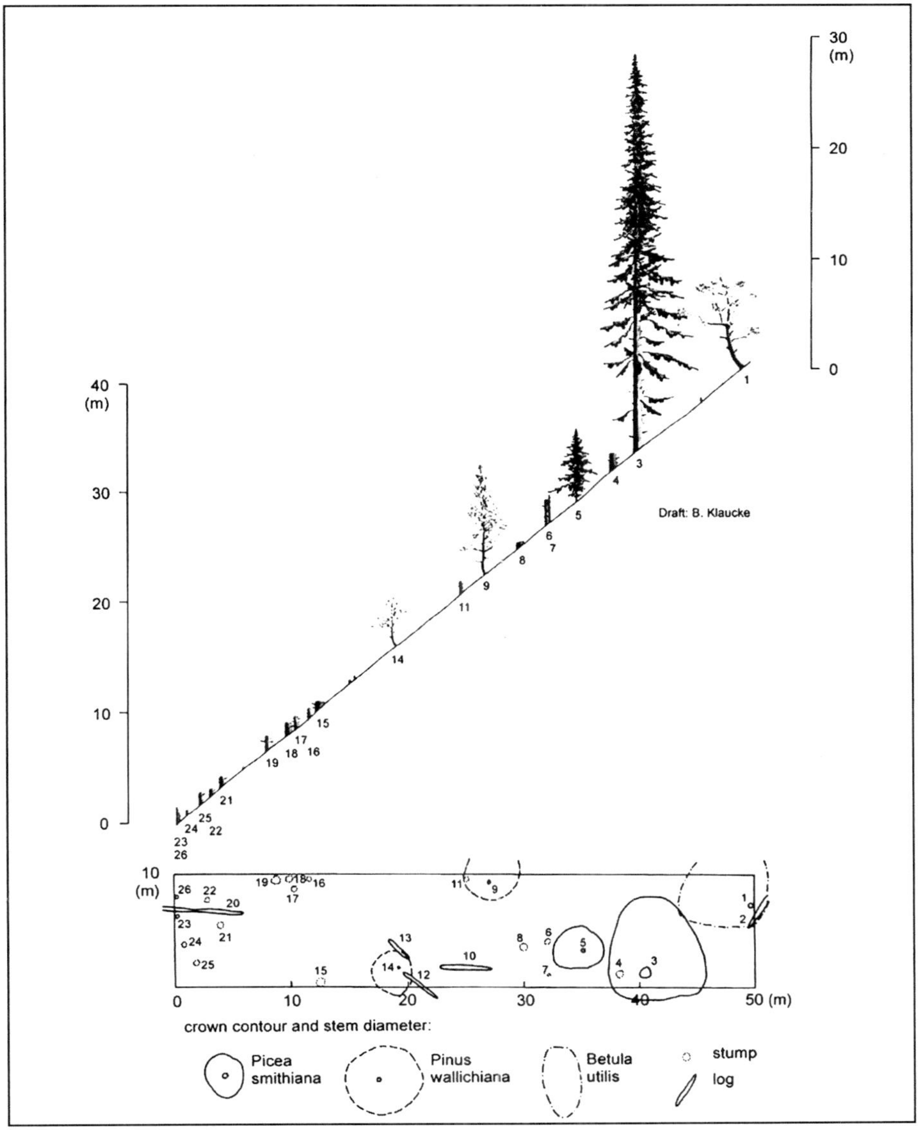

By contrast, highly degraded stands show serious alterations of nearly all structural parameters as exemplified by plot Nilt II (Figure 3). Certain threshold values of ecosystem degradation seem to be exceeded: Large gaps open up within the stands and habitat conditions within the gaps change irreversibly, at least in stands near the drought limit. Structural alterations result in a cascade of change in other ecosystem processes, such as water drainage, physical and chemical soil characteristics, soil erosion, species invasion, and carbon cycling and productivity. Due to extreme low

Figure 3

Profile Diagram and Crown Map of Plot Nilt II

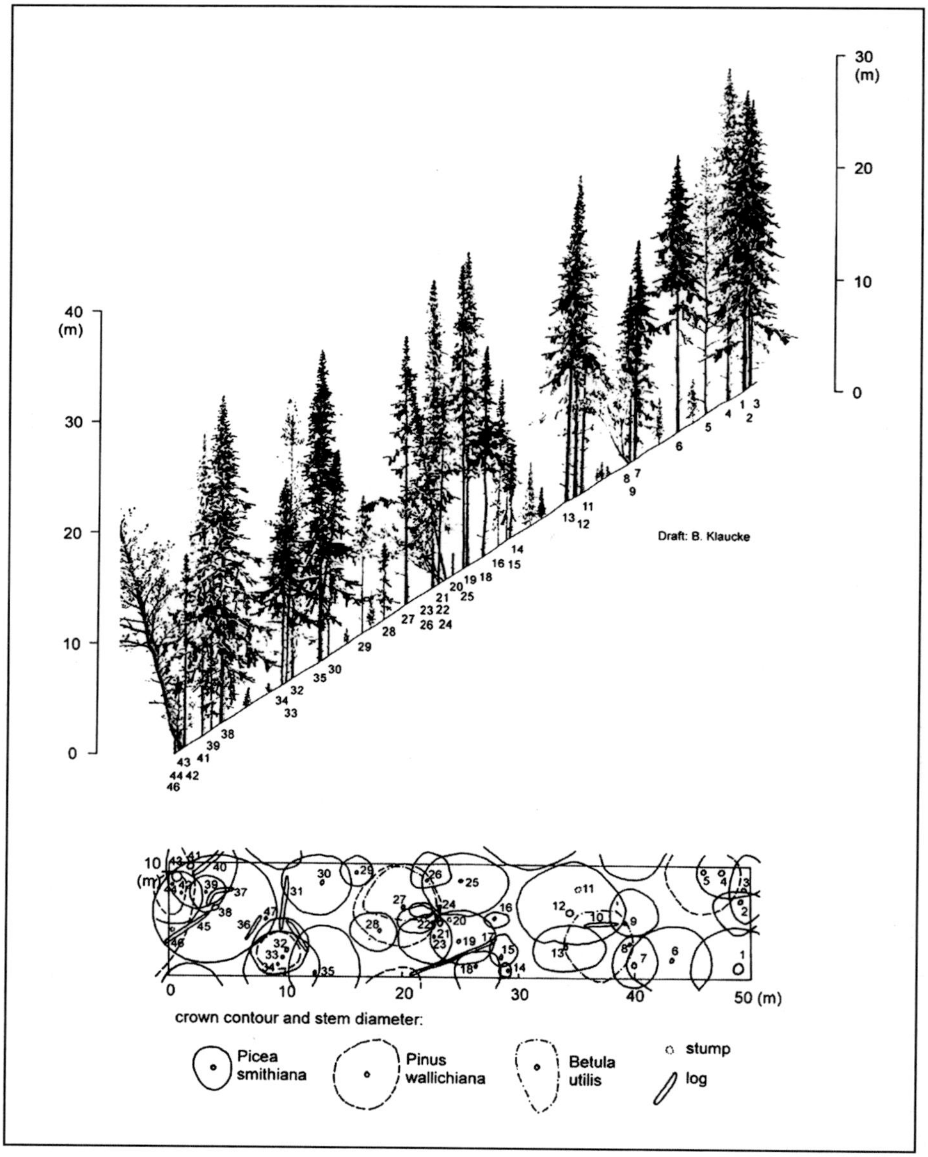

regeneration rates and extreme slow growth processes (Schickhoff, 2000d), stable forest structures would, if at all, only redevelop after very long time periods (several centuries), given that no further disturbances occur. At some sites, particularly at drier, more sun-exposed sites, the regeneration capacity already seems to be exhausted. The irradiation intensity considerably increases, extreme temperatures occur at the soil surface, and evaporation and transpiration also increase. Seedlings and saplings suffer from overheating and dry up.

How can the forest degradation be explained, how can ecological processes be related to the socio-economic system? To answer this question, a basic regional landscape ecosystem model is presented that combines major factors and processes of forest degradation in the Northern Areas of Pakistan in an integrated, spatially-functional perspective (Figure 4). It is a descriptive model of the interrelationships between the anthroposystem and the ecosystem and is based on similar models developed in the Swiss and German MAB-programmes (Messerli and Messerli, 1978; Haber et al., 1983).

Figure 4

Factors and Processes of Forest Degradation within a Regional Landscape Ecosystem Model of Northern Areas

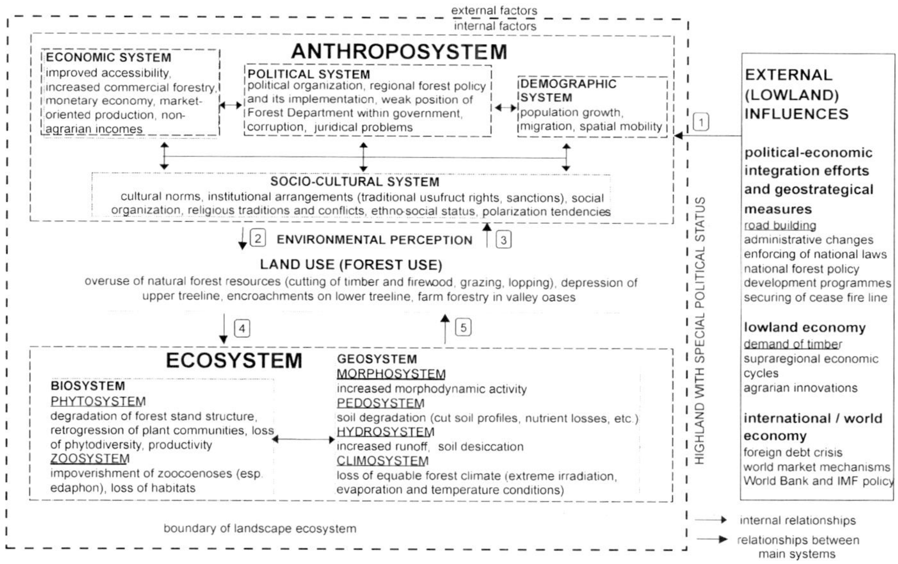

Explanation of Numbers

1. external influences as the major driving forces of socio-economic development;
2. socio-economic development with its concomitant land use changes;
3. repercussions of land use changes on population, economy and society;
4. ecological effects of land use changes;
5. repercussions of ecosystem alterations on human living conditions.

The landscape ecosystem has its regional boundaries, but is open to supraregional, i.e., national and international influences. These external influences are the major driving forces of the regional socio-economic development. They cause changes in the subsystems of the anthroposystem and these changes always affect the ecosystem and its subsystems via the interface land use or forest use respectively. Altered ecosystem conditions likewise have their effects on human living conditions, again modified by the interface land use.

The period under consideration is the period since independence of the subcontinent in 1947. Since then the Northern Areas have a special political status resulting from the partition of the subcontinent and the unsolved Kashmir-conflict (cf. Dani, 1989; Schofield, 2000). The status relating to international law is unsettled, and the Northern Areas are not included in the sovereign territory of Pakistan. Compared to the provinces, they are underprivileged in many respects. This special political status is the key for understanding the problematic development of man-environment relationships. It has marked external influences, e.g., it is the cause for the political and administrative framework of forest utilization. In contrast to the provinces of Pakistan, forest legislation of the Northern Areas virtually favours forest exploitation (cf. Government of Pakistan, 1992). Above all, the Forest Department lacks the power needed to enforce laws and to change forest management systems. Furthermore, it is rendered ineffective by having inadequate staff, infrastructure, funding, and resources to carry out its mandate as well as by being headed by a political administrator rather than a forestry expert. These weaknesses in the political and administrative system are being maintained by influential societal groups in order to continue to exploit the mountain forests (Schickhoff, 2000b).

Among the different external influences that indirectly affect the forest condition, road building must be underlined. As a major factor within the complex of political and economic integration efforts and geo-strategical measures, road building caused far-reaching changes in the anthroposystem and considerably increased the intensity of forest utilization (Schickhoff, 1997). Regarding external influences, the impacts of the lowland economy have to be stressed as well. The integration into supraregional economic cycles and the concomitant change of the traditional village and household economy affects the intensity of forest utilization in a multitude of ways, e.g., as a consequence of the excessive profits in timber business and the high timber demand of the Pakistani lowland, which is highly deficient in wood resources (cf. Jan, 1993), the volume of wood cut is many times higher than in the working schemes prescribed.

As mentioned above, the external influences indirectly affect land/forest use. First, they pass the interposed anthroposystem (cf. Figure 5). Because all subsystems are interconnected, cascades of effects are caused within and between the subsystems at different temporal scales. Changes in one subsystem sooner or later cause transformations other subsystems, e.g., an exogenous improvement of living conditions might cause population growth (change in the demographic system), that would be followed by an increasing need for non-agrarian income (change in the economic system). To meet this demand, traditional usufruct rights of forest use might be modified (change in the socio-cultural system), what in turn would lead to adaptations in the regional forest policy (change in the political system). The development in the anthroposystem immediately causes changes in form and intensity of forest use. The analysis of local constellations of explaining socio-economic variables showed that, regarding utilization intensity, the socio-cultural subsystem is of decisive importance (cf. Schickhoff, 1998; see also Messerli, 1986 for the Swiss Alps). Therefore it is depicted here in a prominent position (cf. Figure 4).

Forms and intensities of land use/forest use are then modified, when overuse of natural resources causes massive ecosystem alterations, and when adverse repercussions on human living conditions are felt. Due to the pronounced interconnectedness all subsystems of the ecosystem are affected by alteration processes. e.g., misuse of forest resources and the resulting loss of phytomass and degradation of stand structures immediately lead to changes in the biotic system. Many processes within both the phytosystem and zoosystem are affected, again affecting abiotic subsystems as well. These processes include e.g., retrogressions of plant communities, loss of phyto and zoo-diversity and productivity, cutting of soil profiles due to increased morphodynamic activity, loss of nutrients, changes in the soil water budget or in irradiation, temperature and moisture conditions. Under present climatic conditions in a subtropical arid high mountain region, these processes may lead to a complete loss of potential forest habitats.

Perspectives of Sustainable Forest Management

Advanced forest degradation and the socio-economic environment of forest utilization raises serious questions about the prospects of economically efficient, socially balanced and environmentally sound forest management. Sustainable forest management has to be seen in the wider framework of an overall sustainable development. According to general consensus, ecological, economic and social dimensions of sustainability have to be taken into consideration simultaneously. Since all dimensions are interconnected and mutually influencing each other, intergenerational sustainability can only be reached, if it is given in each of the three dimensions (cf. Bätzing, 1994; Wilbanks, 1994).

The core problem of the politico-social dimension of sustainable resource use is the aforementioned special political status of Pakistan's northern areas. As long as the administrative framework of resource use and the environmental legislation is not put on the same level as in the provinces of Pakistan, i.e., as long as the forest department's structure and power is not strengthened to enable it to carry out its functions properly, sustainable forest management cannot be achieved. Required institutional changes include upgrading of existing managerial posts, creating new posts and creating a Working Plans Division. The forest department's head should be at par with other provincial and territorial departmental heads and have the status and seniority to assume total government responsibility for the forestry sector in Northern Areas. Strong political support, commitment and urgent attention is especially needed to reform or replace the system of selling standing timber to contractors, which is causing valuable timber forests to be dangerously overexploited, and to bring coniferous forests under controlled management. If these forests are to be saved for future generations, existing management systems need to be completely revamped (Ahmed and Mahmood, 1998). The lack of participation of local identity groups in resource management and the lack of integration of autochthonous institutional management strategies prominently feature among other major obstacles within the politico-social dimension of sustainability.

Apart from high timber demand of the lowland and the integration into supraregional economic cycles, the general monetarization of living conditions that followed road building is a major obstacle within the economic dimension of sustainable resource use. At the regional level, redressment of disparate economic and social structures, and struggle against poverty are important issues. Only if living conditions of all social classes are considerably improved, what requires a reduction of high population growth, the increasing tendency to sell out available forest resources can be stopped. Many problems are associated with the aforementioned system of selling standing timber to contractors (lessees), who are exploiting a valuable national asset for their own personal profit (Schickhoff, 2000b). This is at society's expense and should not be allowed to continue. Total responsibility for forest harvesting and revenues might be given to a Forest Development Corporation under the NWFP (North West Frontier Province) or AJK (Azad Jammu and Kashmir) model. There, these corporations eliminated, at least partially, corrupt and wasteful practices of the previous contractor system (cf. Siddiqui, 1987 and Jan, 1993).

The precondition of sustainablity in the ecological dimension, i.e., to fully maintain functions, services and utilization potentials of ecosystems for future generations, is no longer given in at least 1/3 of all investigated stands. Loss of structural complexity, loss of productivity and reduction of biodiversity indicate that certain threshold values of ecosystem degradation have been exceeded. Those stands cannot return into an initial state without external inputs. Due to low productivity and slow stand development cycles, a long-term appropriate forest utilization is only possible at a very low intensity level. At present, this level is exceeded by far. It is imperative to bring harvesting levels into line with forest capabilities for growth and regeneration.

Conclusion

Advanced degradation of high altitude forests in Pakistan's Northern Areas appears to be a consequence of cumulative effects of disadvantageous factors pertaining to all dimensions of sustainability. At present, sustainability is neither given in the ecological, nor in the economic, nor in the politico-social dimension. Since short-term changes in economic, social and political set-up cannot be expected, further degradation and destruction of forests can hardly be prevented. The future development of high altitude forests in the Northern Areas gives rise to serious concern.

Acknowledgements

I am greatly indebted to Prof. Dr. M. Winiger (University of Bonn) for his encouragement and support while undertaking this research within the framework of the joint Pak-German 'Culture Area Karakorum' (CAK)-Project. Mr. Inamullah Khan (Hamburg) provided invaluable field assistance. The work was funded by the Deutsche Forschungsgemeinschaft (German Research Council). I am also grateful to Pakistani authorities and scientists for their support.

References

Agrawal, A., (2001), "Common Property, Forest Management and the Indian Himalaya", *Contributions to Indian Sociology* 35: 181–212.

Ahmed, J. and F. Mahmood, (1998), "Changing Perspectives on Forest Policy", *Policy that Works for Forests and People*, Series No. 1, IUCN Pakistan and IIED London, Islamabad-London.

Bätzing, W., (1994), "Nachhaltige Naturnutzung im Alpenraum. Erfahrungen aus dem Agrarzeitalter als Grundlage einer nachhaltigen Alpen-Entwicklung in der Dienstleistungsgesellschaft", In Gefährdung und Schutz der Alpen, *Veröffentlichungen der Kommission für Humanökologie* 5, pp. 15–51. Österreichische Akademie der Wissenschaften, Wien.

Blaikie, P. and H. Brookfield, (1987), *Land Degradation and Society*, Methuen & Co., London.

Bryant, R.L. and S. Bailey, (1997), *Third World Political Ecology*, Routledge, London.

Clemens, J. and M. Nüsser, (1997), "Resource Management in Rupal Valley, Northern Pakistan: The Utilization of Forests and Pastures in the Nanga Parbat Area", Stellrecht, I. and M. Winiger (eds.), *Perspectives on History and Change in the Karakorum, Hindukush and Himalaya*. Culture Area Karakorum Scientific Studies 3, pp. 235–63. Köppe, Köln.

Cramer, T., (1997), "Climatic Gradients in the Karakorum and their Effects on the Natural Vegetation", Stellrecht, I. and M. Winiger (eds.), *Perspectives on History and Change in the Karakorum, Hindukush and Himalaya*, Culture Area Karakorum Scientific Studies 3, pp. 265–276. Köppe, Köln.

Dani, A.H., (1989), *History of Northern Areas of Pakistan*, National Institute of Historical and Cultural Research, Islamabad.

Dierschke, H., (1994), *Pflanzensoziologie*, Ulmer, Stuttgart.

Dittmann, A., (ed.), (2000), "Mountain Societies in Transition", *Contributions to the Cultural Geography of the Karakorum*, Culture Area Karakorum Scientific Studies 6, Köppe, Köln.

Dittrich, C., (1998), "High-Mountain Food Systems in Transition—Food Security, Social Vulnerability and Development in Northern Pakistan". Stellrecht, I. and H.G. Bohle (eds.), *Transformation of Social and Economic Relationships in Northern Pakistan*, Culture Area Karakorum Scientific Studies 5, pp. 231–54. Köppe, Köln.

Ehlers, E. and H. Kreutzmann (eds.), (2000), *High Mountain Pastoralism in Northern Pakistan*, Steiner, Stuttgart.

Ehlers, E., (1995), Die Organisation von Raum und Zeit—Bevölkerungswachstum, Ressourcenmanagement und angepaßte Landnutzung im Bagrot/Karakorum, *Petermanns Geogr. Mitt.* 139: 105–120.

FAO (Food and Agriculture Organization of the United Nations), (1999), State of the World's Forests, 1999, FAO, Rome.

Forsyth, T., (1998), "Mountain Myths Revisited: Integrating Natural and Social Environmental Science", *Mountain Research and Development,* 18: 107–116.

Geist, H., (1992), "Die orthodoxe und politisch-ökologische Sichtweise von Umweltdegradierung", *Die Erde*, 123: 283–295.

Government of Pakistan, (1992), *Forestry Sector Master Plan*, Vol. 4, Northern Areas, Prepared by Reid, Collins and Ass./Silviconsult Ltd. under ADB-TA 1170/UNDP PAK/88/018. Ministry of Food, Agriculture & Cooperatives, Islamabad.

Haber, W., Schaller, J., Kerner, H., Spandau, L., Sittard, M., Grossmann, W.D., Richter, U. and A. Kunz, (1983), Ökosystemforschung Berchtesgaden. Durchführung des MAB 6-Projektes "Der Einfluß des Menschen auf Hochgebirgsökosysteme im Alpen-und Nationalpark Berchtesgaden", *Ziele, Fragestellungen und Methoden*, MAB-Mitteilungen 16, Deutsches Nationalkomitee MAB, Bonn.

Hamilton, L.S., Gilmour, D.A. and D.S. Cassells, (1997), "Montane Forests and Forestry". Messerli, B. and J.D. Ives (eds.), *Mountains of the World: A Global Priority*, pp. 281–311. Parthenon, New York-London.

Ives, J.D. and B. Messerli, (1989), *The Himalayan Dilemma: Reconciling Development and Conservation*, Routledge, London, New York.

Ives, J.D., (2004), *Himalayan Perceptions, Environmental Change and the Well-Being of Mountain Peoples*, Routledge, London-New York.

Jan, A., (1993), *Forest Policy, Administration and Management in Pakistan*, Winrock International Institute for Agricultural Development, Islamabad.

Kreutzmann, H., (1989), "Hunza, Ländliche Entwicklung im Karakorum, *Abhandlungen Anthropogeographie 44*, FU Berlin.

Kreutzmann, H., (1995), "Globalization, Spatial Integration, and Sustainable Development in Northern Pakistan", *Mountain Research and Development,* 15: 213–227.

Kreutzmann, H., (1998), From Water Towers of Mankind to Livelihood Strategies of Mountain Dwellers: Approaches and Perspectives for High Mountain Research, Erdkunde 52: 185–200.

Lamprecht H. (1980), Zur Methodik waldkundlicher Untersuchungen in Naturwaldreservaten, *Natur und Landschaft*, 55: 146–147.

Leibundgut, H., (1959), Über Zweck und Methodik der Struktur-und Zuwachsanalyse in Urwäldern. Schweiz. Z. Forstwes. 110: 111–125.

Leibundgut, H., (1993), Europäische Urwälder. Wegweiser zur naturnahen Waldwirtschaft. P. Haupt, Bern-Stuttgart-Wien.

Messerli, B. and P. Messerli, (1978), "Wirtschaftliche Entwicklung und ökologische Belastbarkeit im Berggebiet (MAB Schweiz)", *Geographica Helvetica* 33: 203–210.

Messerli, B., Hofer, T. and S. Wymann, (eds.), (1993), "Himalayan Environment, Pressure-Problems-Processes", *Geographica Bernensia*, G 38. University of Berne.

Messerli, P., (1986), "Modelle und Methoden zur Analyse der Mensch-Welt-Beziehungen im alpinen Lebens-und Erholungsraum". *Schlußbericht zum Schweizerischen MAB-Programm* 25, Bundesamt f. Umweltschutz, Bern.

Miehe, S., Cramer, T., Jacobsen, J.P. and M. Winiger, (1996), Humidity Conditions in the Western Karakorum as Indicated by Climatic Data and Corresponding Distribution Patterns of the Montane and Alpine Vegetation, Erdkunde, 50: 190–204.

Mountain Agenda, (2000), "Mountains of the World". *Mountain Forests and Sustainable Development*, CDE, Berne.

Mueller-Dombois, D. and H. Ellenberg, (1974), *Aims and Methods of Vegetation Ecology*, Wiley & Sons, New York.

Price, M. and N. Butt, (eds.), (2000), *Forests in Sustainable Mountain Development: A State of Knowledge Report for 2000*, Oxon, New York (= IUFRO research series 5).

Rao, K.S., (1997), Natural Resource Management in Himalaya: A Recourse to Issues and Strategies, ENVIS Monograph, G.B. Pant Institute of Himalayan Environment and Development, Kosi Katarmal, Almora, India.

Schickhoff, U., (1994), Die Verbreitung der Vegetation im Kaghan-Tal (Westhimalaya, Pakistan) und ihre kartographische Darstellung im Maßstab 1: 1,50,000. Erdkunde 48: 92–110.

Schickhoff, U., (1995), Verbreitung, Nutzung und Zerstörung der Höhenwälder im Karakorum und angrenzenden Hochgebirgsräumen Nordpakistans. Petermanns Geogr. Mitt. 139: 67–85.

——, (1996), Die Wälder der Nanga-Parbat-Region. Standortsbedingungen, Nutzung, Degradation. In: Kick, W. (ed.), Forschung am Nanga Parbat. Geschichte und Ergebnisse. Beitr. u. Mat. z. Regionalen Geographie 8, pp. 177–189. TU Berlin.

——, (1997), "Ecological Change as a consequence of Recent Road Building: The Case of the High Altitude Forests of the Karakorum", Stellrecht, I. and M. Winiger (eds.), *Perspectives on History and Change in the Karakorum, Hindukush and Himalaya*, Culture Area Karakorum Scientific Studies 3, pp. 277–286. Köppe, Köln.

——, (1998), "Socio-economic Background and Ecological Effects of Forest Destruction in Northern Pakistan", Stellrecht, I. (ed.), *Karakorum-Hindukush- Himalaya: Dynamics of Change*. Culture Area Karakorum Scientific Studies 4/I, pp. 287–302. Köppe, Köln.

——, (2000a), Umweltdegradierung und Biodiversitätsverluste in Hochgebirgsräumen Südasiens. In: Rintelner Symposium VI (Rinteln, 11–14.5. 2000). Berichte der Reinhold-Tüxen-Gesellschaft 12: 153–172. Hannover.

——, (2000b), Politische Ökologie der Gebirgswalddegradierung Nordpakistans. In: Blotevogel, H.H., Ossenbrügge, J. and G. Wood (Hrsg.), Lokal verankert—weltweit vernetzt. Tagungsbericht und wissenschaftliche Abhandlungen/52. Deutscher Geographentag Hamburg, pp. 399–407. Steiner, Stuttgart.

——, (2000c), "The Impact of Asian Summer Monsoon on Forest Distribution Patterns, Ecology and Regeneration North of the Main Himalayan Range (E-Hindukush, Karakorum)", *Phytocoenologia 30*: 633–654.

——, (2000d), "Persistence and Dynamics of Long-Lived Forest Stands in the Karakorum under the Influence of Climate and Man", In: Miehe, G. and Zhang, Y. (eds.), *Environmental Changes in High Asia*, Proceedings of an International Symposium at the University of Marburg, Marburger Geographische Schriften 135: 250–264.

——, (2001), "Juniper Forests in Pakistan: Human Impact on Structure and Dynamics", In: Forest and Walnut Research Institute of National Academy of Sciences/Kyrgyz-Swiss Forestry Support Programme (eds.), Proceedings International Symposium on Problems of Juniper Forests, 6–11 August 2000, Osh, Kyrgyzstan, pp. 215–23. Bishkek.

——, (2002), "Die Degradierung der Gebirgswälder Nordpakistans. Faktoren, Prozesse und Wirkungszusammenhänge in einem regionalen Mensch-Umwelt-System", *Erdwissenschaftliche Forschung* 41. Steiner Verlag, Stuttgart.

Schmidt-Vogt, D., (1990), "High Altitude Forests in the Jugal Himal (Eastern Central Nepal), Forest Types and Human Impact", *Geoecological Research* 6, Steiner, Stuttgart.

Schofield, V., (2000), "Kashmir in Conflict: India, Pakistan and the Unfinished War", *Tauris*, London, New York.

Sheikh, M.I. and A. Aleem, (1975), "Forests and Forestry in Northern Areas", *Pakistan Journal of Forestry,* 25: 197–235 a. 296–324.

Siddiqui, K.M., (1987), "Recent Developments in Timber Harvesting in Pakistan", Sheikh, M.I., Siddiqui, K.M., Strehlke, B. and G. Stöhr (eds.), *Role of Forest Research in Solving Socio-Economic Problems in the Himalayan Region*, pp. 138–145. Pakistan Forest Institute, Peshawar.

Stellrecht, I. and H.G. Bohle, (eds.), (1998), "Transformation of Social and Economic Relationships in Northern Pakistan", *Culture Area Karakorum Scientific Studies 5.* Köppe, Köln.

Stellrecht, I. (ed.), (1998), "Karakorum-Hindukush-Himalaya: Dynamics of Change", *Culture Area Karakorum Scientific Studies 4.* 2 Vols., Köppe, Köln.

Sundriyal, R.C. and E. Sharma, (1996), "Anthropogenic Pressure on Tree Structure and Biomass in the Temperate Forest of Mamlay Watershed in Sikkim", *Forest Ecology and Management,* 81: 113–134.

Troll, C., (1967), Die klimatische und vegetationsgeographische Gliederung des Himalaya-Systems. In: Hellmich, W. (ed.), Khumbu Himal. Ergebnisse des Forschungsunternehmens Nepal Himalaya. Bd. 1, pp. 353–388. Springer, Berlin-Heidelberg, New York.

Wakeel, A., Rao, K.S., Maikhuri, R.K. and K.G. Saxena, (2005), "Forest Management and land use/cover changes in a typical micro watershed in the Mid Elevation Zone of Central Himalaya, India", *Forest Ecology and Management,* 213: 229–242.

Weiers, S., (1998), Wechselwirkungen zwischen sommerlicher Monsunaktivität und außertropischer Westzirkulation in den Hochgebirgsregionen Nordpakistans. Petermanns Geogr. Mitt. 142: 85–104.

Wilbanks, T., (1994), "Sustainable Development in Geographic Perspective", *Annals of the Association of American Geographers,* 84: 541–556.

Winiger, M., (1996), Karakorum im Wandel. Ein methodischer Beitrag zur Erfassung der Landschaftsdynamik in Hochgebirgen. Jahrb. Geogr. Ges. Bern 59: 59–74.

Zurick, D. and P.P. Karan, (1999), *Himalaya: Life on the Edge of the World*, The Johns Hopkins University Press, Baltimore, London.

8 Impact of Floods on Biodiversity in the Brahmaputra Valley

R.B. Singh and M. Bortamuly

Biodiversity maintains the ecological balance and evolutionary process of the environmental resources and has its spiritual, cultural, aesthetic and recreational values. The well-being and survival of human population are dependent on millions of species of plants, animals and micro webs. Particularly, biological diversity is part of human's livelihood and constitutes resources upon which families, communities and future generations depend. In general, terms biodiversity indicates variation and abundance of species and their habitat. It also describes the total availability of life on earth. Biodiversity can be divided into three hierarchical categories—genes, species and ecosystem—that describe quite different aspect of living system. But in this paper only one category that is species diversity is being dealt with, where species diversity refers to the variety of species within a particular region. According to International Union for Conservation of Nature and Natural Resources (IUCN) the global species diversity varies from 2 million to 100 million and 1.4 million have actually been named (Bradbury, 1991). Biological diversity is being eroded as fast as at any time since the dinosaurs died out some 65 million years ago. Some of the important factors, which have a great impact on biodiversity, are certain natural disasters like flood, earthquake, landslide, storm and so on. Being located on the banks of the great rivers most of the forest areas are prone to flood. Particularly, North-East India is one of the worst flood affected areas and yet does not have any proper disaster management policy (Singh, 2004). Flood causes 30 per cent of the total economic loss. Being located on the Brahmaputra floodplain some part of Assam state is severely affected by flood. Domestic cattle, national parks, livestock and valuable human lives are lost. At the same time, India is one of the twelve mega diversity countries of the world. Owing to its varied physical feature, it has about 45,000 species of plants and 75,500 species of animals. Particularly, North-East India is the most important region representing richest zone for communities and species particularly endemic to the region. Wildlife of North-East India enjoys a reputation not only for its variety or numbers but also for the fact that within this region quite a few species of very rare animals and birds are found, which have become extinct from other parts of the world (Assam Remote Sensing Application Centre, 1998).

Much of the North-East's diversity is found in the forest. The average forest cover of North-East India is 69.59 per cent. Highest forest cover is found in the states of Mizoram and Nagaland, which are 86.99 and 85.43 per cent respectively. Assam shows the lowest percentage, which is only 30.20 per cent (Table 1). Again, rapid population growth of the north-eastern states of India poses a tremendous threat to the existing forest cover and the biodiversity of the area. For example, percentage decadal growth of population of the states Assam, Nagaland and Arunachal Pradesh in the year 1991–2001 is 18.85 per cent, 64.41 per cent and 26.21 per cent respectively. It has a great impact directly or indirectly on the biodiversity and occurrence of flood. Moreover, density of population is also high in some of the north-eastern states, for example in Assam it was 286 per sq km in the year 1991 and was increased to 340 per sq km in the year 2001 (Table 2).

Table 1

Forest Cover in Different States of North-East India

States	*Percentage of forest cover*
Arunachal Pradesh	82.21
Assam	30.20
Manipur	77.86
Meghalaya	69.70
Nagaland	85.43
Mizoram	86.99
Tripura	54.79

Source: Forest Survey of India, 1999.

Table 2

Density of Population in Different States of North-East India

States	*Density of population*	
	1991	*2001*
Arunachal Pradesh	10	13
Assam	286	340
Manipur	82	107
Meghalaya	79	103
Nagaland	73	120
Mizoram	33	42
Tripura	263	304

Source: Census of India, 2001.

Whatever may be the existing situation; some protected areas of comparatively high forest cover are found. These are the reserve forest, wildlife sanctuaries and national parks etc. This area is the abode of most of the forest species of wildlife, some of which are on the verge of extinction. Moreover, being located on the southern bank of the river Brahmaputra, Kaziranga National Park is one of the worst flood prone areas. This area gets flooded almost every year, sometimes several times in a year, submerging 80 to 90 per cent of the total landmass. Therefore, this paper deals with the relationship among environment, flood and biodiversity in the north-eastern region taking Kaziranga as the case study.

Study Area and Methodology

Kaziranga National Park, which is 270 km away from Guwahati in the eastern side, is situated in both the districts of Nagaon and Golaghat of Assam. Topographically this area is located between 26°30′ N to 26°45′ N latitude and 98°08′ E to 93°36′ E longitude. The park is bounded by the mighty Brahmaputra on the north and verdant hills of Karbi Anglong on the south. National Highway 37 goes along the southern boundary of the national park. The terrain of the national park is more or less flat with gentle, almost imperceptible slope from east to west. Alluvial deposits of the river Brahmaputra due to recurrent floods form the soil of the national park. The total area of Kaziranga National Park is 430 sq km. The climate of the national park is subtropical monsoon type with high humidity. Rainfall is heavy during summer. The average annual rainfall is 1,320 mm. The maximum temperature is 35°C and the minimum temperature is 8.9°C. The hottest months are July and August. The climatic condition is also somewhat changeable during different seasons of the year. To determine the relation among the environment, flood and the biodiversity in the study area, data has been collected from both the primary and secondary sources. The studied area is visited for two periods, once in the months of July and August that is in the peak of the flood period and once in the month of January that is in the lean season. The primary survey was conducted in four villages, out of which, two villages are located in the transitional zone (that is, villages located between core and the buffer zone) and another two are located in the buffer zone (that is, 1–10 km around the boundary of the national park). A random survey of 120 households that is 30 households from each village of the study area was carried out to collect primary data.

Biodiversity

Flora

There are three broad categories of vegetation cover, which can be readily recognized in Kaziranga National Park (Table 3). All total 400 species of plants have been recorded so far from the national park. The predominant species among the aquatic vegetation is water hyacinth. Other floating and struggling grasses like different *Andropogen spp.*, *Ipomea reptama* (Kalmau), *Pistia straflotes*

(Borpuni), *Lema paucicostala* (Harupuni), (*Nymphia spp.* (water lilies), and so on make the aquatic vegetation. Savannah formation covers nearly 66 per cent of the total area of the park. The most common and widely distributed species of grass in the park is *Erinthus reveneae*, associated with *Saccharum arundinecium*, *Tamarix dioica*, *Saccharum sponteneum*, *Imperata cylendrica*, *Phragmites karka Arindo donex* etc. Woodlands are represented by variety of subtypes or different stages of succession and edaphic variations. Some of the important trees found in the park are *Ajhar*, Mango tree (*Megnifera indica*), Gold mahur (*Delenix regia*), *Hollock* (*Tarminalia myriocarpa*), *Simul* (*Bombax malabaricum*), *Dimaru* (*Ficus spp*), and so on.

Table 3

Different Vegetation Types

Categories	*Percentage to total area*
(a) Aquatic plants	5.58
(b) Eastern wet alluvial savannah or the grassland	66.44
(c) Woodland or the tree forest	27.98

Source: Directorate, Kaziranga National Park.

Fauna

Different types of mammals, birds, reptiles, amphibian and fishes represent fauna of the Kaziranga National Park. The primary mammals found in the

Table 4

Growth of Wildlife Population (Mammals) in the Kaziranga National Park

S.No.	*Species/Year*	*1966*	*1972*	*1978*	*1984*	*1991*	*1993*	*1999*
1.	Rhino	366	658	939	1,080	1,129	1,164	1,552
2.	Elephant	349	422	773	523	515	1,094	882
3.	Wild buffalo	471	555	610	677	1,090	1,034	1,192
4.	Indian bison	1	18	23	36	5	—	—
5.	Swamp deer	213	516	697	756	635	427	398
6.	Samber	120	105	215	150	55	34	58
7.	Hog deer	1,311	4,551	6,855	9,872	2,911	2,048	5,045
8.	Tiger	20	30	40	52	50	72	80
9.	Bear	2	6	4	2	—	2	-
10.	Wild boar	155	522	733	1,645	595	140	431
11.	Capped langur	21	—	19	—	—	21	139
12.	Gibbon	3	—	—	—	—	8	47
13.	Otter	26	95	232	196	—	—	—
14.	Hog badger	1	—	—	—	—	—	—
15.	Barking deer	29	76	95	33	—	—	—

Source: Directorate of Kaziranga National Park, Bokakhat.

national park are Great Indian one horned rhinoceros and so on. Among the mammals, there are 1,552 Indian one horned rhinoceros in Kaziranga, which comprises two-thirds of the total world's population. It has 50 per cent of world's buffalo population and 55 per cent of world's swamp deer population (Dutta, 1991) (Table 4).

Thirteen species of reptiles are reported from Kaziranga. Out of total 13 reptiles, 7 are snakes, 4 lizards and 2 turtle or tortoises. Again, amphibia is a poorly represented group with only 5 records of species under 3 families. Nearly 87 species of fishes have been recorded from Kaziranga. Only five are considered to be rare. The number of fish species is not the only important thing, but the quantity, which is very large, is another important aspect in the park. Moreover there are 133 species of invertebrates in the national park. About 268 species of birds have been recorded so far in Kaziranga National Park.

Flood and National Park

Flood is defined as a relatively higher flow of water in a river than the usual that causes inundation of lowland. During monsoon, these small rivers originating in the Karbi Anglong hills become unbelievably dangerous along with mighty Brahmaputra, which causes flood to the Kaziranga National Park. The flood of the Brahmaputra can generally be attributed to a host of interrelated natural and anthropogenic factors. The natural factors include heavy monsoonal rains and devastating landslides, easy erodibility of rocks of the northern mountains, steep slopes and high seismicity, while the anthropogenic factors include large scale deforestation in the hilly catchments, practice of shifting cultivation, human intervention in the river system including encroachment in the floodplains, destructions of natural wetlands and poorly managed embankment system. The floods of Brahmaputra made a quantum leap, especially after the great 1950 earthquake. The cumulative effect of these factors has aggravated the flood situation on the floodplain of the Brahmaputra as well as the Kaziranga National Park. Usually floodwater rises from the middle of the month of May and it comes in 3–4 waves till October depending on the intensity of the monsoonal rain in the valley. Floodwater reaches a height of 3 to 4 metres at some particular places of the park. Moreover, the floodwater generally remains for 15–20 days or more affecting the wildlife considerably.

Results and Discussion

Flood season is always considered to be a dreaded period for Kaziranga National Park but the role played by flood in maintaining the park ecosystem is also very important. Therefore, both positive and negative impacts of flood are studied to make a proper assessment of the theme.

Positive Impacts of Flood on Biodiversity

This national park developed naturally with its numerous biodiversity. Therefore, regulated flood and its quick recession is no doubt beneficial for the ecosystem as well as related biodiversity of the park in many ways.

Maintenance of Vegetation Status and Soil Formation

The annual flood coupled with the annual burning of the grassland are the two most important and major factors, which influence the vegetation of the park. Flood is responsible for creating different site conditions and soil formations, preferred by different types of vegetation. Flood in Kaziranga National Park is one of the most important factors for maintaining the present vegetation status and for arresting any further progress in the process of plant succession. It is observed that fire and annual flood play an important role in soil profile dynamics and development by erosion of the topsoil and by fresh deposition of silt/sand. The process is repeated every year in the park, which helps in formation of fresh vegetation.

Breeding Ground of Fish

Various streams and beels of the Kaziranga National Park serve the purpose of breeding grounds and nursery for large fish population. As soon as the floodwater enters the park, a vast area gets submerged and the fishes found in the numerous beels and nallahs come out to lay eggs in the current of the floodwater. Some fishes from these areas also go into the Brahmaputra River along with the receding flood. Therefore, during flood or just after receding of the flood, fishes are found in abundant quantity in the river Brahmaputra and its tributaries.

Cleaning of Water Hyacinth and Better Conditions for Birds

The beels and other waterlogged areas are generally carpeted with water hyacinth, which is unpalatable for herbivores of the park. The cover of water hyacinth helps water bodies becoming shallow and harms the existing wetland ecosystem. The floodwater also helps in washing away the water hyacinth from beels, streams and nallahs, which grow into a very thick, impenetrable cover. After the flood, the fresh water with very few water hyacinth gives way to the water birds like ducks, herons, pelicans and so on. After floods, the healthy stock of fish in the beels or other water bodies helps in providing food to the bird community of the park.

Replenishment of Beels and Other Water Bodies

Kaziranga National Park is a suitable habitat for Indian one horned rhinoceros due to its wetland ecosystem and presence of necessary conditions. The water of the various beels of Kaziranga National Park becomes very shallow during dry season, which affect the habitat suitability of rhino as well as other water loving animals. Annual flood replenishes the dry water bodies and streams with high amount of water and makes it again suitable for these animals. Flood also maintains the water quality through cycling.

Negative Impacts of Flood on Biodiversity

In recent years, flood has become a common feature in the park with its increasing magnitude. Flood has its both short-term as well as long-term negative impact on its biodiversity and both results into severe destruction to the whole ecosystem.

Casualties of Animals

The worst and the most pathetic situation is that some animals especially the calves and the aged ones and the deer species are sometime washed away by the current of the flood of turbulent river Brahmaputra. In the year 1998 the flood of river Brahmaputra had inundated more than 95 per cent of the park area causing tremendous devastation by its waves. The animal mortality during the flood went up to 629, where numbers of rhinos being 39 and the worst affected hog deer being 506 (Tables 5, 6, 7).

Table 5

Death of Rhino during Flood from 1980–1998

Year	*Adult*	*Sub-adult*	*Calf*	*Total*
1980	6	1	6	13
1981	2	—	1	3
1982	1	—	—	1
1983	—	—	—	—
1984	—	1	—	1
1985	—	—	1	1
1986	1	—	—	1
1987	1	—	2	3
1988	9	10	19	38
1989	1	—	—	1
1990	—	—	1	1
1991	3	—	3	6
1992	—	—	—	-
1993	—	—	3	3
1994	—	—	2	2
1995	3	—	2	5
1996	1	—	1	2
1997	—	—	1	1
1998	18	5	16	39
Total	46	17	58	121

Source: Status report, Kaziranga National Park, 1998.

Table 6

Animal Mortality During 1998 Flood in Kaziranga National Park

	Animals	Death due to drowning	Death due to knock down	Total	Animal rescued
1.	Rhinoceros	39	—	39	1(Calf)
2.	Elephant	5	1	6	1
3.	Buffalo	20	—	20	-
4.	Hog deer	478	28	506	41
5.	Swamp deer	8	—	8	2
6.	Samber	10	—	10	—
7.	Wild boar	14	3	17	2
8.	Hog badger	5	2	7	—
9.	Bear	1	—	1	—
10.	Porcupine	9	—	9	—
11.	Fishing cat	—	1	1	—
12.	Cobra	—	2	2	—
13.	Python	—	1	1	—
14.	Civet	—	2	2	—
	Total	589	40	629	47

Source: Status Report on Kaziranga National Park, 1998.

Table 7

Flood Related Casualties in Kaziranga National Park in the Year 2002

	Animals	Drowning	Knocked down	Killed by people	Rescued and died	Total
1.	Hog deer	10	16	2	—	30
2.	Rhino	6	—	—	—	6
3.	Wild boar	5	1	—	—	6
4.	Buffalo	3	—	—	—	3
5.	Samber	1	1	—	—	2
6.	Swamp deer	1	—	—	2	1
7.	Hog deer	1	1	—	—	2
8.	Python	1	—	—	—	1
9.	Civet	—	4	—	—	4
10.	Capped langur	—	6	—	—	6
	Total	28	29	2	2	61

Source: Directorate, Kaziranga National Park.

There is a correlation between the magnitude of flood and the quantity of biodiversity of the area. Flood causes animal as well as floral mortality in the national park. Moreover, the stated relation between flood and quantity of biodiversity can be emphasized with the comparison of the percentage increase/decrease in some selected animal population (Figure 1) and flood level taken in different areas of the park in different years (Figure 2). It also correlates with the percentage area submerged by flood in different years.

Figure 1

Increase/decrease in Selected Wildlife Population Taking 1966 as the Base Year

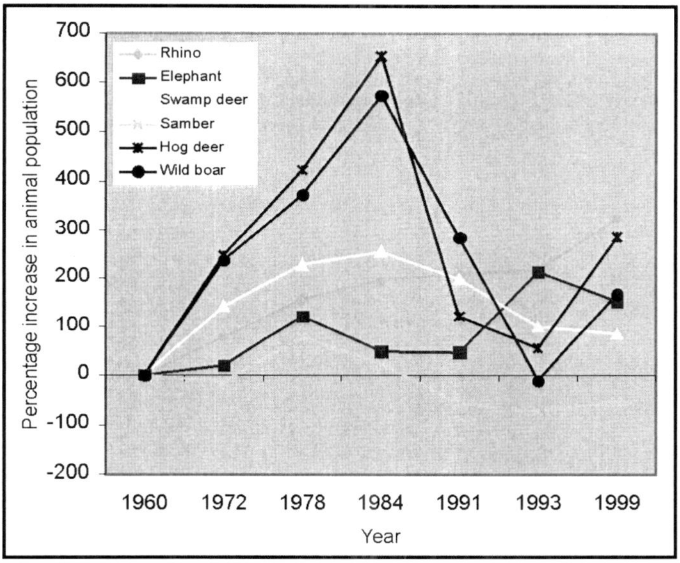

Figure 2

Flood Level at Number 2 Mihi Tower and Holalpath Camp

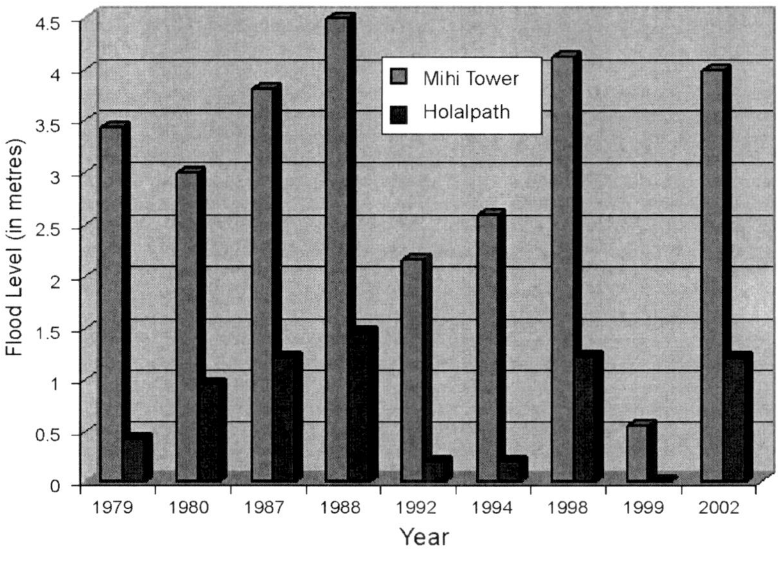

The flood level in different years of Kaziranga National Park shows that, flood of 1987, 1988, 1998 and 2002 were the most dangerous among the stated years, where there is more than three metres of height in number 2 Mihi Tower. The animals, most affected by flood are rhinos, swamp deer, samber, hog deer, wild boar etc. Percentage increase of these animals taking 1966 as the base year shows a stable growth up to 1984 except elephant and samber. After that there is a steep fall in percentage increase of almost all the animals. The basic reason of steep decreasing slope in the percentage increase lines of the animal is that the devastating flood of 1988 destabilized the whole ecosystem of the park. They could not keep pace with earlier increasing status. The main cause of this decrease is considered to be flood. Population of samber also shows a negative trend due to the 1988 flood. After 1988, there was another great flood in the park in the year 1998, which also correlates with the decreasing trend of the animal population like rhinos, samber, swamp deer and so on. Once again, the devastating flood of 1998 has disturbed the stable growth of animal population with its direct or other indirect short-term and long-term impacts. Moreover, every year, flood deposits a thick layer of silt on the ground of the park and the some smaller species of plants along with its seeds are buried under metres of thick silt deposition and thus these species have disappeared from the park.

Migration of Animals and Related Problems

As soon as the park is submerged by the incoming floodwater the animals of low lying areas of the park start migrating to the Karbi Anglong hills. As the surrounding areas of the park are populated, protection of animals during the period of migration from and back to the park becomes an uphill task. During their migration from the park to Karbi Anglong hills, they have to cross the National Highway 37 where heavy vehicles ply round the clock. During floods a considerable number of wildlife are lost due to knock down by fast moving vehicles.

Shortage of Fodder and Malnutrition

The grazing animals in the Kaziranga National Park severely suffer from shortage of fodder during flood situation. During floods, all the grasslands are submerged under water. Even, sometimes undergrowth of the forested areas also gets submerged under water during high floods. As a result, grazing animals have to suffer from shortage of fodder and malnutrition.

Erosion Due to Flood and Impact on Habitat

Soil erosion is a serious problem in Kaziranga National Park during summer floods. Erosion is primarily attributed to the instability of the river in this region. This can be understood through the changing bank line from 1967–68 to1998 as seen on base map and IRS–1A and IRS–1D data. Considering the retreating level, the park has lost 28.87 sq km of area due to erosion during

1967 to 1998. Again, a number of floral species have disappeared in the river water with the eroded land (SAC, 1999).

Siltation and Impact on Habitat Suitability

Beels and other perennial water bodies inside the national park are extremely natural without human interference. But with increasing flood of the river Brahmaputra siltation occurs, which results into fragmentation of beels. This again results in the overall reduction of the area under water bodies. In 1967–68, wetland occupied 36.6 sq km (8.5% of the total area) of area, which decreased to 27.4 sq km (6.7% of the total area) by 1997, thus showing an overall decrease in the area under wetlands. Thus, soil erosion and sedimentation, which directly or indirectly influence the water regime changes, result in rapid alteration of community structure (IIRS, 1997).

Damage to Infrastructure and Disruption in Communication

The floods generally cause considerable damage to the anti-poaching infrastructure, namely, roads, bridges, patrolling path, guard camps of the park. During the flood season, communication between the various camps and with the range headquarters becomes very difficult.

Suggestions

The park environment provides an ideal habitat for majority of the herbivores, which constitute 98 per cent, as well as the carnivores, which constitute 2 per cent of the biodiversity. The present study has highlighted the environment and other problems related to the flood disaster of the area. Some suggestions are worth mentioning to achieve sustainability in the park ecosystem (Directorate, Kaziranga National Park, 2002).

1. Effort should be made to increase the area under beels and short grasses wherever possible. Therefore, artificial but ecologically viable depressions should be created. Moreover, desiltation programmes should be carried out in a more scientific way to make the existing water bodies more suitable for the park animals.
2. Though tree species should be increased in some particular areas but succession upon the grassland should be prevented. Tall grasses should be burnt for this purpose. But the way of burning should be proper and scientific such as there is no loss of any endangered species.
3. The provision of providing shelter to wild animals during flood season has to be made inside the park itself. For these, elevated platforms of the size of about one hectare or more have to be created, planted and maintained.
4. Addition of proposed areas to the present park territory will not only provide additional land but also corridors needed for animal movement and migration. It will also legally prevent the encroachment by the people on the park boundary.

5. Attempts should be made to minimize the impact of river current along the northern boundary of the park by adopting proper ecologically viable engineering measures. This will prevent the habitat loss and siltation during floods to a large extent.
6. The park should be monitored annually on a regular basis to detect the changes in the landmass and water bodies due to erosion, accretion and siltation. Remote sensing could serve as an effective tool to accomplish this job timely and cost effectively (Singh and Kumar, 2004).
7. People's participation and government initiative are two necessities for fulfilling every step for the well-being of the national park. Moreover, awareness programmes and environmental education should be carried out so that the villagers feel themselves as the proper custodian of the park.

References

Assam Remote Sensing Application Centre (ARSAC), (1998), "Satellite Survey of Kaziranga's Vegetation Cover", A Report by ARSAC, Guwahati.

Bradbury, I.K., (1991), *The Biosphere*, John Wiley and Sons, London.

Directorate, Kaziranga National Park, (2002), Status Report, Kaziranga National Park.

—, (2002), "Project Proposal of the Action Plan to Combat Flood Damage in Kaziranga National Park", Directorate, Kaziranga National Park, Assam.

Dutta, A.K., (1991), *Unicornis*, Konark Publisher, New Delhi.

Indian Institute of Remote Sensing, Dehradun, (1997), "Landmass Dynamic and Rhino Habitat Suitability Analysis in Kaziranga National Park", Assam, Report by Directorate, Kaziranga National Park, Bokakhat.

Singh, R.B., (2004), "Hydroinformatics for Integrated Watershed Management for Mitigating Floods and Droughts", In *Flood and Drought Management*, (eds.) G.N. Mathur and D.K. Jain, Central Board of Irrigation and Power, New Delhi, pp. 19–26.

Singh, R.B. and Kumar, Dilip, (2004), "Monitoring, Mapping and Mitigation of Flood Disaster in India Using Remote Sensing and GIS: A Case Study", In *Monitoring, Prediction and Mitigation of Disasters by Satellite Remote Sensing*, (eds.) K. Takara, Y. Tachikawa and T. Kojima, DPRI, Kyoto University, pp. 53–59.

Space Application Centre, Ahmedabad and Brahmaputra Board, Guwahati, (1999), *Erosion at Kaziranga National Park, Assam*, Print Vision, Ahmedabad.

9 Nature of Plant Diversity on Tropical-Monsoon Highland Slopes

A Micro-Environmental Study in Suryamaninagar of Tripura, North-East India

Sudip Dey, Jimmi Debbarma, Rina Debbarma, Susmita Choudhuri and Bulti Roy

Functional effect of environmental factors (physical, chemical, biological and anthropological) in a small area can be defined as 'micro-environment'. Highlands have a specific micro-environmental significance as they have some specific micro-level environmental characters due to elevation, slope, soil condition, groundwater conditions etc., which influence the local ecology. Tripura, a peripheral state of North-East India is marked by its geomorphic diversity especially highlands and lowlands. Folded hills, dissected highlands and lowlands of this area are strong evidences of geo-tectonic and geomorphic process dynamics through the geological history of Tertiary and Quaternary periods (Dutta, 1974; Nandi, 1974). The dynamic nature of the climatic condition through the geological period influences the development of landform of Tripura along with tectonic processes (Nandi, 1985). According to Alam et al. (2003) this is a part of eastern geo-tectonic province of the Bengal Basin. The highlands of Tripura, which developed by sediment deposition and thereafter uplifted by tectonic process during the Tertiary-Quaternary period (Nandi, 1985; Dey, 2005), strongly support the above view. The highlands consist of 'Dupitilla' type of sediments, which are estuarine brown clayey sand, mottled silty clay, white to grey sandrock with silt bands and white clay of late Pliocene epoch. The present soil conditions over highlands are sandy and sandy loam (Geological Survey of India, 2004). The top of the highlands are covered by grasses.

It has been observed the highlands (locally known as *tillas*) slopes are covered by dense forest in Tripura. Sal, Bamboo, Gamar, different types of grasses etc., grow on these slopes which are also economically very important species. From the ancient period shifting cultivation has been done on the slope of the highlands and hills by the tribal people of this state. In most of the places of the adjacent low lands, which are locally known as *lungas*, are used for food crop production and the highlands tops are used as residential purpose in the rural Tripura. This is a very common traditional practice in this state. Recently the tilla tops and slopes are also used for plantation of rubber and tea. Considering the physical and ecological importance we decided to perform a micro-environmental study on the plant diversity of highland slopes of Suryamaninagar area of Tripura to assess the nature of plant diversity of the highlands. This study area is located only 20 km south to Agartala, the capital city of Agartala. Geomorphologically this area consists of highland (tillas) and adjacent lowlands (lungas). The tilla slopes are normally 25° to 57°, but at some places the slopes are recorded more than 80°. The slopes of the highland of this area are covered by dense tropical-monsoon mixed types of vegetation which made the present authors interested to conduct the study. The main objectives of the study are: (a) to identify the morphological slope and plant relationship and (b) the range of plant diversity in this area.

Methodology

Pre-Field Study

Prior to visiting the field area, the present authors surveyed the existing literature such as reports of Geological Survey of India, Department of Environment of Government of Tripura etc., and recent research papers published in different journals and presented in different seminars, congresses etc., to prepare a complete fieldwork programme. Historical literatures often preserve some valuable evidences of physical environmental change as historical events are influenced by environmental conditions. To assess the early environmental change through historical period this author used some existing relevant historical records of this area available from different literatures.

Methods of Fieldwork

The fieldwork was conducted from the January to August of 2006. Different techniques have been applied and instrumental survey works (simple dumpy level and clinometers etc.) have been done during fieldwork to find out the geomorphologic evolution of this area. Angles of slopes of different parts of tills lands were measured. Micro-level angle of slopes were measured by clinometer. Pattern and characters of natural vegetation were studied systematically. Plant coverage on the bottom, slopes and top of the tilla lands were measured on the basis of their height.

$$\textit{Plant coverage} = \textit{Number of plants/25 sq metre}$$

Besides that different plant species and their families were identified during the field work.

Post-Field Data Analysis

Geomorphic and botanical data collected through field investigation have been used for statistical testing. Micro-relief variations of the 5 selected slopes were calculated from the data of micro-slope angles. Relationship between micro-angle slopes of this area and the plant coverage has been analyzed by sophisticated statistical software. Every 25 sq m area has been considered as land unit or 'squire' on the selected profiles and maximum micro-slope angle has been taken for the measurement of relation between with plant coverage. Maximum micro-slope angle is considered as the indication of undulation range at micro-level. Some antic maps and literature were consulted along with field data for detection of environmental changes. Geological and geomorphological map (Scale–1: 50,000), Current landuse map published by Settlement office, toposheets (Scales: 1: 63,360) etc., were also consulted.

Results and Discussion

Micro-Relief Variation on Highland Slopes

The present study area has a remarkable micro-relief variation which also influences surface runoff and growth of small and medium plants. On the basis of the field data of five selected profiles (Table 1) of the highland (station distance is 5 m) it has been observed that maximum micro-relief variation is 0.14m and minimum–0.01 m while maximum total elevation change observed is 0.7 m within 25 m length of the profile. So, this landform is very undulating in small scale.

Table 1

Micro-Relief Variation of Selected Profiles

Distance from the bottom (in metre)	*Micro-Relief variation*				
	Profile 1	*Profile 2*	*Profile 3*	*Profile 4*	*Profile 5*
0–5	0.14 m	0.05 m	–0.1 m	0.04 m	0.03 m
5–10	0.22 m	0.13 m	0.14 m	0.16 m	0.02 m
10–15	0.13 m	0.07 m	0.07 m	0.02 m	0.05 m
15–20	0.03 m	0.03 m	0.07 m	0.07 m	0.02 m
20–25	0.13 m	—	0.08 m	—	—
25–30	0.05 m	—	—	—	—

Source: Field Data Generation.

Relationship Between Micro-Slope and Plant Coverage

During the fieldwork it has been observed that the micro-level slopes influence the distribution of plant and plant coverage (number of plants/25 sq m) on the land. Examination of field data by sophisticated software also strongly supports that hypothesis. Distance is also very important to identify the zoning of high and low plant coverage. Table 2 is showing the nature of plant coverage on five selected slopes.

Table 2

Plant Coverage on the Slopes of the Highland

Profiles	*Distance of the stations*	*Maximum Micro-level slope angles (in degrees)*	*Plant coverage according to height (in n/25 m²)*			*General plant coverage (in n/25 m²)*
			< 1 m	*1–4 m*	*> 4 m*	
1	0–5	16	125	8	2	135
	5–10	30	75	7	—	82
	10–15	15	100	20	—	120
	15–20	2	63	25	—	88
	20–25	16	75	30	—	105
	25–30	7	88	20	—	108
2	0–5	8	75	13	3	91
	5–10	16	50	19	4	73
	10–15	11	100	24	3	127
	15–20	9	63	20	6	89
3	0–5	8	63	21	3	87
	5–10	20	38	35	3	76
	10–15	10	50	40	5	95
	15–20	10	38	35	—	73
	20–25	15	75	45	4	124
4	0–5	11	75	20	15	130
	5–10	21	13	15	2	30
	10–15	4	0 (Grasses)	12	1	13
	15–20	10	25	14	—	39
5	0–5	4	50	10	35	95
	5–10	7	63	35	15	113
	10–15	10	13	8	38	59
	15–20	2	13	20	58	91

n = Number of plants

Source: Field Data Generation.

From the above table it is clear that maximum number of plant coverage found in those squires (25 m²) where maximum micro-slopes are 5° to 20°. It can be explained as the most suitable slope condition for plant on the highland. There is a general tendency of decrease of plant coverage when the maximum slope crosses 20° and it has been observed from the field that in squire where maximum micro-slope is 30° the plant coverage is little decreased than the earlier squire. On the top of the highland the plants are very scattered and most of the area is covered by grasses.

On the basis of field observation we divided the plants into three groups according to their heights, namely Small plants (<1 m), Medium plants

(1–4 m) and Tall trees (>4 m). Maximum coverage of the first group or <1 m high plants found in those squires on the slope where the maximum micro-slope angles are 10° to 20°. On that squire where the maximum micro-slope is 30° the plant coverage is medium. The plant coverage character of this group is very similar to the general plant coverage character. (Fig. 1 to 4)

Figure 1

General Plant Coverage Status on Different Micro-Angle Slopes of Squires

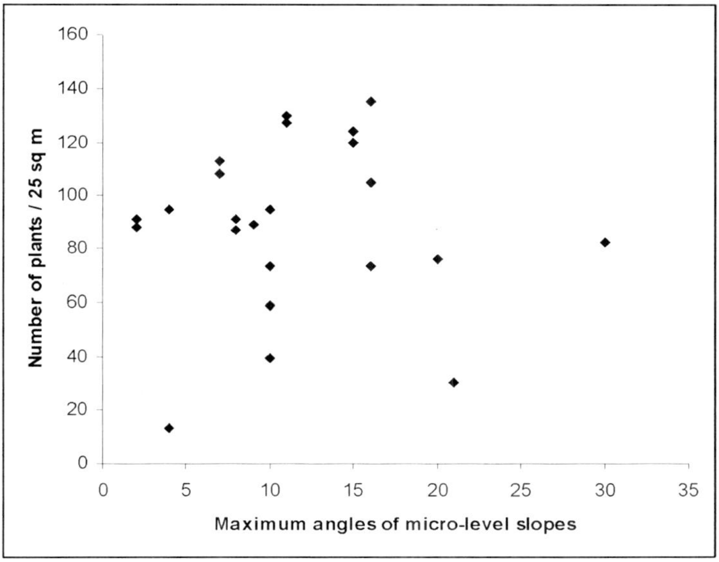

Figure 2

Small Plant Coverage Status on Different Micro-Angle Slopes of Squires

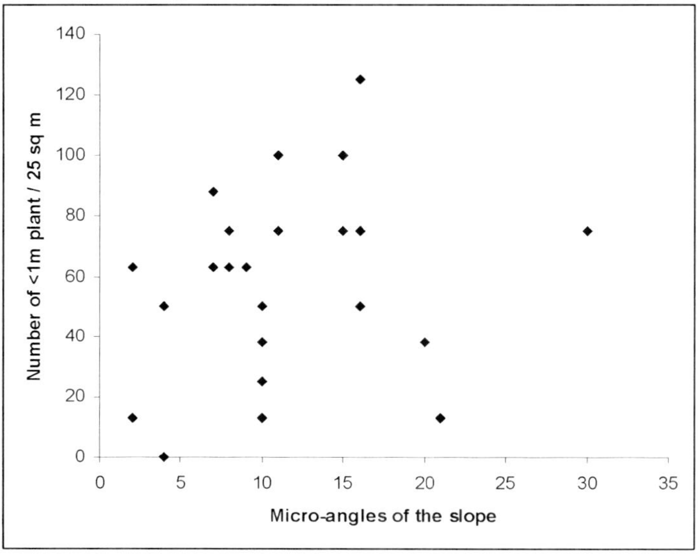

Figure 3

Medium Plant Coverage Status on Different Micro-Angle Slopes of Squires

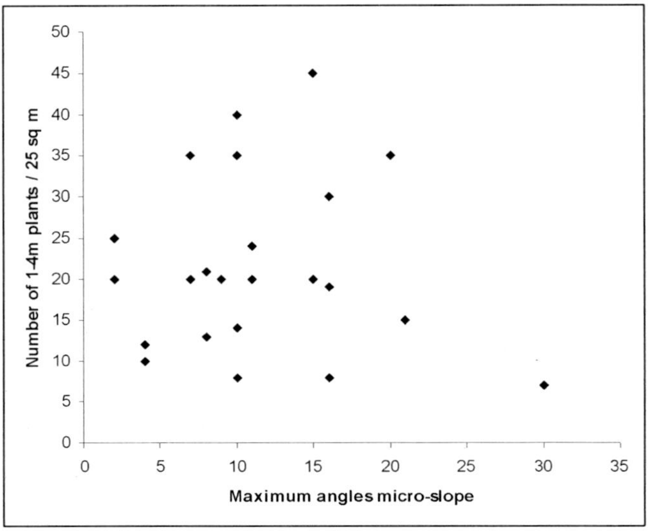

Figure 4

Tall Plant Coverage Status on Different Micro-Angle Slopes of Squires

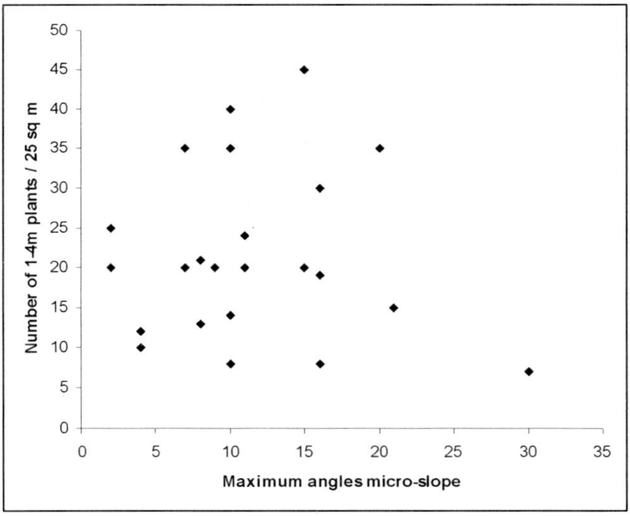

The second group (1–4 m) also follows the character of the first group. But in the squire of maximum slope 30° plant coverage is very low (only 7). A remarkable difference has been observed in case of third group which represents tall trees of >4 m height. High level plant coverage of the tall trees is found where maximum micro-slope angles are less than 5°. On 5°–10° maximum slopes plant coverage is medium. Very few tall trees are found where maximum slopes are more than 20°.

Types of Plants

This high land is very rich in valuable plant resources. In this place total 34 families are found and total 73 types of species of plants are observed by the present authors. Table 3 is showing the plant species and the families where they are belonging.

Table 3

Name of the Plant Species and their Families

	Local names (Bengali)	*Scientific names*	*Name of the family*
1.	Arjun	*Terminalia arjuna*	Combuaetaceae
2.	Acanda	*Calotropis gigantea*	Aselepiadaceae
3.	Dudhilata	*Pergularia daemia*	
4.	Pargacha	*Hoya parasitica*	
5.	Akhashmani	*Acacia moniliformic*	Fabaceae or Lequminosae
6.	Krishnachura	*Delonix regia*	
7.	Tetul	*Tamarindus indica*	
8.	Badar lathi	*Cassia fristula*	
9.	Lajjabati	*Mimosa pudica*	
10.	Ban atasi	*Crotolaria rethsa*	
11.	Kal-kasunda	*Cassia sophera*	
12.	Babla	*Acacia Arabia*	
13.	Ranjan	*Adenanthera pavonina*	
14.	Shan	*Crotalaria junacea*	
15.	Aam	*Mangifera indica*	Anacardiaceae
16.	Amalaci	*Phyllanthus emblica*	Euphorbiaceae
17.	Muktajhuri	*Acalypha indica*	
18.	Jetropha	*Jatropha gossypifolia*	
19.	Croton	*Croton bonplandianum*	
20.	Ban kakrol	*Momordica chochinensis*	Cucurbitaceae
21.	Dhela kucha	*Cephalandra indica*	
22.	Kumra	*Cacurbita pepo*	
23.	Karala	*Momordica charantia*	
24.	Kathal	*Artocarpus intergrifolida*	Moraceae
25.	Dumur	*Ficus cunea*	
26.	Bout	*Ficus benghalensis*	
27.	Rubber	*Ficus elastica*	
28.	Kamranga	*Averhopa carambola*	Oxalidaceae
29.	Kamini	*Murraya exotica*	Rutaceae
30.	Ashaseora	*Glycosmic pentaphylla*	
31.	Bel	*Aegle marmelos*	
32.	Kul	*Ziziphus jujuba*	Rhamnaceae
33.	Chudhara	*Lantana camara*	Verbenaceae
34.	Segun	*Tactona grandis*	
35.	Gamai\Gamar	*Gamelina arborea*	
36.	Churkanta	*Andropogon aciculaths Cynodon*	Gramineae Malvaceae
37.	Durbaghas	*dactylon*	
38.	Khus	*Eragrostis cynosuroides*	
39.	Jaba	*Hibiscus rosasinensis*	
40.	Barela	*Sida frondosa*	

Cont'd...

Cont'd...

41.	Kathali chapa	*Artobotrys odoratissimus*	Annonaceae
42.	Aata	*Annona paradisiaca*	
43.	Nayantara	*Catharantus rosea*	Apocynaceae
44.	Chatim	Alstonia scholaris	
45.	Kurchi	*Holarrhena antidysentaerica*	
46.	Kala	*Musa balbisiana*	
47.	Karabi	*Nerium odorum*	
48.	Narkel	*Cocos nucifera*	Palmae (arecaceae)
49.	Neem	*Melia azardirachta*	Meliaceae
50.	Peara	*Psidium guyava*	Myrtaceae
51.	Jam	Eugenia tambolana	
52.	Dalim	*Punica granatum*	Lithareceae
53.	Seto drun	*Leucas easpera*	Labiatae or Lamiaceae
54.	Tulsi	*Ocimum sanctum*	
55.	Thankuni	*Centella asiatici*	Umbelliferae (Apiaceae)
56.	Ban Begun	*Solanum torvun*	Solanaceae
57.	Kakmachi	*Solanum nigrun*	
58.	Banhana	*Cestrun nocturnum*	
59.	Aloo	*Solanum tuberosun*	
60.	Tomato	*Lycopersicum esculenta*	
61.	Kachu	*Colocasia antiquorum*	Araceae
62.	Bet	*Calamus tennis*	
63.	Shathi	*Curcuma zedoaria*	Zingiberaceae
64.	Khet papra	*Oldendandia corymbosa*	Rubiaceae
65.	Polyporus	*Polyporus gilvus*	Polyporaceae
66.	Kolmi	*Iponoea reptans*	Convolvulaceae
67.	Riccia	*Riccia robusta*	Ricciaceae
68.	Teris	*Pteris vittata*	Polypodiaceae
69.	Sushni	*Marsilea minuta*	Marcileaceae
70.	Agaricus	*Agraricus campstris*	Agaricaceae
71.	Ascobolus	*Ascobolus scatigenus*	Ascobolaceae
72.	Ban data	*Amaranthus gangeticus*	Amaranthaceae
73.	Hiradukhi or Dramina	*Dracaena spicata*	Liliaceae

Source: Field Data Generation, 2006.

It is very remarkable that among the total 73 types of species, all are found together within a very small area of the highland (only within 4 km^2 area). It has also been observed that the number of types of plant species are the same on the total area of the highland (about 12 km^2). So, it can be said that the level of plant diversity is very high in this area.

Another interesting feature is that among the 34 families 19 families have only one type species, 5 families have 2 types of species, 3 families have 3 types of species, 3 families have 4 types of species and 1 family has 5 types of plant species. Only one family (Fabaceae or Lequminosae) has 10 types of plant species. This is also a strong evidence of high level plant diversity on the present study area (Table 4 and Figure 5).

Table 4

Level of Species Diversity

S.No.	*Name of the family*	*Types of plant species*
1.	Combuaetaceae	1
2.	Aselepiadaceae	3
3.	Fabaceae or Lequminosae	10
4.	Anacardiaceae	1
5.	Euphorbiaceae	4
6.	Cucurbitaceae	4
7.	Moraceae	4
8.	Oxalidaceae	1
9.	Rutaceae	3
10.	Rhamnaceae	1
11.	Verbenaceae	3
12.	Gramineae	3
13.	Malvaceae	2
14.	Annonacea	2
15.	Apocynaceae	5
16.	Palmae (arecaceae)	1
17.	Meliaceae	1
18.	Myrtaceae	2
19.	Lithareceae	1
20.	Labiatae or Lamiaceae	2
21.	Umbelliferae (Apiaceae)	1
22.	Solanaceae	5
23.	Araceae	2
24.	Zingiberaceae	1
25.	Rubiaceae	1
26.	Polyporaceae	1
27.	Convolvulaceae	1
28.	Ricciaceae	1
29.	Polypodiaceae	1
30.	Marcileaceae	1
31.	Agaricaceae	1
32.	Ascobolaceae	1
33.	Amaranthaceae	1
34.	Liliaceae	1

Source: Field Data Generation, 2006.

Figure 5

Level of Plant Species Diversity

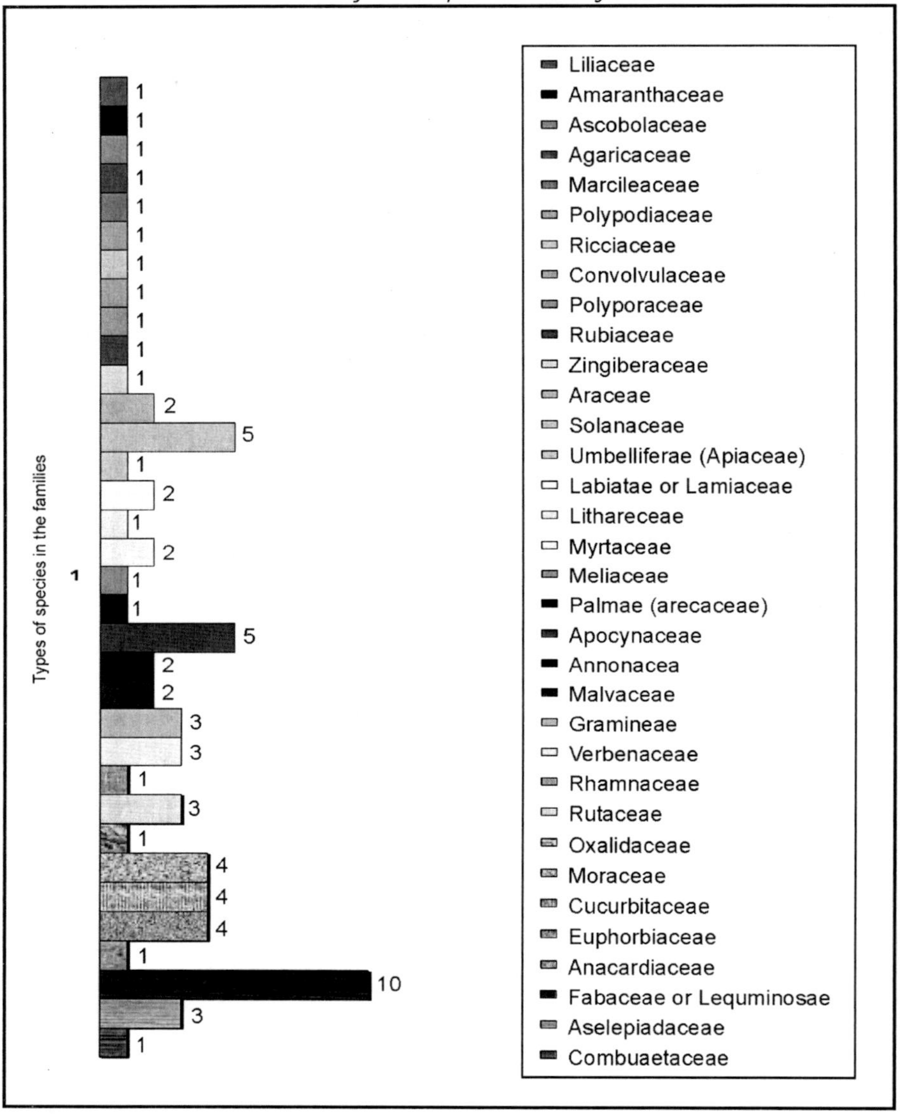

Penetration Zones of Plant Species

The penetration zones of different plant species is also assessed on the basis of maximum slope area. It has been observed most of the species grow where the maximum slopes are 2° to 20°. Only some species grow where the maximum slope crosses 21°. Among the small plants *Cynodon dactylon* and *Lantana camara* are very prominent who can penetrate even on high angle and highly undulating areas. Among the medium plants *Dracaena spicata* is the only species which can grow under this environmental condition. One remarkable thing is that these small and medium plants can penetrate everywhere, from low micro-angle area to high micro-angle area. It is also observed that no tall tree grows on high angle (>21°) and highly rugged area. On the basis of that observation the plants of this area can be defined in to two groups: (a) everywhere penetrating plants and (b) low to medium angle area penetrating plants. Table 5 shows the different species and their penetration zones with different maximum micro-slope angles.

Table 5

Plant Penetration Zones on the Profiles

Profile No.	Distance from bottom to top	Maximum angle of micro-slopes	Plant group according to height		
			Small (< 1 m)	Medium (1–4 m)	Tall trees (>4 m)
1	0–5	16	Cynodon dactylon, Lantana camara.	Calamus tennis, Acacia Arabia. Dracaena spicata,etc	Ziziphus jujuba, Eugenia tambolana.
	5–10	30	-do-	Dracaena spicata,etc	
	10–15	15	Cynodon dactylon, Leucas easpera,	Ziziphus jujube Dracaena spicata,etc	
	15–20	2	Leucas easpera, Andropogon aciculaths	Dracaena spicata,etc	
	20–25	16	Centella asiatica, Pteris vittata, Marsilea minuta, Cynodon dactylo Andropogon aciculaths,etc	Dracaena spicata,etc	
	25–30	7	Cynodon dactylon, Centella asiatic Pteris vittata,Leucas easpera		
2	0–5	8	Cynodon dactylon, Marsilea minuta Leucas easpera, etc	Dracaena spicata,etc	Ziziphus jujuba, Glycosmic pentaphylla,etc.
	5–10	16	Cynodon dactylon, Leucas easpera, Lantana camara, Andropogon aciculaths	Cestrun nocturnum, Dracaena spicata,etc	Eugenia tambolana, Glycosmic pentaphylla
	10–15	11	Cynodon dactylon Andropogon aciculaths,	Dracaena spicata,	Ziziphus jujube, Glycosmic pentaphylla,
	15–20	9	Cynodon dactylon, Marsilea minuta, Lantana camara, Leucas easpera	Dracaena spicata,etc	Glycosmic pentaphylla, Calamus tennis
3	0–5	8	Cynodon dactylon, Leucas easpera,	Dracaena spicata,etc	Cassia sophera
	5–10	20	Pteris vittata, Cynodon dactylon, Mimosa pudica	Dracaena spicata,etc	Cassia sophera, Acacia moniliformic

Cont'd...

Cont'd...

	10–15	10	Cynodon dactylon, Mimosa pudica, Lantana camara	Dracaena spicata,etc	Cassia sophera
	15–20	10	Cynodon dactylon, Mimosa pudica, Marsilea minuta	Dracaena spicata,etc	
	20–25	15	Cynodon dactylon, Mimosa pudica, Marsilea minuta	Calamus tennis, Alstonia scholaris	Ziziphus jujuba, Glycosmic pentaphylla Eugenia tambolana
4	0–5	11	Cynodon dactylon, Mimosa pudica, Marsilea minuta Leucas easpera,	Cestrun nocturnum, etc	Glycosmic pentaphylal Eugenia tambolana,etc
	5–10	21	Cynodon dactylon, Mimosa pudica	Dracaena spicata,etc	Eugenia tambolana
	10–15	4	Cynodon dactylon, Andropogon aciculaths	Dracaena spicata,etc	Glycosmic pentaphylla
	15–20	10	Cynodon dactylon, Andropogon aciculaths, Mimosa pudica, Marsilea minuta	Banhalud, bit	
5	0–5	4	Cynodon dactylon, Mimosa pudica	Dracaena spicata Cestrun nocturnum etc,	Acacia moniliformic, Calamus tennis, Mangifera indica
	5–10	7	Cynodon dactylon, Mimosa pudica	Dracaena spicata Cestrun nocturnum etc	Acacia moniliformic, Calamus tennis, Mangifera indica
	10–15	10	Cynodon dactylon, Mimosa pudica	Dracaena spicata Cestrun nocturnum etc	Acacia moniliformic, Calamus tennis, Mangifera indica
	15–20	2	Andropogon aciculaths, Pteris vittata, Centella asiatica, Marsilea minuta,	Dracaena spicata etc	Calamus tennis, Glycosmic pentaphylla, Eugenia tambolana

Source: Field Investigation, 2006.

General Assessments

The highlands of this area can be marked as very important micro-environmental unit since they are very rich in natural and ecological resources. This study shows that the selected highland is also very important from ecological point of view. From this study the following facts come out which area as follows:

1. The slopes of the highland are very rich in plant diversity and complicated adjustment with the micro-geomorphic condition.

2. The micro-relief variation shows that the slopes of the highland are very rugged from micro-geomorphic point of view.
3. Micro-slopes are not ignorable since the study shows that they have a relationship with the plant coverage as well as pattern of plant distribution on the slope.
4. The general plant coverage shows that the plants are found up to maximum micro-slope of 30°.
5. General plant coverage also shows that high level plant growth occurs on the 5°–20° maximum slope areas.
6. Small (< 1 m) and medium (1 m–4 m) plant distribution almost follow the general plant distribution nature and they have a similar type of relationship with micro-level slopes.
7. Tall trees normally grow where maximum micro-slope angle is less than 5°. They have a different distribution pattern and relation with micro-slopes than the other height groups and general distribution.
8. The level of plant diversity is very high as maximum numbers of species and numbers of families are found in a very small area.

References

Alam, M. et al. (2003), "An Overview of the Sedimentary Geology of the Bengal Basin in Relation to the Regional Tectonic Framework and Basin-fill History", *Sedimentary Geology*, 155; Elsevier, pp. 179–208

Dey, S., (2005), "Conceptual Models for the Assessment of Tertiary-Quaternary Geomorphic Evolution of Paleo-coastal Tripura", *Annals of the National Association of Geographers*, India; Vol. XXV, No. 1, June, pp. 73–80.

Dutta, G.N., (1974), "Some Geomorphological Aspects of Tripura State", Abstracts of Proceedings of Indian Science Congress, p. 130.

Geological Survey of India, (2004), Report on Melaghar Block (Unpublished); pp. 1–18.

Menon, K.D., (1975), *Tripura District Gazetteer*, Government of Tripura, Agartala, pp. 1–27.

Nandi, M., (1974), "The Relation of Agartala Floods to Haora River Catchments, Tripura State", *Records of Geological Survey of India*, Vol. 108; Pt–2; pp. 149–152. Nandi, M. (1977), "Physiography of Tripura State", *Geog. Rev., Ind. Geog. Soc. Ind.*, Vol. 39, No. 4, pp. 367–373.

Nandi, M., (1985), "Geographical Account of Tripura", *An Anthology of Tripura*, J. Gan-Chowdhui (ed.), Inter-Indian Publication; New Delhi; pp. 1–9.

10 Land Use and Resource Utilization Pattern

A Case for the Pindar Basin of Garhwal Himalaya

Vishwambhar Prasad Sati

The management of natural resources in the mountain environments has emerged as one of the most significant challenges to human understanding and organizational ability in the current century. Although the mountains and uplands constitute about 20 per cent of the surface of the earth, it is difficult to find an area not affected by their environmental characteristics. The upper parts of the mountains are always covered by snow and they are the sources of major rivers in the world. The vital cultural and ecological importance of these mountain waters further enhance their significance in satisfying the needs of all living creatures. Of somewhat less significance ecologically, but nonetheless of great economic importance, are the uses of the mountains for forestry, agro-horticulture, mineral extraction, livestock rearing, tourism, and recreation.

Pindar basin that constitutes the major part of the Alaknanda basin in the Garhwal Himalaya, is the most ecologically fragile zone. The series of landslides, which washed away the thousand tonnes of soil per year, reduce the rate of soil fertility. Thus, the whole basin is landslide prone area. Every year, one or more landslides occur. The basin can be divided into three-physiographic zones according to their elevation. The highly mountainous area, which is always covered by snow is the major sources of rivers, like Pindar, Kail, and its tributaries. Alpine meadows and small patches of field terraces characterize the second zone. Where, mainly herbs in the high altitude are grown, and agriculture in the middle zone is practised quite extensively. This zone is called 'Danda'.[1] Along these two practices, horticulture is slowly but steady being practised throughout the basin. Even some quality fruits are grown. In the low-lying area, which is the third physiographic zone, traditional subsidence cereal farming is being practised extensively. But the production of the crops is very low. This zone is called 'Gangar'.[2] The topography of the basin prevents the developmental process and optimum utilization of resources. Besides, the main valley route, from Karanprayag to Gwaldom and Mundoll, the entire basin lags behind, in terms of, transportation facilities. In spite of these huge impediments, there are many fertile patches in terms of agricultural practices, horticulture, herbs culture, potato production, and various other food grains; pulses, oilseeds etc., are intensively and extensively grown. Only, the need is of developing a trend for sustainable development of these huge resources. The present study discusses on the resources available in the region and their analyses to enhance the economic status of the region.

Study Area

The Pindar basin comprising of 1,826 km^2 extends from 30° N to 18° N latitude and 79°13′ to 80° E longitude. It consists of the eastern part of Garhwal Himalaya with height ranging from 800 m to 6,800 m. River Pindar originates from the 'Pindari Glacier' in district Almora (32 km) and flowing an approximate 100 km with its numerous tributaries, confluences in to the Alaknanda River at Karanprayag in Chamoli district (68 km). The watersheds of Ramganga in the south, the Saryu in the east, Mandakini in the north and Alaknanda in the north-west, giving it a distinct socio-geographical identity, separate the study region. This basin consists of the four communities developmental blocks of the Chamoli district, viz., Karanprayag, Narainbagar, Tharali, and Deval in ascending order from lower to higher altitude and a part of Kapkot block in Almora district (Figure 1).

Methodology

The present study is based on primary and secondary data collected through questionnaire and schedules and from various sources respectively. District Statistical Handbook of district Charmoli is also used. Maps have also been drawn for interpretation of data.

Pattern of Resource Utilization

The Pindar basin is characterized by heterogeneity in terms of landscape, resources and their utilization patterns. Right from the valley region to the highly elevated dueling zones, the climatic conditions regulate the all types of utilization pattern. While, the valley region is considered for developing service centres and the wide patches of the cultivated fields are characterized by growing cereal crops; the middle and upper belts are known for growing varieties of crops, vegetables, fruits, herbs, rearing livestock, goats, sheep; manufacturing wooden, Ringal, and Bamboo goods. The production of potato, apple and citrus fruits, milk and milk made goods is gaining impressive mark since the decade of nineties. In this regard, some patches of Ghaish-Blan, Kurer-Kwarad, Sol-Dungri, Shubhtal Jhaltall have the progressive position. The cash crops are taking the place of cereal farming. Consequently, the economic level of the populace is undoubtedly being developed. The resource utilization pattern is varied from place to place, based on infrastructural facilities, particularly transportation and extension services. The undeveloped agricultural practices are also due to the remoteness of the villages from the

Figure 1

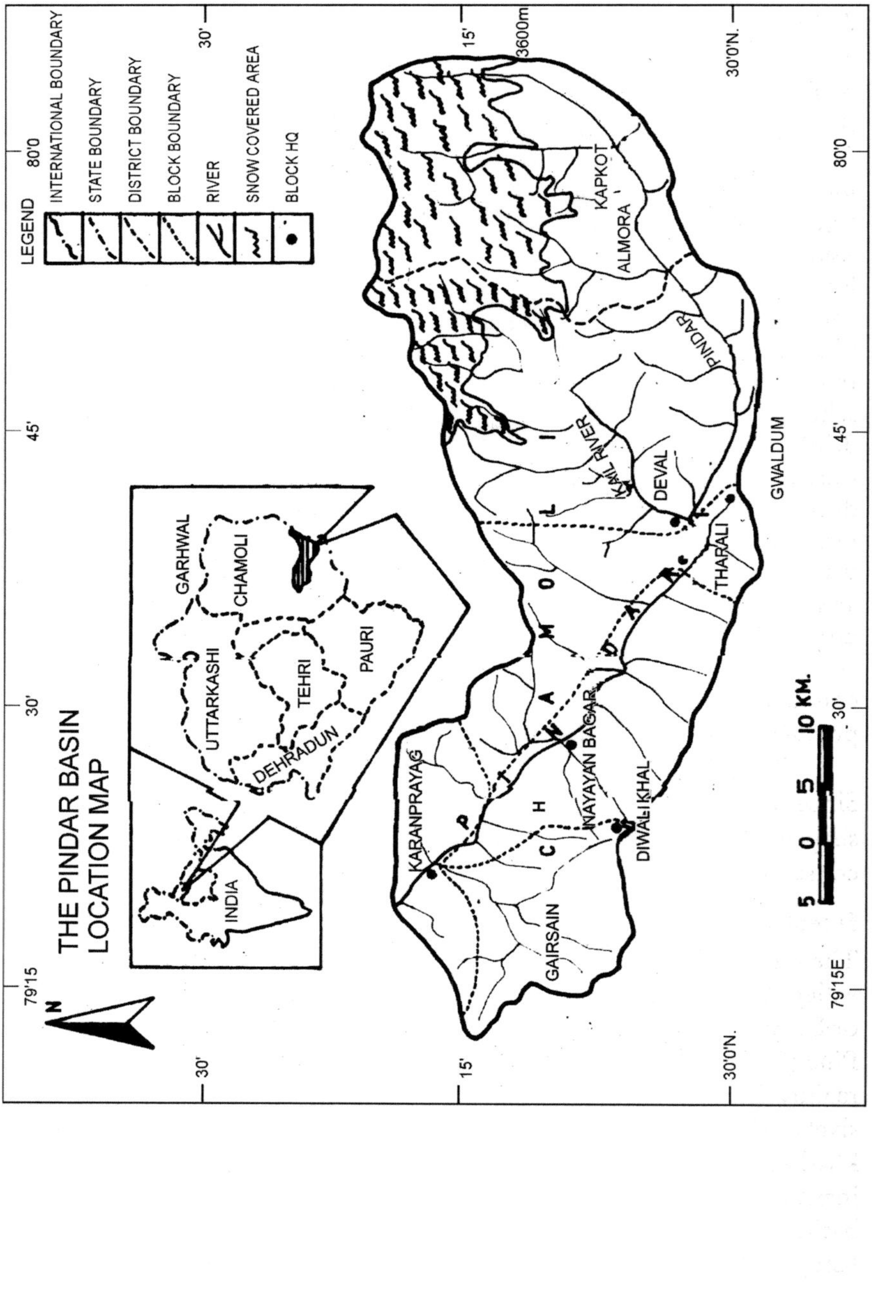
THE PINDAR BASIN
LOCATION MAP
LEGEND
INTERNATIONAL BOUNDARY
STATE BOUNDARY
DISTRICT BOUNDARY
BLOCK BOUNDARY
RIVER
SNOW COVERED AREA
BLOCK HQ
INDIA
UTTARKASHI
GARHWAL
CHAMOLI
TEHRI
PAURI
DEHRADUN
KARANPRAYAG
GAIRSAIN
NARAYAN BAGAR
DIWALI KHAL
THARALI
DEVAL
GWALDUM
KAIL RIVER
PINDAR
ALMORA
KAPKOT
CHAMOLI
PINDAR
3600m
5 0 5 10 KM.
79'15
79'15E
30'
45'
80'0
15'
30'0'N.
N

main stream. Settlements are sparsely distributed. But the highest population is found in the middle altitude. In the high and low altitude, the population is very low. Because, the high altitudinal zone has very low temperature, while, in the low-lying areas the land for settlement and agricultural practices is less. The study is being done on the following manner.

Water Resource

The Pindar River is the main drainage system and the biggest river of the region. There are many other perennial sources, which are the tributaries of the Pindar. The main tributaries are Kail Ganga, Pranmati Gad, Meing Gadhera, Kewar Gadhera, Chopta Gad, and Ata Gad. Besides these there are so many small perennial streams, and natural springs. But due to uneven and rugged topography, water crises prevailed in most of the villages, during summers in particular and winters in general. The Pindar River has many ideal locations for constructing dams. The two best locations are near the Bangoli and Simli. These two sides have already been proposed for dam construction and many times work is started during the winter and summer but during the rainy season, the finished works are swept away by the river. Micro-hydle schemes can be introduced, on the other hand, to mitigate the problem of drinking water and electric supply, and even for irrigation purposes. The old practice of water mills which locally known as 'Gharat system' can be reintroduced as a form of floor mill and electricity generation. Still these are in practice in some of the areas. Even with the dominance of electric mills the water mills are also functioning in almost all the small tributaries. In some areas, the water mills come into practice only during the rainy season because during the other seasons the volume of water remains less to meet the need for running-water mills. The author during counting of these water mills has conducted a survey and it was found that more than 200 water mills are still in practice in the entire basin.

There are many diversities in the course of the river from outlet to inlet. Since, the river flows with its youthful stage the streams make water fall, V shaped valleys, gorges etc. The wide and stretched valleys of these streams constitute the fertility of land.

Forest Resource

The Pindar basin is very rich in terms of forest and its diversity. Right from the valley region to the highly elevated Alpine meadows, locally known as Kharak or Bugyal, the diversities in plants are found extensively. In the middle altitude, Pine (Chir) are found while in the upper reaches, temperate coniferous forest mainly Banj (oak), Buransh, Tilonj, Devadar, Surain etc., are found extensively. Except these forest types, many other fodder plants like, bhimal, khadik, are also grown along with the agricultural fields. Oak is the predominant flora of the moist temperate forest growing on the elevated region of the basin. Kharsu (quercussemicarpifolia), Tilonj (Q. Dilitata), Rianj (Q. lanuginose) and Banj oak (Q. leucotricophora) are the four main species, which have a traditional edge over the other species in this region. Out of these four species Banj oak (Q. leu), which being a very rich fodder and ideal fuel has been an inseparable part of folk life in hills, is the one, which starts growing from 1,676 m with Rhododendron trees. It is found up to 2,286 m mixed with the other species of the same genera. The oak forest of the Pindar basin appears to be thriving due to the imposing presence of pine trees. A note in *The Hindustan Times* by HT correspondence, "will oak disappear from Nainital" is a real story in the hills of Uttranchal.[3] According to the altitude and forest types, the basin can be divided into the following zones:

1. *Low-lying river valley [below 1,000 m elevation from mean sea level (msl)].* Mainly along the valley of Pindar River, right from Karanprayag to Tharall, where the height does not reach above 1,000 m. Presently, afforestation is carried out by the department of forest in this region to avoid landslides. These plants are of eucalyptus, popular, bamboo etc.
2. *Middle altitude (between 1,000 to 1,600 m elevation from msl).* Pine is found extensively. There are many patches, where dense pine forests are found. They start from Bangoli to the valley of Pindar along with both sides of its tributaries. Such as; the slope of Kaver gadhera, Mieng gadhera, Pranmati gad, etc. The present trend of pine trees distribution is that they are invading the coniferous regime. Pine trees invaded most of the oak dominated areas and environmentalists advocating that this process exists due to temperature incensement in the atmosphere.
3. *Temperate belt (between 1,600 to 2,000 m elevation from msl).* The area constitutes Love-Kush top, watershed of Pindar and Ramganga, Kanpur Garhi, Kankhrakhet, Shubhtal and Chhaltal, Kurur Kwarad, Sol-Dungri etc. The region is very rich in terms of soft forest wood, which is very useful for many purposes. Main trees found in this area are oak, buransh, tilonj, surat, devdar, etc. This region is inaccessible from reach. Thus, the woods are unutilized.
4. *Highly elevated region (between 2,000 m to snow line).* This region is accounted for herbs. Extensive fields of herbs are found. These are locally called Kharak or Bugyal The main bugyals are; Khankhrakhet, Bedni bugyal, Roopkund, Shall Samunder etc. Bedni Bugyal and Roopkund are world famous spots and they can be developed as the tourist places. Different kinds of herbs are found in these Bugyal.

Forest based small industries can be developed in these areas. Particularly, where these forest are found. They are not used properly in terms of small-scale industries. While every year thousands of tonnes of timber is used in form of fuel-wood as well as building materials they are also used for fodder need.

Agricultural Resources

Subsistence agriculture is the main stay of the people inhabited in the Pindar basin. Most of the workable force is engaged in the agriculture and its allied practices. But this practice does not fulfil the food need; therefore, to cape with food scarcity, the people of the region have out-migrated to the other parts of India. The production and productivity of the crops are very low. Cost benefit analysis reveals that the output by the agricultural production is low, even, below the level of input. Cereal crops are dominant in the cropping pattern. The production of fruits is getting to impressive level, but the land devoted to fruit plants is very less. The climatic conditions for production of fruits are congenial. There are two fruit belts demarcated. These are Kalimati Janlechatti and Gwaldom-Lolti. Many other small fruit belts are also located there, which can be developed for intensive fruit production. (Table 1)

Table 1

Area Under Different Crops in the Pindar Basin

Crops	*Block (area in ha.)*							
	Karanprayag	*%*	*Narain Bagar*	*%*	*Tharali*	*%*	*Deval*	*%*
Paddy	2238	47.7	1632	43.72	1594	61.2	570	34.3
Wheat	2778	51.7	1733	46.4	780	29.9	777	46.8
Barley	166	3.0	190	5.1	61	2.3	62	3.7
Masur	7	0.1	10	0.3	6	0.2	5	0.3
Pea	—	—	4	0.1	4	0.2	—	—
Arhar	9	0.2	15	0.4	6	0.2	6	0.3
Pulses	24	0.4	73	1.9	58	2.2	17	1.0
Oilseeds	14	0.3	31	0.8	30	1.2	12	0.7
Potato	130	2.4	45	1.2	67	2.7	211	12.7

Source: District Statistical Diary of Chamoli District

The table reveals that most of the crops grown in the entire basin are mainly cereals, oilseeds, and pulses. Presently, the production of potato is being increased. There are ideal conditions for production of cash crops like potato, pulses, oilseeds and different types of vegetables; therefore, these cash crops can be grown extensively. Though paddy and wheat are leading crops of the area (more than 80% of cropped area) their yield is very low owing to some localized factors (traditional views and old methods of ploughing and harvesting). The harvesting period of the paddy is variable according to elevation but the ploughing period is the same everywhere. In the 'Gangarh', the paddy harvesting begins one month earlier than the 'Danda'. The plouging period is March-April both in the Gangarh and the 'Danda' and the harvesting period is the early September in the 'Gangarh' and October in the 'Danda'.

Land Use Pattern in the Pindar Basin

Land is one of the most components of ecosystem and it constitutes main part of economy of the Himalayan environment. Land use pattern is varied according to relief and slope factors. The following table reveals the land use pattern in the Pindar basin (Table 2 and Figure 2).

Table 2

Land Use Pattern in the Pindar Basin (Percentage of Geographical Area)

Block	*Forest land*	*Pasture land*	*Agricultural land*	*Horticultural land*	*Cultivable land*
Karanprayag	52.2	3.3	28.4	8.5	0.7
Gairsain	74.0	2.4	12.6	6.9	2.9
Narain Bagar	67.2	8.9	18.5	7.1	0.3
Tharali	25.2	0.5	4.8	0.2	0.4
Deval	80.8	1.6	5.7	6.5	1.4

Sources: District Statistical Handbook of Chamoli.

Figure 2

The Pindar Basin Land Use

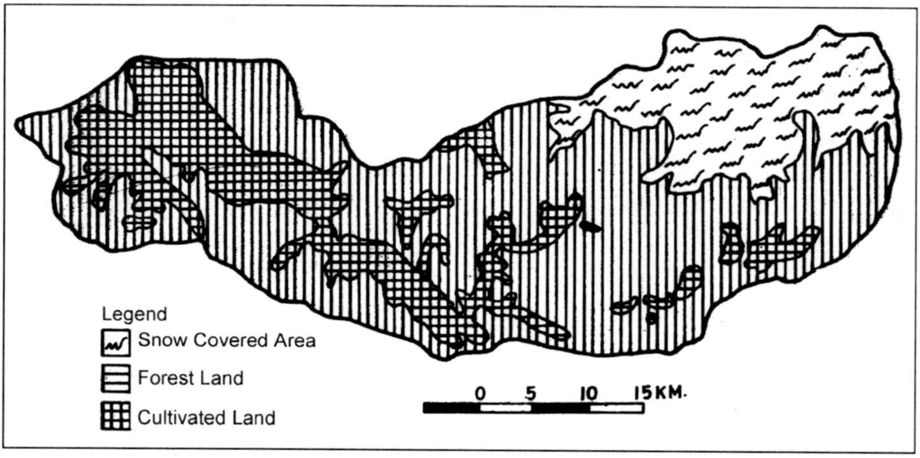

The table reveals that forest covers the highest percentage of the total geographical land followed by agricultural land. The percentage varies from one block to another. There are three crop seasons. Rabi, Kharif, and Jayad. Under the Rabi crops, wheat, oilseeds, barley, masure etc., are grown, while, under the Kharif crops, rice, koni Jhangora etc., are grown. Pulses and vegetables are grown with the Jayad season.

Cropping Pattern

"The kind and sequence of crops grown over a period of time on a given area of soil can be described as the cropping system". The overall cropping pattern of the Pindar basin is typical of and underdeveloped agricultural economy. Here, nearly 90 per cent of the total cropped area is devoted to subsistence food crops mainly grown for domestic consumption and local market. Commercial or cash crops occupy a very negligible portion of the cropped area. Due to spatial differences in the agronomic, economic, and cultural conditions, variations in the cropping pattern in the region can be seen, but on the whole these variations are of very little significance. Rice and wheat dominate the agricultural realm from one corner of the region to the other, obviating the culture of the other crops. The influence of the monsoon on the cropping pattern is very dominant, with the result that the total cropped area about 70.75 per cent is under Kharif or rainy season crops. In the region, whatever may be the type of soil or the amount of rainfall, the dominance of food grains in the cropping pattern is everywhere obvious. A case study of cluster of five villages has been done for interpretation of cropping pattern. These cluster of villages are located almost in the centre of the basin. Agriculture is the main occupation of the villagers. Of the total population, 49 per cent people are engaged in the primary works, mostly in subsistence agriculture. The method of agriculture is traditional. Agricultural implements are mainly made of wood and oxen plough the fields. The seeds are sown using spray method. The main crops are paddy, wheat, manduwa, pulses and millets, and in the vegetables, potato,

Table 3

Cropping Pattern

A. 2000

Crops	*Sar (in ha) 45.6*	*Period of sowing and harvesting*		*Crops*	*Sar (in ha) 45.6*	*Period of sowing and harvesting*	
Name of Crop	*First sar (Mally Sar)*	*Sowing Period*	*Harvesting Period*	*Name of Crop*	*Second (Mally Sar)*	*Sowing*	*Harvesting*
Manduwa (Kodo)	23.1			Paddy	15.03	March	September
Jwar-Bajra (Junoru-Bajuro)	3.5			Koni-Jhangora	10.7	April	September
Urd	8.0	June-July		Paddy-Koni-Jhangora	30.7		
Soyabeen	7.4						
Taur	0.5						
Kulthi	0.1						
Bhatt	1.0						
Ramdana	0.5						
Potato	1.5						
Crop	Second Sar (Tally Sar)	Land Near and around the settlement					
Wheat	39.2	November-December	May-June	Maize	4.7	June	September
Barley	8.4			Different type of vegetable in rainy season	5.3		
Oil seeds	8.8						
Total	56.4				10.0		

Source: Primary data collected by the author.

B. 2001

Crops	Sar (Area in ha) 41.6	Period of sowing and harvesting		Crops	Sar (Area in ha) 45.8	Period of sowing and harvesting	
Name of Crop	First sar (Mally Sar)	Sowing Period	Harvesting Period	Name of Crop	Second sar (Mally Sar)	Showing period	Harvesting period
Paddy	32.8	March	September	Wheat	32.6	November-December	May-June
Koni-Jhangora	3.4	April	October	Oilseed	6.5		
Paddy-Koni-Jhangora	5.4						
Crop		Second Sar (Tally Sar)		Land near and around settlements			
Mandua	29.0	June-July	October-November	Maize	5.2	May	October
Jwar-Bajra	7.5			Vegetables	3.8		
Urd	5.2				9.0		
Soyabeen	5.1						
Tuar	2.7						
Kulthi	4.1						
Bhatt	0.3						
Ramadana	0.5						
Potato	2.0						

Source: Surveyed by the author.

pumpkin, cucumber, brinzal, etc. The production of potato has been recently increasing. The people are self-reliant for their food grain needs. Agriculture is totally dependent upon rainfall, because of inadequate facilities of irrigation. Under the cropping pattern, the total agricultural land is divided into two parts; Tally and Mally Sar, which are called 'Sar System'. Since March-April till September-October, there are grown the Kharif crops in the Tally Sar. The millets are grown in the Mally Sar in the same period under the Kharif season. Table 3 reveals the cropping pattern in 2000 and 2001 under Rabi and Kharif Crops.

The cropping pattern of this village is systematically characterized by mixed cropping. The crop rotation, which is locally known as 'Sar System', has been giving good yield of crops. Within a year, the land is given-up fallow for six months. The cropping pattern of this village represents the entire Uttranchal state except the plains of Dun, Dwar, and Tarai and Bhaber regions, where the extensive cultivated land extended and the cropping capacity, intensity of crops, crops production and per ha yield of crops is very high.

Horticultural, Herbicultural and Tea Gardens Practices

The entire region is suitable for growing fruits, vegetables, herbs, and tea. The production of fruits is limited only on some patches of Gwaldom-Tharali, Khun-Kothali, and Kalmati-Jangle Chatti, while the climatic conditions are similar in the basin. Among the fruits, apple, orange, nuts, and pears are grown widely. But their quality and quantity are very low. They are produced and utilized locally.

Herbs are naturally grown in the meadows, which are locally known as 'Kharaks'. The main areas where herbs found extensively are Khankharakhet, Bedani Bugyal, Kanpur Garhi, Beni Tal, Shuvtal and Jhal tal. These herbs are locally utilized for medicinal purposes and have the positive results following the use by local people.

Simultaneously, a tea garden practice is centuries old but presently its existence is very less. Following the creation of Uttranchal state, the government has launched the scheme for establishing tea gardens in the different geographically ideal locations. In the Pindar Basin, there are above five suitable belts for tea production. Those included are: Gwaldom-Lolti belt, which is still in existence, Shuvtal-Chhaltal belt, Beni tal belt, Kalimati-Janglechatti belt, Khakharakhet-Kanpurgarhi belt and many others which are located in the middle latitudes along with very gentle slopes.

These practices can help to boost the economy of the entire basin. Similarly, they can check the problems of environmental hazards, because, this region is landslide prone.

Cattle Farming

Raising livestock is an integral component of mountain farming systems. In the mixed crop livestock farming system, there is a dynamic relationship between common property resources, livestock, and crops. Livestock depend to a certain extent on fodder and grass grown of common property resources as well as on crops residue; at the same time, the animals return fodder, grass, and crop residue to common property resources and fields in the form of manure, and provide much needed draught power. Livestock are thus integral to the sustainability of hills and mountain farming.

Dairy farming is an integral part of mountain agriculture. In the Uttranchal state of India, small holders who make up an overwhelming majority, have evolved two major systems of livestock management in mixed crop livestock farming system: sedentary and migratory. Cattle and buffaloes are the only dairying species throughout the basin. Use of goats and sheep as dairy animals is extremely rare. The production of cattle over the last three decades has decreased or remained stagnant, while that of buffaloes has recorded a remarkable increase. Cattle make up about 49 per cent and buffaloes 51 per cent of the 61,404 dairy animals in the Pindar basin. The number of dairy animals in the Uttranchal state is 11,00,839. Out of them 52 per cent are cattle while buffaloes are 48 per cent. All the buffaloes in the region are indigenous. This milch species is largely an admixture of improved type Murrah and Bhadwari. The overall average composition of indigenous cattle is 99.2 per cent. The Table 4 shows per day production of milk and need of fodder in the three different altitudinal village of the similar longitude:

Table 4

Per Day Production of Milk and Need of Fodder in The Pindar Basin

Name of Village (elevation)	*Accessibility from the road head*	*Milk production by cow (indigenous)*	*Milk production by buffalo (indigeous)*	*Fodder used oak/others (in kg) Wint./Summ*	*Crops residue whole year*
Prethi (1,654)	5 km away	50 kg	150 kg	600/700 kg	400 kg
Ali (1,200)	0.5 km away	25 kg	20 kg	Nil/200 kg	400 kg
Kimoli (1,905)	10 km	80 kg	300 kg	800/900 kg	300 kg

Source: Data collected by the author.

The table denotes that per day production of milk and need of fodder in the Pindar basin varies from one place to another according to the height. The production of milk is increased according to increase in the elevation. The milk obtained from buffaloes is much more than cows. Similarly, altitude and oak fodder play a vital role in the production of milk. The table reveals that comparatively with the increasing altitude, milk production increased. The fodder need is also met at a large scale in the high altitude. The forest mostly meets the fodder need because there are large grassland and forestland in or surrounding of the villages. The excessive dependency on the forest for fodder and fuel need threatens the existence of environment. The growth rate of plants in comparison to per day exploitation is very low. About 32 quintal fodder is being utilized per day in these three villages as the table reflects. Even though, the production of milk is not desirable as it is in the other part of the nation. Though the climatic conditions are very feasible for production of milk and rearing of high variety animals, they are yet to be introduced. Atleast the accessible areas, which are connected by the roads, may be used for introducing cross-breed animals. This process will boost up the economic condition of the villages.

A comparative case study of the three villages located in the different altitudes and their varied accessibility from the road head has been carried out by the present author. The Table 5, reveals the some of the aspect of animal rearing and their economic significance:

Table 5

Village Name	*Height*	*Accessibility*	*Cows*	*Buff-aloes*	*Sheep*	*Goats*	*Oxen*	*Households*	*Animal/ Family ratio*
Wan	2,100 m	10 km up from main road head	62	35	135	106	21 pairs	104	Average 4 animal/family
Meing Gadhera	1,010 m	On the road	116	67	nil	nil	32 pairs	234	Average 1 animal/family
Prethi	1,600 m	5 km away	67	65	nil	nil	16 pairs	105	Comparatively higher than previous one

The table denotes that the village that is located in the high altitude has the high ratio of animal even the number of goats and sheep are more than in the villages located in comparatively low altitude. Whereas in the villages of low altitude, the number of cows and buffaloes are more while the ratio is less. The Table 4 shows per day production of milk and need of fodder in the three different altitudinal villages of the similar longitude.

Primitive agriculture is the main occupation of the Himalayan people. Therefore, the economy rests on agriculture and its allied practices. The motivation of government and non-governmental organization towards and application of modern technology in the fields of agriculture in the Pindar basin did not turn an appreciable mode of development due to its remoteness from the means of transportation, rugged terrain and unwillingness of the people towards development. The natural resources are still unexploited. The problems of soil erosion and landslides are the major issues to be solved at the right time. This basin is known for rapid landslides, throughout its course. Consequently, every year, thousands tonnes of soil is eroded by the river and heavy loses of lives take place.

Tourism

Tourism has a vast scope in the entire basin. There are many places of tourist's interests. Roopkund, which is known as untold mystery, Bedani Bugyal, Gwaldom and Karanprayag, are the major sites for tourism and its development. The rich cultural heritage in the forms of fairs, festivals, and customs can help in the development of tourism. The presence of Great Himalayan Range and its peaks exhibits the potential for tourism.

Ecology and Environment

The Pindar basin consists of very high mountains and steep precipitous slopes. Throughout the area, marginal land farmers have attempted to squeeze out a living from lands not suited for crop production, and have only ruined the entire landscape in the process, but have created hazards for floods and sedimentation in the area down below. These steep mountainous areas have their own characteristics response to erosion. Avalanches, land slides, and land slips, etc., are fairly common. High intensity run off of base land surface would not usually permit any trace of soil on these hills. However, there are few redeeming features, which make the slopes sufficiently stable to allow some kind of land use. Poor impact of raindrops coerces skeletal soil and fast vegetation recovery is some of the conditions that favour soil protection. The pressure of population, except, in areas where soil is easily workable, is too little. Soil working is restricted to hand digging. Holdings are small, usually confined near the habitation. Poor accessibility of the area would not provide for major technological changes in near future. On the basis of these assumptions, it may not be necessary to think of major land use changes. The strategy for watershed control would mostly confine to improving practices within major land use regions. The survey aspect should therefore aim at locating areas whose potential is not being fully utilized. For example, if marginal land is suited for agriculture, it should ecologically, be treated through reducing its pressure of overgrazing and developing plantation agriculture on it.

Ecological problems dealing with environmental degradation consequently result in low productivity in the Pindar Basin. Natural hazards, i.e., cloud bursting, soil erosion, mainly in rainy season, and unrationalized human interferences like unscientific quarrying and blasting for road and dam construction, over grazing, over lopping, over fallowing, over ploughing, and dam deforestation, consequently, give birth to low fertility rate and low production and productivity. Generally, landslides in the valley occur when hillside or valley side slopes fail due to certain geological, climatic, and biotic factors.

The Pindar River is perennial. During the winter, the river flows far below the danger mark and in summer mostly on the average level and sometimes above the level due to the snow melting in the Himalayan ranges, but during the rainy season, it flows above danger mark. A heavy rainfall occurred in the valley on 20 July, 1970, consequently, the two steep slopes slides and blocks the water of the river for more than 15 minutes; afterwards, flood situation occurred and swept away human and cattle life and damaged the low lying areas of the valley. The river Alaknanda was influenced by the flood and even the low-lying areas of Srinagar town were silted. During the blockade people made a good collection of fishes. Beside this, there were series of landslides which took place during the different rainy seasons. The cloud burst of 1972 in Paithani, natural calamity of Shilodi in 1889, calamity of Bainoly, cloudburst of Gadini Gar in 1992, and cloud burst of Musudiar in 1993 were among the other natural calamities that took place in the course of the Pindar River.

A landmark anomaly is found in the basin. Mainly the basin is divided into two climatic zones; namely, 'Danda' a highly mountainous zone and 'Gangar' a low-lying area of the valley. The word 'Danda' is characterized by highly precipitous zone, predominant with humidity in the atmosphere, which is useful in healthy production of crops without irrigation. The effect of drought of 1987–88 was negligible in this zone (Danda). The corps which are grown in this zone are mostly cash crops like ramdana, potato, soyabean, beans, kulthi, tore, etc., in high elevation of the zone (more than 1,500 m height), and rice, wheat, barley, manduwa, koni, jhangora, jawr, bazaar, corn and various types of pulses grown between the 1,200–1,500 m altitude. Among the millets; jawar, bajara, koni, jhangora, and among pulses; tore and leguminous plants are sown in the steep slopes and between the ridges of the fields. Due to long roots, the plants protect the soil erosion. The term 'Gangarh' points to the low-lying areas below 1,200 m and is characterized by the low rainfall and a dry hot zone. Irrigation is essential for high production and productivity in these zones. The main means of irrigation are canals and lift irrigation schemes. The zone is more productive than the Danda zone. The main crops are paddy, wheat, barley, pea, lentil (masur) and gram.

The land resources of the valley like forests, water, mineral, soil, and physical landforms have been deteriorated frequently due to excessive erosion and frequent occurrence of mass wasting which is a direct consequences of deforestation, overgrazing, over lopping, over fallowing, over ploughing, unscientific mining and quarrying, over blasting for construction of highways and high dams, and the neglect of the conservation measures. The over lopping is generally taken into illegal practices by the graziers in the whole of the Garhwal (Saxena, 1988).[4]

The heterogeneity in climate and high anomalies in the landforms make a critical variation in the potentiality and capability of crops. Owing to the inaccessibility of the villages, due to lack of means of transportation, the use of fertilizers, insecticides are negligible. The traditional methods of ploughing and small size of fields lead to the low capability of yields.

Conclusion

What appears to be lacking in the approaches to development planning in the Pindar basin is integration between aspects of environment and economic development, on the one hand, and between development of infrastructure and economic activities and among different activities, on the other. Such an integrated approach is feasible only if it is based on area-level planning; decentralized implementation of plans made at the central and state levels has so far proved to be inadequate. No doubt, certain environmental aspects need to be centrally regulated, but development plans need to be prepared and implemented at the decentralized area level within rationally and clearly defined regulations and guidelines in the context of the use of environmentally sensitive resources. This alone can ensure that the development patterns of the region are based on its specific resource endowments and relative advantages and are sustainable both economically and ecologically.

The rough, rugged and uneven landscape, has given birth to a low living standard of people who depend, directly or indirectly on agriculture. This is the fate of the area, which is unable to fulfil the needs of people to meet just a square meal. Hardly, one-third of the annual need is met by the local agricultural production. Obviously, the inaccessibility of the region, outdated modes of cultivation, near-total absence of fertilizers and insecticides, and new seeds account for the low yield of crops. There is a redeeming feature too. There is a rich fertile linear belt on both sides of the river from Gwaldom to Karanprayag covering more than 95 per cent cultivated land. Easy accessibility, use of fertilizers and insecticides, HYV seeds, sufficient irrigational facilities would definitely raise the production in these river terraces many times higher than anywhere else in the basin. From the poor Danda the legendary term 'Ukharh' (dry land) still reflects the poor cultivation. The 'Gangarhs' are fortunate enough to have enough patches of 'Talouns' (irrigated land), which provide a scope for experimental farming. However, the developmental agencies have yet to venture into such experiments in the area.

To protect the valley from the geographical hazards following preventive measures should be introduced:

1. High power explosives should not be used for road construction.
2. Underground drainage system should be adopted for small channels and streamlets.
3. To protect roads from erosion, high embankments enforced with iron concrete should be constructed.
4. Large-scale deforestation, over lopping, over grazing, and bare fallowing should be checked.
5. Unscientific farming on the valley slopes should also be prevented.
6. Afforestation and fruit gardens should be encouraged on the valley sides and spurs.
7. The soil and water loss caused by gully formation and fast flow of seasonal and perennial rivers, must be checked and conserved by scientific measures.

Notes

1. Sati, V.P.: 1993: "Problems of Agricultural Ecology and their Management in the Pindar Basin", *National Geographer*, Vol. XXVIII, No. 2, pp. 143–148.
2. Sati, V.P., ibid.
3. *The Hindustan Times*, 12 November 2000.
4. Saxena, P.B.: 1998: "Alaknanda Basin, A Modern Approach in Geography".

11 Landslides Hazard and Its Influences on Biodiversity in Nanda Devi Biosphere Reserve

R.B. Singh and Suraj Mal

Any sudden and rapid down slope movement of bedrock, regolith or a mixture of these two as a result of pull of gravity is commonly known as landslide (Crozier, 1986). The constant pull of gravity makes all hill-slopes and high mountains susceptible to landslides. When it occurs, earth's material is transported down-slope until a stable slope is re-established. Landslides have negative implications on mountain biodiversity. It sometimes alters the evolutionary process of ecosystem. It leads to loss of biodiversity in terms of habitat loss, loss of flora and fauna, causalities of animals etc. Landslides pose massive threats to floral diversity in terms of devastation of forests. It is equally a great threat to faunal diversity as the accidents of wild and domestic animals and human population are very common phenomena in mountain regions. Studies show that landslides have increased in the Himalayas (Singh and Pandey, 1996) particularly in Nanda Devi Biosphere Reserve (NDBR). Study shows that landslides have increased in NDBR due to which landslide induced biodiversity loss has also increased. Therefore, present paper aims at following objectives:

- To assess the trend of landslides occurrences in the study area.
- To assess the linkages between the landslides and loss of biodiversity.

Present research paper is based upon primary and secondary data sources. The secondary source of information comprises of archival records, academic works, library collection and research publications. The institutions such as Wadia Institute of Himalayan Geology, Dehradun; Wildlife Institute of India, Dehradun; Forest Survey of India, Dehradun; Indian Institute of Remote Sensing, Dehradun; various departments of geology, geography etc., and researchers have been consulted for secondary data.

Apart from this, study area was visited twice in September and October, 2005 in order to collect primary data. Primary data was collected through questionnaire and interview methods (group and individual) to gather the information regarding the general perception of the villagers about landslides and its influence on biodiversity. Villages for survey were selected on the basis of purposive random sampling. A total of 200 respondents from 10 villages were surveyed. Four villages near the core zones, four villages inside the outer boundary of buffer zone and two villages in the transition zone were surveyed.

The data was analyzed and synthesized by using various Geographical Information System (GIS), cartographic, statistical method, tabulation, diagram, graphical and other techniques. Various GIS software such as Arc View 3.2a and Erdas Imagine 8.7 were also used in present study.

Study Area

The NDBR is located in districts of Chamoli (Garhwal), Pithoragarh and Almora (Kumaon). It is situated in the Great Himalaya in the upper watershed of the river Alaknanda, Rishi Ganga, Dhauli Ganga and Pindar River. It lies between 79°13′ E to 80°17′ E longitude and 31°04′ N to 30°06′ N latitude. For the first time, Eric Shipton and W.H. Tilman made their way to the Nanda Devi basin successfully in 1934. An area of 182.63 sq km was then declared as Nanda Devi Sanctuary in the year 1939 with the influence of Shipton and Tilman. The sanctuary was converted into National Park in 1982 covering an area of 624.62 sq km. Further the conservation efforts were accelerated with the upgradation of National Park to Biosphere Reserve in 1988, increasing the area up to 2,236.74 sq km. Subsequently in 2002, the area of Biosphere Reserve was increased up to 5,860.69 sq km including the Valley of Flowers National Park and surrounding area in it. An area of 524 sq km was further added as transition zone, bringing NDBR to a total area of about 6384 sq km. The area of core zones is about 712 sq km while buffer zone covers about 5,148.57 sq km. It surrounds the core zone on all sides.

Presently there are 47 villages in the buffer zone of the Reserve and 54 in the transition zone, while both the core zones are free from the human habitation. Of the 47 villages of the buffer zone, 34 are in the Chamoli sector of Reserve while 10 and 3 are in the Pithoragarh and Bageshwar sectors respectively.

The NDBR was one of the 14 established biosphere reserves throughout India to represent country's vast biological and geographic diversity under the Man and Biosphere programme of UNESCO. The entire Reserve has been divided into three zones viz., the core zone, the buffer zone and the transition zone. There are two core zones i.e., Nanda Devi National Park and Valley of Flowers National Park both designated as world heritage sites.

Physiographically, NDBR is a vast glacial basin with high altitudinal variations. It consists of many snow laden lofty peaks, large and small glaciers, glacial torrents, waterfalls, alpine meadows, hillocks, colourful boulders, moraines, steep gorges and overhanging cliffs, besides the thick forest down the valley. The boundary of NDBR runs along prominent natural features.

The climate is monsoon type with an average annual precipitation of 900 mm. It varies from 700 mm to over 2,000 mm, which is restricted to short period of two months (July and August). The months of September-November are dry. Winter rain occurs occasionally in January with very low temperature. Thunderstorm with hail may occur in April-May with a short summer season. The maximum temperature ranges between 32–25°C in July, while minimum temperature goes below 0°C.

Landslides in Himalaya

Landslides are very common in Himalaya especially during monsoon season. Himalayan landslides can be attributed to the fact that, Himalaya lies at the convergence zone of two lithospheric plates i.e., Indian Plate in the south and Eurasian Plate in the north. Thus geologically, it is considered very active as the height of the Himalaya is still increasing. It is made up of complex geology, geomorphology and geo-hydrology. There are several big thrusts viz., Main Boundary Thrust (MBT), Main Central Thrust (MCT), Almora Thrust, Vaikrita Thrust, Krol Nappe, Chail Nappe etc., dividing Himalaya into number of fragments. Presence of large number of faults and lineaments make the region geologically very fragile and susceptible to landslides. Little imbalance in elements of above stated shear stress factors might trigger landslides. During rainy season, additional water is added to the existing factors, which accelerate shear stress factors. Thus, condition of the bedrock become fluid and landslides take place. Crozier (1986) suggested that landslides do not occur in isolation, there are number of factors that trigger landslides and in return landslides also influence their physical environment. As per field investigations, it was found that most of the landslides are natural and anthropogenic activities such as construction of roads and dam are triggering events.

Increasing anthropogenic activities particularly construction of roads and dams have further contributed to the process of slope failures. Introduction of road has become a prime cause of landslide occurrence in Himalayan region (Singh and Ghai, 1996). Similarly, Gongxian (1996) reported that railway tunnels were an important cause of landslides in mountains of China.

Salient Features of Himalayan Landslides

Some salient features of Himalayan landslides are described below:

- Landslides in Himalayan region are not of very huge scale. In Rockies and other huge mountains, there are large numbers of landslides involving hundreds to thousands of million cubic metres of rock material but numbers of huge landslides are very less in Himalaya.
- Repetitive landslides are significantly more in Himalaya. These are noticed particularly in highly fractured and sheared rock-mass and also in soft-hard rock alterations. In the intervening period, climatic factors become operative until sufficient weathered and disintegrated materials are brought down by heavy downpour.
- Though in Himalaya, summer rains start in month of June, devastating landslides mostly takes place in later part of the monsoon i.e., August and

September. Probably, this is because initially the subsoil can accommodate lot of percolation. But when the subsoil becomes saturated after a few months, it cannot any more absorb water. Thus, the surface run-off increases, which causes landslides. Another factor may be that earlier rain weakens the slope forming materials while later rains remove them.

- It is observed in number of cases that along the river valleys when the main fissility in competent rock dips into the hills relief, fractures open up but when the same fissility dips towards the valley, fractures rarely develop. Many failures are known to occur due to slip along such fractures during construction, sometimes killing people. A large number of Himalayan landslides are of shallow depth.
- Most of the landslides in the Himalaya are caused by anthropogenic interferences like road and dam construction.

Landslides in Nanda Devi Biosphere Reserve

NDBR lies at the interface of glacial and non-glacial environment and peri-glacial lies between the two. So the mechanisms and types of mass wasting change according to prevailing environment. At the lower altitudes in non-glacial i.e., fluvial environment, landslides are common phenomena. In the peri-glacial environment, *geleflucion* (freezing-thawing process) is most common resulting in disintegration of bedrock that forms the pattered ground, while avalanches are found at high altitude in peri-glacial and glacial environment. Field investigations, discussion with concerned researchers working in Reserve and extensive literature survey suggests that landslides triggering events can be categorized into two groups, (1) natural factors and (2) human activities particular slope modification. (Figure 1)

Figure 1

Hot Spring Near Tapoban

Natural Driving Forces

It includes geomorphology, geology, excessive rainfall, triggering events etc. Slope of geologically active NDBR are made up of complicated geo-structural and topographic set-up. Reserve is located north of the MCT-II, while the Vaikrita thrust i.e., MCT-I passes across the area. Presence of number of faults and lineaments like Malari-Debrugheta fault provide a very fertile ground for landslide occurrences. A total of four hot springs are reported from the Chamoli sector of the Reserve.

Presence of numbers of nick points in river courses of Alaknanda, Bhundyar Ganga, Sarashwati, Dhuali and Rishi Ganga as reported by Shah (1991), Bisht (2002) and by researcher himself during the field survey in 2005 prove in rejuvenation and fragile geology of the region. Shah (1991) and Bisht (2002) reported about five knick points in Alaknanda River, from Satopanth Lake to Vishnuprayag indicating rejuvenation and channel sedimentation phases. The longitudinal profile of the upper reaches of Alaknanda (near Satopanth Lake) is steeper and of glacial erosion, while channel gradient is quite low in downstream of Satopanth Lake that represents glacial depositional phase. The profile shows sudden steepness downstream of Badrinathpuri (Figure 2). The channel gradient is low at Mana-Badrinath basin and Lambagarh where Alaknanda was blocked for some time during the past. Gradient of river is steep from Lambagad up to Govind Ghat, then again slopes lose its steepness. Comparing the longitudinal profiles of Alaknanda, Bhundyar Ganga and Saraswati rivers, a distinct knick point is observed at 3400 m, which clearly represent a dynamic phase of rejuvenation in the area. There is an occurrence of riverbed scours and plunge pool marks on highly fractured and jointed granitic rocks up to 200 m from the present riverbed in

Figure 2

Longitudinal Profile of Alknanda, Saraswati and Bhundyar Ganga Showing Knick-Points

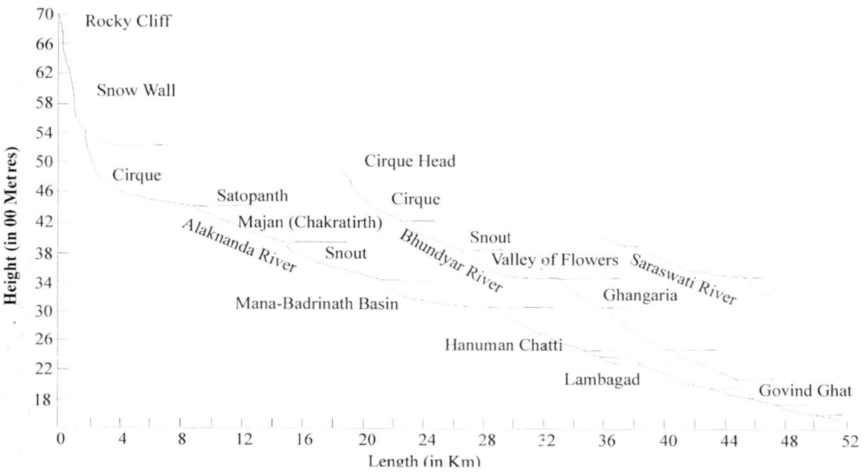

Source: Shah, 1991.

Saraswati at Mana, high channel slope of Bhundyar Ganga after 3,400 m formation of Mana-Badrinath basin. Below this altitude the deep gorge downstream of Govind Ghat are clear indicators of rejuvenation in the area.

Dhauli Ganga has most tortuous course than any other river of the region. There are three sudden falls in the river of Dhauli Ganga between the villages of Malari and Tapoban. The fall (knick point), which is about 6 miles above the village of Tapoban, is the maximum. It falls about 150 m in the distance of 250 yards only. Two knick points observed in the river are near village of Jumma and 3 kms downstream of village of Malari. Most of the course of the river is under huge boulders. Thus the river is hardly visible most of the time. Whole course of the river can be said to be through a narrow pass with almost perpendicular cliffs on either side, several thousand feet high and wild in the extreme. All the above stated factors implies that the region has fragile geology and experienced rejuvenation in past. It provides a fertile ground for landslides occurrence. Wherever excessive rainfall, triggering events and fertile ground for landslide occurrence are found, it results in landslides.

Human Driving Forces

Figure 3a

Stage-I Natural Slopes: Pre-Modified Stage

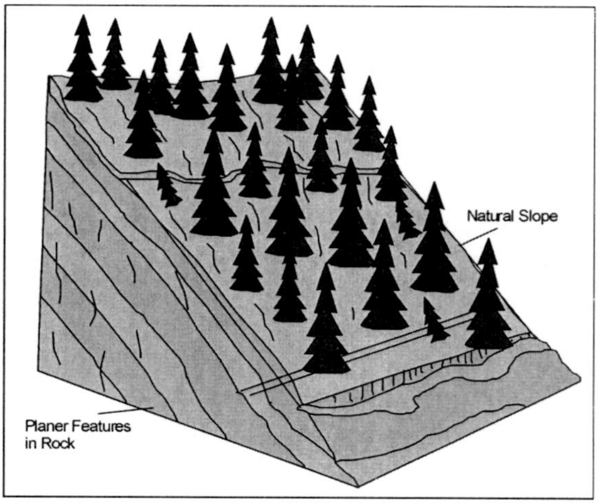

Human factors include activities like road construction, dam construction, hotel-resorts etc. These have strong bearing on the landslides. First road in Himalayan region was introduced way back in British colonial period (Singh and Ghai, 1996). But it was only after the Indo-China War of 1962 that the road construction was intensified in the Himalaya. Thereafter, Indian engineers blasted gigantic networks of road and communication facilities deep into hills of the Himalaya. In NDBR, the history of road construction goes back to 1960s. In 1964, for the first time, roads were constructed in the Reserve. Local people reported that initially about 135 kms of roads were constructed in the region. About 50 kms of road were further added to existing network of roads in 2000. Now the road in Mana valley has been declared as national highway (Haridwar-Badrinath National Highway) due to which slopes are now in the process of being over-modified. Thus, shear stress factors are exceeding shear strength factors and landslides are increasing. After the introduction of roads, landslides have become very frequent in the Reserve. The correlation between construction of roads and landslides is depicted in Figure 3a–c. It has a three-stage relation.

First stage, which is the pre-modified stage of natural slope, shows the steady-state condition (Figure 3a). Then the natural slopes are modified or undercut for the construction of road and dam etc. Thus, steady-state condition is disabled (Figure 3b). Then landslides take place so that steady-state condition is re-established (Figure 3c). Therefore, landslides can be said to be nature's rule of equilibrium.

Figure 3b

Stage-II Disabled Steady-State Condition

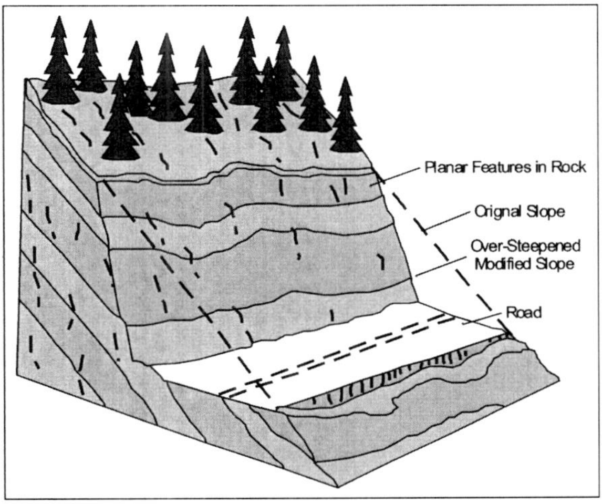

Figure 3c

Stage-III Nature Re-establishing Steady-State Condition

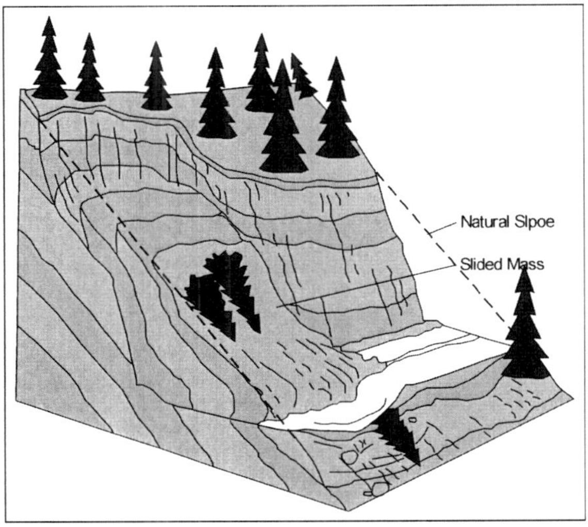

Built-up land has also increased in recent past due to which natural slopes have been over-modified. It has also led to deforestation in the Reserve. Thus, in many areas, steady-state condition has been disabled and have eventually resulted in increased frequency of landslides.

Strong positive correlation between occurrence of landslides and construction of roads were observed during the field investigations. Villagers reported that earlier landslides were not very common but after the introduction of road, this has become a common phenomenon. About 80 per cent of the landslides occurring now are result of road construction.

According to field survey, most of the villagers consider that road construction, excessive rainfall and seismicity are main causes of landslides. About 83.5 per cent, 88 per cent, and 88.5 per cent respondents reported road construction, excessive rainfall and seismicity respectively as the most common causes of landslides occurrence in the region. Whereas declining forest cover, grazing and agricultural land expansions are other causes of landslides (Figure 4). Agricultural land expansion, grazing and deforestation are not main causes of landslides in the region as they are in other Himalayan region, which is the result of ban on utilization of local resources. Road construction has the recent origin as the prime cause of landslide in the region.

Figure 4

People's Perception Regarding Causes of Landslides

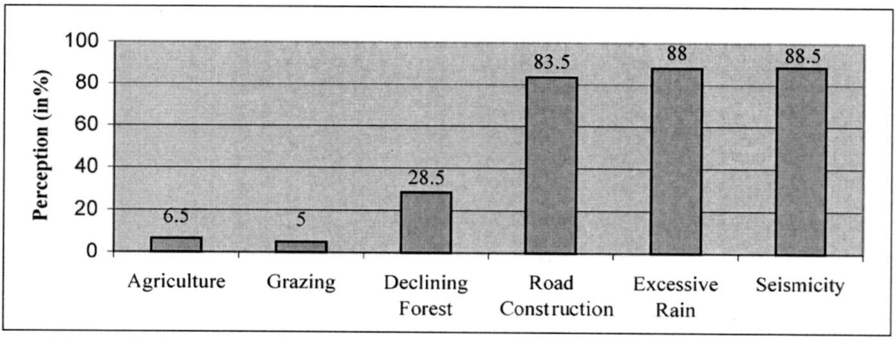

Most of the region is the peri-glacial and glacial so ideally there should be much snow fall rather than rainfall. But local people reported that snowfall has decreased and rainfall has increased in last few decades. This can be attributed to global warning due to which rainfall regions are shifting upwards and taking place of snowfall in the Reserve. Thus, excessive rainfall in first half of monsoon saturates the bedrock and in later half removes it in the form of landslides.

Road construction and rainfall have increased in the Reserve thus landslides have also become very frequent. Survey results show that the landslides have been increasing rapidly in region in last three decades, as about 51 per cent of the respondents reported the same, while 33 per cent reported that landslides are increasing slowly and only 16 per cent stated that trend of landslide is almost static (Figure 5).

Figure 5

People's Perception Regarding Trend of Landslides

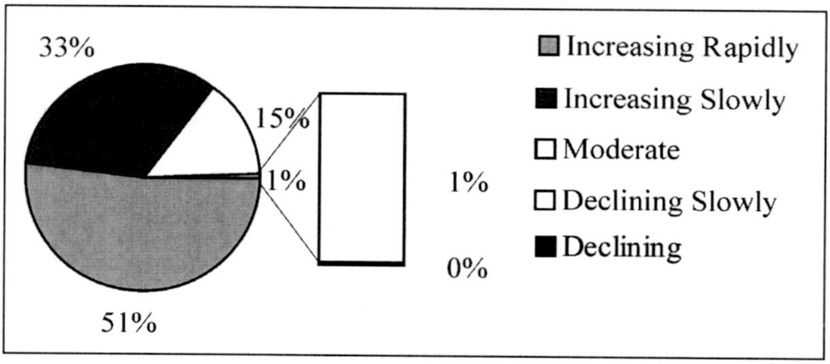

On the basis of literature and field survey, history of major landslides has been constructed. A total of 10 major landslides have been noticed. There must be some more huge landslides in the region. But these are not documented properly due to high inaccessibility of the region to the outer world. A very interesting fact can be seen (Table 1) that almost all the huge landslides are natural. About 30 landslides were noticed during field survey (Figure 6). Of the total landslides, 21 were human induced and 7 were natural. All the human induced landslides have recent origin. Thus, it can be said that human activities have aggravated the problem of landslides in recent past.

Table 1

History of Landslides in NDBR

Year	*Causes*	*Description*
1939	—	Mana village was partially abandoned due to extensive rockfalls.
1978	Avalanches	Bamni village near Badrinath was completely washed away.
1968	Related to Flash Floods	Rishi Ganga in Garhwal was blocked up to a height of 40 m due to a slide at Reni village. The dam breached in 1970, caused extensive damage.
July, 1970	Exceptionally heavy rainfall	Alaknanda river caused considerable loss of life among pilgrims. Many bridges, houses and an entire village were washed away.
Sep., 1970	Exceptionally heavy rainfall	Landslide and house collapses killed 223 people.
July, 1970	Landslide due to dams	Floods in Rishi Ganga created 40 m high blockades near village of Reni. Lake silted up by May, 1970 and eventually blockade breached in July, 1970.
Sep., 1999	Earthquake induced landslide	North of Joshimath.

Source: Literature Survey.

Figure 6

Landslides and Its Influences on Biodiversity

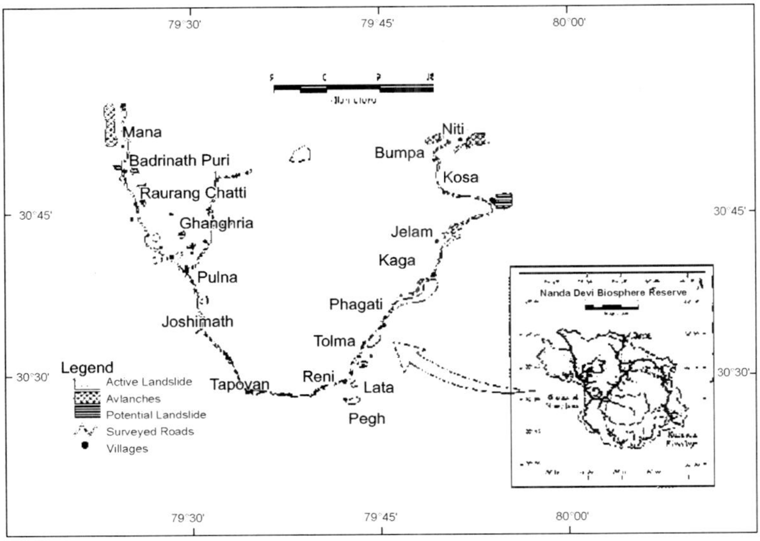

Presently, intensity of landslides occurrence is high in NDBR as about 87 per cent of the respondents indicated the same, whereas only 13 per cent indicated moderate, low and very low intensity. People who reported moderate low and very low intensity of landslides inhabit upper reaches of river valley and practice seasonal migration. During monsoon, these areas receive less rainfall. Thus, landslides are not as common as in lower reaches of river valley. During winter when avalanches take place, they migrate to low altitude. Thus, they do not witness frequent landslides. People's perception about present status of landslide is presented in Figure 7.

Figure 7

People's Perception Regarding Present Status of Landslides

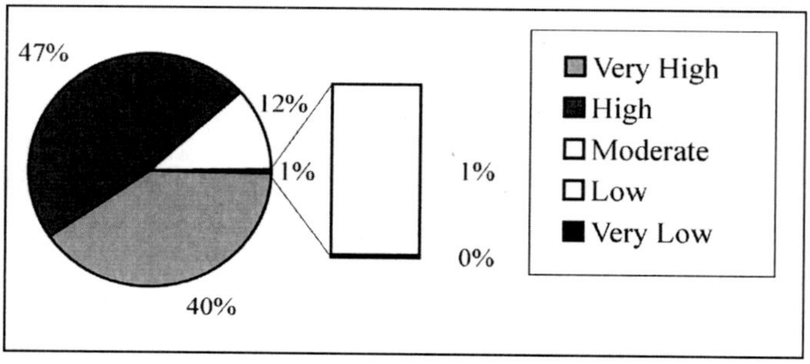

Landslides and Its Influences on Biodiversity

Landslides are closely related with loss of biodiversity. According to Crozier (1986) landslides are product of their environment and in return they also influence their environment. Landslides are caused by combination of natural and human factors; in return landslides devastate large areas under forests, grassland or bare land, which is the loss of habitat of number of endangered and threatened animals. Devastation of forests and grasslands are the loss of flora (Figure 8). Some time animals meet with accidents due to landslide, which is loss of fauna.

Figure 8

Landslide Causing Habitat Loss

During last three decades, landslides and avalanches have increased significantly due to which biodiversity loss has also increased in the Reserve. On the basis of field investigations and discussion with concerned scientists and villagers, it was established that it is not the occurrence of landslides only that influence the biodiversity, but biodiversity itself influences the landslides pattern too. These complex relations are listed below:

- It was observed that there was a strong correlation between species of pine family and occurrence of landslides. Wherever the pine trees are found, the area is severely affected by landslides. This statement may be little critical because forests are considered as major determinant of landslides. It is often said that area, which is under forest cover, would be least exposed to landslides and vice versa. But as far as pine species are concerned, on the basis of field investigations and discussion with scientists and local villagers, above stated relation can be established. Discussion with scientists suggests that the head-weight of pine tree is very high and wherever the pine is found on steep slopes, landslides are likely to take place during monsoon, if other conditions are favourable.

- Based on field investigations it was also found that there was strong correlation among Buransh tree, schist (geology) and landslides. It was noticed that wherever Buransh are found on schists, the area would be more prone to landslides. Field investigations show that these two species of flora and landslides have increased in the region. Thus, above stated relation can be established among them.

Two valleys viz., Mana Valley (Upper Alaknanda Valley) and Niti Valley (Dhuali Ganga catchments) have been approached for study, thus relations of landslides and biodiversity loss are initially classified in two classes i.e., case of Mana Valley and case of Niti Valley. Further, these are classified in natural and human induced landslides. The criteria to classify landslides as natural and human induced were the degree of influence of human activities. The landslides, which are near the man made structures such as roads, villages and dams, were called human induced landslides. The landslides, which are far away from every human activity and in natural environment, were said to be natural landslides. About 40 landslides and about 17 avalanches sites were encountered during the field investigations, while rock fall is common phenomena along the roads and rivers.

Case of Mana Valley

The Mana Valley consists of two major centres of religious tourism i.e., Badrinathpuri and Govind Dham. It also harbours Valley of Flowers National Park (VoFNP). Thus, the valley faces heavy inflow of tourists. The road of the valley has been declared as national highway, due to which slope modification is very intense. This has resulted in increased frequency of landslides. About 16 landslides were encountered in this valley, of which only 3 were found to be natural and rest of the 13 landslides triggered by human activities. Extensive field survey resulted in 9 of the landslides to be active and 7 to be dead which can be active any time, hence termed as Potential Landslides (Table 2). Landslides along riverside are

Table 2
Landslides in Mana Valley

Factors	*Active Landslide*	*Potential Landslide*
Natural	Valley of Flowers (2)	Pulna (1)
Human Induced	Chain, Govind Ghat, Hanuman Chatti (a), Rorang Chatti (a), Hathi Pahar, Baldoda, Lambagad (a)	Lambagad (a & b), Vinayak Chatti, Hanuman Chatti (b), Rorang Chatti (b)

Source: Primary Survey.

not numbered since it is very difficult to observe them because of deep canyons and rough terrain. But they are very common. Rock fall is also very common along roads. In almost all the cases of human induced landslides, modification of natural slope for road construction was found to be prime cause of landslide.

It is depicted from Table 2 that most of the landslides are human induced. Most of the human induced landslides are triggered by slope modification for the construction of road, except landslide near Lambagad. Landslides near Lambagad are a result of the construction of dam and tunnels. For the construction of tunnels, the blasting technique is used beneath the mountains. It impacts in many ways like blasting results in expansion of cracks and fragile condition of regolith and bedrock, which can any time slide down the slope. Construction of tunnels results in vacuum under the mountain, which results in subsidence of tunnels if the faults, lineaments etc., are reactivated. There are three landslides near Lambagad of which two are very active and one is presently dead. It can reactivate any time as one of the dead landslides activated before two years. Landslide near the Hathi Parvat is a recent one. It came into existence after the introduction of Vishnu Prayag Hydal Project. Natural slopes near the Hathi Parvat are overmodified and undercut for efficient transport network for fulfilling the demands of project. Now this landslide is a perennial one. All the landslides are resulting in vegetation loss, which is also loss of habitat (Figure 9) and ultimately affects the fauna of the region.

Figure 9

Road Constructions, Landslide and Vegetation Loss

Landslides occurring due to natural steepening of slopes and reactivated geology are called natural landslides. Landslides, which were dead from long time, reactivated after Chamoli earthquake in 1999. Landslide near Pulna village and both the landslides of VoFNP are such slides (Figure 10). Natural steepening of slopes as a result of undercutting action of river along its bank is a common phenomenon in study area. Most of the river courses are severely affected by this problem.

Figure 10
Landslide in Valley of Flowers

At the higher altitudes avalanches are most common. They can be seen 2,100 m above mean sea level. About 3 km of the area above Badrinathpuri to village of Mana, along the left bank of river Saraswati, severely suffers from avalanches every year (Figure 11). There are about 10 avalanches sites. About 7 avalanches sites were seen in VoFNP (5 km away of Ghanghria) and from the way to Govind Dham (4 km away of Ghanghria) (Figure 12). Avalanches sites were also noticed below 1 km of Ghanghria.

Figure 11
Avalanches Sites in Badrinathpuri

(Photo by Manish Mehta)

Figure 12
Avalanches Near Goving Dhan

Case of Niti Valley

Altogether, landslides are less common in Niti Valley than that of Mana Valley. It is due to the fact that construction activities are strictly prohibited in this valley. Road transport is not very intense. About 11 landslides were encountered in the valley of which about 6 were human accelerated and rest 5 natural.

In this valley, human induced landslides are not much because road construction is not very intense. All the six human induced landslides were active. One of the human induced landslides is very huge and intense (unnamed place), which is about 6 km ahead of Suraithota and rest 5 are big (Table 3). This landslide is human induced and natural as well. The river of Dhuali Ganga widens its course and undercuts the base of the natural slopes. Border Road Organization (BRO) does slope modification for the construction of road. Dhauli Ganga washes away the road at least twice in a year, and BRO try to construct it. Hence this landslide is very active. This site is of about 2 km.

Table 3
Landslides in Niti Valley

	Active Landslide	Potential Landslide
Natural	Daunigiri, Malari, Saigiri	Peng, Lata, Tapoban
Human Induced	A.T. Nala, Suraithota, Ravigram, Tapoban, Gadinala.	

Source: Primary Survey.

Landslide near Saigiri village is a good example of natural landslide. This site was activated for the first time in 1999 due to heavy rainfall. Earlier it was a small perennial drain and lifeline of Saigiri village. But in 1999, it activated at night and took only two hours to get its present shape (Figure 13). It devastated whole of the village and killed about 10 people. The village is now uninhabited. This landslide is active throughout the year.

Figure 13
Debris Flow Near Siagari Village

Landslide near Lata village is also a good example of natural landslides, which first time occurred after Chamoli earthquake of 1999. Local people reported that initially it was like a small volcano. Lava and hot water burst out of earth. Thereafter, it has become a rockfall zone (Figure 14).

Figure 14
Landslide in Lata Village (1999)

Landslide near Dronagiri village at Daunagiri Parvat is, perhaps, the result of activation of Malari-Debrugheta fault. Presences of huge sediment load in the stream make this area more prone to landslides (Figure 15). This landslide caused huge loss of forests and wildlife.

Figure 15

Landslide Near Daunagiri Village Causing Habitat Loss (1999)

A large number of avalanche sites were also noticed in this valley. Avalanches sites are very common to observe after Lata village and 2,200 m above sea level. Avalanches sites were observed near Peng-Reni, Suraithota, Malari, Gamshali and Niti village. In almost all the cases, avalanches have become major threat to biodiversity as they cause significant loss of flora (Figure 16) and fauna. Loss of vegetation due to avalanches was very common but no evidence of loss of animals was found. However people reported that it was very common in the Reserve.

Figure 16

Avalanches and Loss of Vegetation

Survey results show that accidents of wild animals due to landslides are common phenomena in study area but it is very difficult to observe because the terrain of the region is very rough and undulating. Number of tourist guides, porters and local villagers reported the same. During the monsoon season, injured animals can be observed in interior regions of Himalaya. High degree of inaccessibility on the one hand and accidents of wild animals in extremely inaccessible area on the other make very difficult to observe landslides affected animals. Fortunately, I came across of one landslide victim animal i.e., Bharal during field survey, which was brought for treatment by Reserve authority (Figure 17).

Figure 17

Bharal Affected by Landsides

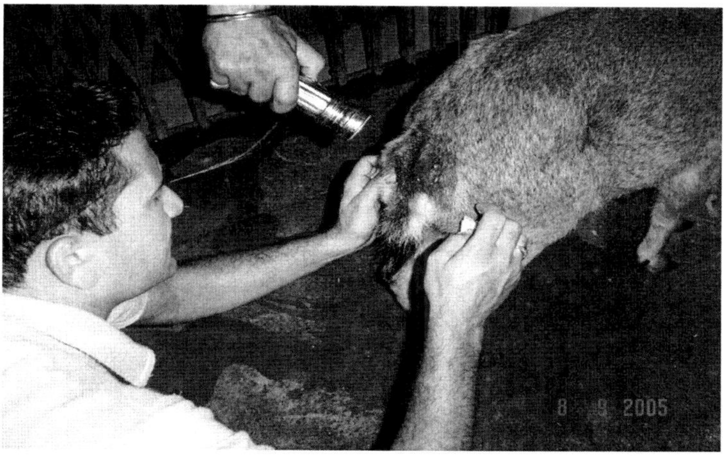

Conclusion

NDBR is geologically very dynamic and fragile. It is made up of complicated geology and geomorphological structures. Complex geo-hydrology coupled with intense human induced slope modifications has made this region very fertile for landslide occurrence. Landslides have increased in recent past particularly after the introduction of roads. Earlier, landslides were not very common. They were caused by natural factors mainly, but in last two decades landslides have increased tremendously. The increase in landslides is resulted mainly of human activities like road and dam construction. Though, natural landslides have also increased but human induced landslides have increased rapidly. Upward movement of rainfall regions has also influenced the landslides occurrence up to some extent. This often results in devastation of flora and fauna of the region. The loss of flora and fauna is likely to increase in near future as the landslides will increase in the Reserve.

Select References

Anonymous, (2001), "Atlas: Landslide Hazard Zonation in the Himalaya of Uttaranchal and Himachal State Using Remote Sensing and GIS Techniques", NRSA, Hyderabad.

Bisht, M.P.S. Mehta M. and Nautiyal, S.K., (2002), "A Report on Geomorphic Hazards Around Badrinath Area (Uttaranchal), and Innovative Control Measures Proposed", Department of Geology, H.N.B. Garhwal University.

Bisht, M.P.S., Mehta M. and Nautiyal, S.K., (2003), "Geology of Nanda Devi National Park", In *Biodiversity Monitoring Expedition: Nanda Devi 2003*, Uttaranchal State Forest Department, Wildlife Institute of India, GB Pant Institute of Himalayan Environment and Development, H.N.B. Garhwal University, India, pp. 55–62.

Crozier, M.J., (1986), *Landslides: Causes, Consequences & Environment*, Croom Helm Ltd., London.

Gongxian, W., (1996), "Relationship Between Railway Tunnel Damages and Mountain Landslides", In, *Landslides: Glissements De Terrain*, (ed.) Kaare Sennest, Vol. 1 Rotterdam, Netherlands, pp. 463–466.

Kamal Kumar, S. and Virdi, N.S., (2003), *Geotechnical and Seismotectonic Feasibility of Vishnuprayag Hydel Project, District Chamoli, Uttaranchal'*, Wadia Institute of Himalayan Geology, Dehradun.

Kumar, K., Tolia, D.S. and Kumar, S., (1996), "Landslide Hazard Evaluation in a Part of Himalaya", In, *Landslides: Glissements De Terrain*, (ed.) Kaare Sennest, Vol. 1 Rotterdam, Netherlands, pp. 257–262.

Mehrotra, G.S., Prakesh, S. and Dharmaraju, R., (1996), "Landslide Hazard Assessment–A Need for Environmental Management of Himalaya", In, *Landslides: Glissements De Terrain*, (ed.) Kaare Sennest, Vol. 1 Rotterdam, Netherlands, pp. 309–314.

Mullick, B.B., (1996), "Landslides in Himalaya: Causes, Types and Management", In, *Landslides: Glissements De Terrain*, (ed.) Kaare Sennest, Vol. 1 Rotterdam, Netherlands, pp. 327–332.

Naithani, A.K., Prasad, C., Bisht, M.P.S. and Kumari, G., (1997), "Landslide Zonation and Geo-environmental Appraisal along Main Central Thrust Zone in Mandakini Valley, Garhwal Himalaya, India", *Himalayan Geology*, Vol. 18, pp. 135–143.

Shah, M.P., (1991), "Some Geomorphic Observation on Badrinath-Satopanth Area, Chamoli District, Garhwal Himalaya", *Journal of Himalayan Geology*, Vol. 2, pp. 185–195.

Singh, R.B. and Ghai, S., (2000), "Human Induced Hazards and Development in a Part of Upper Yamuna Watershed, Garhwal Himalaya", In, *Contributions to Environmental Geoscience* (eds.) Pathan, A.M. and Thigale, S.S., Aravali Books International (P) Ltd., pp. 151–160.

—, (1996), "Environmental Disasters in Marginal Areas: A Study of Road Construction Induced Impacts in a Himalayan Yamuna Catchment", In, *Development Issues in Marginal Regions* (eds.) Singh, R.B. and Majoral, R., Oxford & IBH Publication, Delhi, pp. 311–320.

Singh, R.B. and Pandey, B.W., (1996), "Landslide Hazard in Indian Himalaya and Canadian Rockies: A Comparative Analysis", In, *Landslides* (eds.) Chacon et al., Balkema, Rotterdam, pp. 63–69.

—, (2000), "Landslide Hazard in Himachal Himalaya", In, Proceeding of the International Workshop on Geoenvironmental Hazard, pp. 71–84.

Yuji Maruo, (1979), "Geology and Metamorphism of the Nanda Devi Region Kumaon Higher Himalaya, India", *Himalayan Geology*, Vol. 9 (1), pp. 3–17.

12 Forest Biomass Withdrawal from the Middle Himalayas

A Case Study of Bajarni Village

Anuradha Sharma

The forest, "if properly managed, can enrich human life in variety of ways both material and psychological. Poorly managed they can be a source of disruption of the Environment of an entire region". These sources of disruption, which man created unknowingly in the past, resulted in the extinction of many well developed and flourishing civilizations. F. Engels (1947) is of the view that "people who rotted out forests in Mesopotamia, Greece, Asia Minor and other places if they get to plough land never dreamt that they were thereby laying the ground for the present desolation of these countries by depriving them of centre for collection and conservation of moisture".

M. Shafi (1950) has expressed similar views on the subject when he says, "History tells us that the decay of a large number of prospering nations in ancient times has been due to deforestation. In Greece, Anatolia, Spain and Iraq the destruction of forests has seriously interfered with their climate and with the moisture content of the soil on which, in the ultimate end, every nation depends. It will not be too much to say that deforestation has done more damage than any war and has resulted in the destruction of the greatest empires. In India the excavations of Harappa and Mohan Jo-Daro refer to the great civilizations which flourished in the Indus Valley, but gradual destruction of forests, allowed the rolling sea of sand to engulf the colony". Human beings have been interacting with the forests since time immemorial. Interaction is the mutual action between two biologically active systems, which receives energy from the sun, undergoes photosynthesis, converting this energy into biomass and the other that thrives on the energy so produced i.e., man and animals. The animals (domesticated ones) interact with the forest directly as well as indirectly. When these animals are allowed by man to graze in the forest they interact directly but when they consume lopped branches of the forest species or grasses cut and brought from the forested areas by men they can be taken as interacting with forest rather indirectly.

Thus, we see that forests constitute one of the prized assets of man which not only moderate climate, reduce soil erosion, regulate stream flow providing biomass to be consumed by its native dwellers, but this resource is a backbone of the modern economy too, as it provides values as fuel, building material, industrial raw materials and many by-products. Unfortunately Himalayas, a storehouse of this resource are experiencing serious ecological disorders. All these ecological and environmental problems of the Himalayas emerge from so-called economic development and anthropogenic transformation of the space. In addition, in the process of economic development, forests of the Himalayan region would be the obvious victims, as forest products are exploited for diverse purposes of economic transformation, more particularly in the plain areas. But unfortunately all this is done without any consideration of overall natural productivity, sustainability and stability of natural processes.

The importance of forests has a special significance for India where forests roughly occupy 23 per cent of the total area with still different percentages arrived at by the various departments and research organizations, not withstanding out of India's total forested area the share of Jammu and Kashmir state is only 4.5 per cent. In Kashmir region 56.56 per cent of its area is forested. The corresponding figure of Jammu is 45.51 per cent. We can say that more than half of the Kashmir region is occupied by the forests while little less than half of the Jammu region has forest cover. Moreover, the forests of the state are second to none in quality and value. The Jammu region, our study area, attracts out attention because the entire region enjoys considerable amount of climatic and altitudinal variations.

Plight of Himalayas is very serious in respect of vanishing forest cover. The Jammu region, the area of present study, lies between 32°17′ North to 34°18′ North latitude and 73°36′ E to 76°48′ East longitude. The entire area under study is situated in Himalayas where vanishing forest cover is posing serious threat due to this depleting resource. The entire Jammu region extends from 300 m to 5,000 m in the middle Himalayas. The impact of large scale exploitation of forests in the state as a whole has resulted in degradation of forests. Resultantly, poverty and associated ills have gradually crept into the lives of the people. Mountain ecosystem once densely forested has been drastically altered. In the process the inhabitants are facing a major ecological crisis and economic hardships.

It is in this context that the present study has been conceived and an attempt has been made to study the withdrawal pattern of forest resources which results in this interactive pattern. No doubt the study concerns only about one village but taking the location, climate relief and other environmental conditions of this village into consideration, it can be argued that it adequately represents the middle Himalayan region in which it is situated.

Study Area

Village Bajarni lies in district Doda, tehsil Doda and is 15 km away from the main city of Doda tehsil. The approach to the village is by bus till Bhagwa after which one has to climb up from the bus stop for reaching the village. There is no other approach to the village except animal transport. People go to Doda city for purchasing their requirements. The village is situated on a steep slope. There is a big nallah flowing in the village called Laddar Nallah. It is perennial. The source of the Nallah is at 'Lal Draman' and finally this nallah joins Chenab River at the base of the hill on which village Bajarni is situated. There are six other non-perennial nallahs within the village. There are 10 springs and no dead spring has been recorded, which indicates a balanced hydrological cycle prevalent in the area. At the top of the hill on which Bajarni is situated there is a beautiful meadow named as 'Lal Draman'. It is a beautiful spot at the height of 2,700 m a.m.s.l above which there is a very dense timberline of conifers. The climate of this village is largely temperate.

Methodology

The entire study is based on primary data collected through extensive and intensive field work. The aim of the study is (i) To analyze the primary data as to peep into the ecological and economic scenario of the people of this (representative) village lying in middle Himalayas of Jammu region; (ii) to assess the biomass withdrawal and interaction of animal population with forests and the importance of this interaction on the overall ecological condition of the village; and (iii) to draw meaningful conclusions, so that the study may be helpful in the developmental planning of the village in particular, the surrounding area in general.

The village under study is a very big village with 757 households. Due to constraints imposed by time and resources it was not possible to survey each and every household, hence 70 households during random sampling method were selected. The basis of the data was thoughtfully prepared questionnaire in which the questions were designed to bring out the withdrawal during interaction of man and animals in the forests. Our empirical study is, therefore, focused on generating data pertaining to the parameters identified i.e., biomass withdrawn from the forests by man in the form of fuel wood, lopped branches of trees and the grasses cut from the precipitous slopes as well as from the enclosed areas especially kept for the purpose. However, in doing so care was taken to consider the environmental, social and economic conditions influencing the interaction with the forests. The study area enjoys two dry and two wet spells in a year. March to June is the first dry spell which is followed by a wet season from July to September in which rainfall is caused by the monsoon winds. This rainy season is then followed by a dry spell from the late Sept to early December. A wet season prevails from December to March due to western disturbances.

The biomass withdrawn from the forests includes the lopping of the trees for fodder and the cutting of the grasses for precipitous slopes as well as from the enclosed areas specially kept for the purposes. This cut grass is stall fed to animals during the winter season (mid-November to mid-March) when the grazing is not possible, due to precipitation. The present study has been conceived to assess the biomass withdrawn for the forests from these two accounts as well. It is pertinent to add that the grass cutting process starts from mid-August and continues up to mid-October. However, the grass is cut from the enclosed areas only after it attains its maximum growth in the month of October and yet the people continue to store the grasses cut from the precipitous slopes from August onwards. An assessment of the grass biomass of so withdrawn from the forests has been attempted in the present study. The assessment of the lopped biomass was made by actually weighing the lopped branches of trees by people, selected as sample cases, for seven days and then working out an average per household per day. However, the samples were selected randomly. On the basis of the average, so worked to, the assessment of the total biomass withdrawn by loping was made possible for the entire village under study.

Grasses withdrawn from August to October are meant for stall feeding to the cattle during the intense winters Although the requirement of stall feeding varies depending upon the number and type of animals kept by a household and yet everybody wants to store as much grass as possible because it finds readily available local market too. Thus, the storage of grass biomass is independent of requirement and assumes an almost uniform pattern. In such a situation, the assessment was made by surveying the green grass withdrawn by single household. While surveying care was taken to assess as accurately as possible, the volume amount of grass stored.

Thereafter, an average for a household was worked out which when multiplied by total number of households would give us the total biomass withdrawn in the shape of grasses.

The data regarding the consumption of fuel wood was generated by conducting intensive field work in the village. The consumptional patterns of wood were observed. Enquires were made and actual wood consumed in different months was actually weighted before being released for the use in the selected households. All this was made possible by taking into consideration certain households representing different social and economic strata and were thus made sample households in the village under consideration. On the basis an average consumption of fuel wood per household was arrived at which was then multiplies by total number of households to get total consumption or conversely total withdrawal of biomass from the forests as fuel wood. On the basis of the above discussion, particularly with respect to climatic condition, a working schedule for biomass withdrawal was worked out which is elaborated in the Table 1.

Table 1

Schedule for Biomass Withdrawal

Forest biomass	*Period of Collection*	*No. of days in a year*
Fuel wood	January to December	365
Lopping	February to June and September to November	244
Grasses	August, September and October	92

Analysis and Discussion

Fuel Wood

Wood is the principal source of fuel in the village, rather the only source. Presently when the forests have been cut for the commercial purposes on an extensive basis, the broad leaved species have been heavily lopped for fodder, the pressure exerted by ever-growing population on the demand of fuel and timber has created a new situation. It is pertinent to state here that the wood in the selected village is also used for heating the houses as it lies above an altitude of 2,200 m from the mean sea level. As the temperature in the village drops to 0°C during the winter month, this therefore necessitates raising the temperature inside the house by using wood as a fuel. This phenomenon leads to higher level of fuel consumption due to the influence of climatic conditions prevailing in the different months of the year.

Before analyzing the withdrawal of fuel wood from the forests its has to be understood that there is a marked variation in the consumption of wood in different months of the year. It has been found that the consumption is lowest (16 kg/day/household) from May to September when fuel is not used for heating purposes. But in April, October to November when heating during the nights, mornings and evenings becomes essential the consumptional level increases from 16 kg/day/household to 24 kg/day/household. This figure reaches its peak of 40 kgs/day/household during the intense cold from December to March. Thus, the entire consumption per household from May to September (153 days) comes to 2,448 kg while the corresponding figures for 121 days of December, January, February stand at 4,840 kg which is almost two times the figures for 153 days of summer months (May to September). Similarly, total consumption per household in April, October and November reaches, 218 kg for only 91 days (Table 2).

Table 2

Average Fuel Wood Consumption in Village Bajarni: Fuel Wood used/day/household (in kgs)

April, Oct. & Nov.	*Total in 91 days*	*May to Sept*	*Total in 153 days*	*Dec. to Mar.*	*Total in 121 days*	*Average per day*
24	2184	16	2448	40	4840	25.95

The entire fuel wood used in the village under study is obtained from the forest. Females and children are the main fuel wood collectors but adults males also venture in the time of distress. The summer season is the optimum time for the collection of dry dead wood lying on the forest ground. But this, being insufficient to meet the requirements, the lopped wood for fodder is added to it and yet, as revealed by the villagers even this fails to meet the entire demand. Resultantly, it becomes necessary for the adults to cut branches of such valuable trees like deodar (*Cedrue* deodara), Kail (*Pinus* wallichiana), Fir (*abiespinsow*) and some of their common broad leaved associates. The villagers sometimes and when in need, may not hesitate to even cut a very healthy and valuable tree of any species prematurely, they can lay hand an, to meet their demand, contrary to the law established both by nature and man.

An assessment of the total fuel wood and in the village under study and seasonal variations in the consumptional behaviour as indicated in the earlier discussion has been assessed in the following table:

Table 3

Total Fuel Wood Consumption in Village Bajarni (in kg)

No. of Households	Apr, Oct. & Nov/day/ household	Total in 91 days by the village May to Sept.	May to Sept./day/ household	Total in 153 days	Dec. to March/day/ household	Total in 121 days
757	24	16,53,288	16	18,53,136	40	36,63,880

The summer and rainy season extend from May to September, with high temperature and mild rainfall. Fuel wood used during this period for the whole village accounts for 18,53,136 kg. Severe winters that start with the advent of December and extend up to the end of March demand still higher level of wood consumption and during this period, the village consumes a total of 36,63,880. Thus the total fuel wood consumed by the whole village in a year stands at 7,170.30 m tonnes (Table 3).

Lopping

Lopping is a common practice in the village under study. The experienced ones make certain consideration for lopping: (i) Leaf content of branch lopped, (ii) Length of the branch and (iii) Age of the branch. While in the field it was gathered that with the increase in the age of the branch, its leaf content and overall weight increases appreciably and so does its length and hence is preferred.

The intensive fieldwork has revealed that this practice is not carried out throughout the year and is restricted to 244 days i.e., eight months. The head loads of lopped tree branches were weighted to assess the actual biomass withdrawn. The length of the branch measured revealed the variation between 70 cm to 1 m. Every household indulges in lopping activity to provide better diet to the livestock and also to supplement the total requirement of fodder. Further, it was also observed that the head load of individuals varied appreciably. The data was then collected separately for women (young, middle-aged and old) and men (young, middle-aged and old). It was further observed that only one person for the family was spared for the collection of lopped biomass. The most common tress lopped are oaks of different shades, which are social associates of conifer forests (Table 4).

Table 4

Biomass Withdrawn By Lopping in Village Bajarni (in kgs)

No. of Household	Lopping/day/household	Lopping in 244 days	Total lopping in the village (in m tonnes)
757	25	6100	4617.700

The average head load per household calculated reveals that 25 kg of lopped biomass is withdrawn per day and in 244 days each household collects 6,100 kg. The figure for the whole village shoots to 4,617.700 m tonnes. This quantity of biomass lopped feeds on an average, the total livestock population of 7,534; this staggering figure therefore is reduced to only 1.75 kg per head.

Grass Cutting

Grass cutting is another important activity of the village under study. The total number of livestock is 7,534 in the village. The climatic condition allows the animals to graze only for 182 days in a year at the best and for the rest of the year these livestock are stall fed. Though the crop residue in also used as fodder for stall-feeding but is far less than required. The grasses are not available throughout the year, therefore the villagers look forward towards forests for lopping trees and cutting the grasses available on the slopes of the mountains. However, the grass cutting in this village begins in the month of August and ends in October. Working on the basis of detailed methodology, the quantity of grasses withdrawn from the forest has been estimated to 30 kg/day/household (Table 5).

Table 5

Withdrawal of Grass Biomass from Forests (in kgs)

No. of Household	Grass withdrawn/day/ Household	Grass withdrawn from Aug. to Oct./Household (92 days)	Total biomass (Grasses) withdrawn in the village (92 days) in m tonnes
757	30	2760	2,089.32

In 92 days (the grass withdrawing season in a year) each household would, therefore, withdraw 27.60 quintals of forest grass. The total withdrawal of the grass biomass of the village under study amounts to 2,089.32 m tonnes. This quantity of biomass is fed to 7,534 livestock heads of the village under study for the entire period of stall feeding (183 days–when grazing is not possible because grazing period has been taken as 182 days in a year) Then per head consumption per day of grass biomass so withdrawn from the forest comes to only 1.51 kg. It, therefore follows that it is only partially that the farmer has to depend on the forests for providing fodder to his livestock wealth and has to depend upon other sources like grass cutting from his individual land, kept specifically for the purpose, residue of agricultural crops and the green leaves lopped from the forests.

Finally it has been analyzed that the total forest biomass withdrawal from the village under study under three parameters alone has been 13,874.32 m tonnes the share of fuel wood, lopping and grass cutting has been 7,170.30 m tonnes, 4,617.70 m tonnes and 2,089.32 m tonnes respectively from the total forest biomass withdrawal. In conclusion it can be said that 13,874.32 m tonnes of forest biomass withdrawal by a single village under limited parameters certainly points to immediate attention. The principal objective of the present study was to understand the material/resource movement in the fragile Himalayan environment where already damage had been done. The village does not enjoy the resource availability, though the requirement of the people and livestock is on the higher side due to prevailing climatic and economic conditions. Hence, it is pertinent to note that, despite its location within the forest, the degraded Himalayan forest covers bound the Himalayan dwellers to survive in poverty-ridden environment and are deprived of the facility that should flow to them naturally if the forests are properly managed, conserved and used.

Reference

Engles, F., (1947) "The Dialectics of Nature", *Gospolitizdat*, Moscow, p. 143

Shafi, M., (1950) "Forests in The National Economy of India", *The Geographer*, Vol. 3 No. 1, p. 9.

13 Loss of Forest Cover and Land Degradation in Jhum in India's North-East

A Case Study

Abha Lakshmi Singh and Md. Bahar-Ud-Din Shah

Jhum cultivation involves slashing down of trees and bushes over the forest areas, drying and burning, sowing of seeds of host of crops including paddy by using stick, dibbler or by hand before the onset of monsoon. Crops are raised for a few seasons then the area is abandoned once in 2 or 3 years because of loss of soil fertility and erosion. The farmers called 'jhumias' then shift over to other lands and resort to similar practice with cutting and burning down the forest. Leaching, erosion and loss of fertility takes place rapidly and the field per unit of land becomes progressively lower. Land-water system which is basic life supporting factor and a prime mover of socio-economic development has already fallen into the clutch of the law of diminishing returns with the reduction of productivity vis-à-vis inputs and gross physical degradation of the system (El Moursi, 1984; Christanty, 1986).

On the other hand, high hills offer low returns and restrict alternatives. Difficult terrain encourages isolation of small communities. To such isolated communities, jhum satisfies their minimum food and other basic requirements. In fact, it is a way of life evolved as a reflex to the physiographical character of land under unique geo-climatic conditions. There are other convincing reasons behind the practice of jhuming as it is the genera de vie of the tribal societies. The climate, the terrain, the food habits of the people, their socio-economic and cultural needs, their self-reliance and their desire to be close to nature all have a say in jhum. In other words, their way of life, social organization, economic institutions, political systems, ceremonies, feasts, festivals and the entire gamut of life are the products of jhuming system. The tropical countries of Africa, Latin America and South-East Asia accounts for over 98 per cent of the total area under shifting cultivation. More than 525 million people live and farm on tropical hillsides which covers 12.9 million sq km and forms 9 per cent of the earth's land mass, out of which Africa has 40 per cent (Consultative Group of International Agriculture Research, 2001). In

Asia, about 53 per cent *of* the total land mass is in upper watersheds, which is the home to 65 per cent of the 1.6 million rural population (Centre for Science and Environment, 2001). In India 4.9 million hectares in the 11 states of Assam, Arunachal Pradesh, Manipur, Meghalaya, Mizoram, Nagaland, Tripura, Sikkim, Orissa, Madhya Pradesh and Andhra Pradesh is affected by the problem of jhum and about 4.44 million families are engaged in this.

North-east comprises of the seven sister states of Assam, Arunachal Pradesh, Manipur, Meghalaya, Mizoram, Nagaland and Tripura Out of the total geographical area of north-eastern region (25.5 million hectares), 2.7 million hectares is under jhum of which 17 per cent is under use at any given time (Manipur Remote Sensing Application Centre, 1995).

The North-Eastern region, Western Ghats, Western Himalayas and Andaman and Nicobar Islands constitutes 8.73 per cent of the total forest cover of India. Of this, the north-eastern region comprises 5 per cent of the total forest cover. In India's North-East, jhum is causing large scale loss of forest cover and land degradation. Due to jhuming the loss of forest cover in Manipur was 60.3 thousand hectares, in Nagaland it was 57.3 thousand hectares, in Mizoram it was 29.2 thousand hectare, in Assam it was 25.7 thousand hectares, in Arunachal Pradesh it was 7.5 thousand hectares, in Meghalaya it was 7.5 thousand hectares, and for Tripura the data was not available. The total loss of forest cover in the North-East was estimated to be 187.5 thousand hectares (Table 1) (Ranjan and Upadhya, 1999) and soil erosion was estimated to be 170 tonnes per hectare (Indian Council of Agricultural Research, 1999). Increasing pressure on available land and consequent short jhum cycle have caused large scale land degradation. The density of population of north-eastern region has

Table 1

Loss of Forest Cover in India's North-East (in thousand hectares)

States	*Due to jhuming*		*Other reasons*		*Total loss*	
	1993–95	*1995–97*	*1993–95*	*1995–97*	*1993–95*	*1995–97*
Arunachal Pradesh	16.9	7.5	NA	NA	16.9	7.5
Assam	22.4	25.7	37.7	15.9	60.1	41.6
Manipur	6.5	60.3	NA	NA	6.5	60.3
Meghalaya	21.8	7.5	NA	6.2	21.8	7.7
Mizoram	79.2	29.2	NA	NA	79.2	29.2
Nagaland	5.8	57.3	NA	NA	5.8	57.3
Tripura	NA	NA	NA	0.3	NA	0.3
Total	152.6	187.5	37.7	76.4	190.3	203.9

NA: Data Not Available.

Source: Ranjan, R. and Upadhya V.P. 1999, "Ecological Problems due to Shifting Cultivation", *Current Science*, Indian Academy of Sciences, Bangalore, Vol. 77 No. 10, November 25.

increased from 118 in 1991 to 147 in 2001. The state of Manipur reported 30.02 per cent growth rate in population from 1991 to 2001, while the hill districts of Manipur reported a growth rate of 50.97 per cent. The hill districts includes Senapati (81.96%), Chandel (72.80%), Churachandpur (29.81%), Tamenglong (29.23%) and Ukhrul (28.98%).

Thus, indiscriminate cutting and destruction of forest trees and vegetation for the jhuming purpose causes denudation of forest (World Resource Institute, 1990). Burning down of dried vegetation affects the soil properties. Not only that, in treeless hill slopes which had been cleared by slash-and-burn method, occurs soil erosion, rill wash, gully wash etc., due to certain exogenetic agents like wind and rain (Cramb, 1989; Tobing, 1991). In this paper an attempt has been made to study the loss of forest cover and land degradation due to jhum. Manipur has been selected as the study area because this state has suffered the maximum loss of forest cover due to jhuming in the north-eastern region. Five villages of Ukhrul district of Manipur were selected for in-depth investigation of the problem.

Data Base and Methodology

Data were collected mainly from primary sources through field surveys, surveys of the selected villages and selected households. The survey was conducted on the basis of questionnaire interviews. Field work was done during the years 2001 and 2002. For getting accurate information the sampled villages and households were visited frequently.

The following methods have been used in this study:

1. Data for assessing the loss of forest cover, loss of soil fertility and soil erosion were drawn from a comprehensive survey of the 5 selected villages, each lying in different tribal blocks and inhabited by different tribes. Out of the total 222 villages in Ukhrul district only 5 villages were selected and sampled for in-depth study because of inaccessibility, bad or no roads, lack of transport, difficult terrain, tribal population and insurgency. Mostly the survey was conducted by foot.
2. In the 5 sampled villages 10 to 100 per cent jhumia households were sampled and information regarding the number of households practicing jhuming, area under jhuming, loss of forest cover and loss of soil fertility etc., were collected.
3. A questionnaire was developed with the help of questionnaires used in similar studies (e.g., Christanty, 1986; Cramb, 1989; El Moursi, 1984; Tobing, 1991; Singh, 1978; Singh and Singh 1978; Shah, 2003). The questionnaire was designed precisely and vividly so that the respondents could answer without suspicion and hesitation.
4. For analyzing the slope of study area a simple formula was used i.e., Tan ø = VI/HE. The tangent on an angle (theta) is arrived by vertical interval (VI) divided with the horizontal equivalent (HE). The slope categories

thus derived were grouped into different class intervals like nearly level (<0% slope), very gentle slope (1–3% slope), gentle sloping (3–5%), moderately sloping (5–10%), strongly sloping (10–15%), moderately steep to steep slope (15–35%) and very steep slope (>35%).

5. For assessing the loss of soil fertility soil samples were collected from the jhum fields in the 5 sampled villages before and after burning and tested at the Soil Testing Laboratory of Indian Council of Agricultural Research, Imphal to assess the pH value, organic carbonic carbon, phosphorus and potash.

Study Area

Ukhrul district (94°0′ to 94°45′ east longitudes and 24°15′ to 25°45′ north latitudes) is the easternmost district of Manipur. The district is located at an elevation ranging between 388 and 2,834 metres above the mean sea level. It is bounded by the Imphal district on the south, Nagaland on the north, Senapati district on the west and Myanmar on the east. The total area of the district is 4,544 sq km and the total population is 1,40,946. Ukhrul district reported 28.98 per cent growth rate of population from 1991 to 2001 (Census of India, 2001). Being a rural district, most of the population is rural in character and comprises of tribals (94 per cent of the total population). The most predominant tribal communities living here are the Tangkhul Nagas, Thadou and Vaiphei. The district is further subdivided into five tribal development blocks spread over 222 villages. Five villages selected for sampling are situated in the different tribal development blocks. The whole district is having monsoon type of climate. It receives very heavy rainfall for seven months (April to October). The district is rich in forest resources like tropical hardwood, timber, superior varieties of bamboo, cane etc. About 61 per cent of the total area is under forest (1993–94). Thus, mountainous topography, undulating slopes and wet weather conditions in Ukhrul district provides ideal conditions in which tribals practice jhum cultivation as their way of life.

Discussion

Five villages lying in the different tribal development blocks of Ukhrul district were selected for in-depth study. About 8 to 21 households were sampled from every village (Table 2). The villages sampled are Lungphu (located in Phungyar—Phalsat tribal development block), Yeasom (located in Kasom Khullen tribal development block), Nungbi Khullen (located in Chinga tribal development block), Mongkot Chepu (located in Ukhrul tribal development Block) and Maku Kuki (located in Kamjong Chassad Tribal Development Block). All the sampled villages are located on hills having steep slopes.

Lungphu village is about 50 km away from Manipur's capital Imphal. This village has a total population of 477 and 73 households (6.5 persons per household) of Tangkhul Nagas tribes. Of the total households (73), 67 households were practising jhum. About 8 households (10.95 per cent) having a population of 67 persons were sampled. Of the total sampled households, 7 households are practising jhum. Of the total land under cultivation (18.48 acres) belonging to the sampled households, 14.86 acres of land was under jhum.

Yeasom village is about 50 km away from Imphal via Thoubal and 20 km away from the nearest market centre at Yairipok. This village has a total population of 239 and 46 households (5.2 persons per household) of Tangkhul Naga tribes. Of the total households (46), 40 households were practising jhum. About 8 households (17.3 9%) having a population of 48 persons were sampled. Of the total sampled households, 7 households are practicing jhum. Of the total land under cultivation (15.20 acres) belonging to the sampled households, 10. 81 acres of land was under jhum.

Nungbi Khullen village is about 150 km away from Imphal. This village has about 985 people and 186 households (5.3 persons per household) of Tangkhul Naga tribes. Of the total households (186), 176 households were practicing jhum. About 21 households (11.29 per cent) having a population of 130 persons were sampled. Of the total sampled households 20 households are practising jhum and of the total land under cultivation (19.62 acres) belonging to the sampled households, 16.31 acres of land was under jhum.

Mongkot Chepu village is about 35 km away from Imphal. This village has about 689 people and 104 households (6.6 persons per household) of Kuki Simte tribes. Of the total households (104), 95 households were practicing jhum. About 12 households (11.53 per cent) having a population of 75 persons were sampled. Of the total sampled households, I I households are practising jhum. Of the total cultivated land (11.90 acres) belonging to the sampled households, 11. 02 acres of land was under jhum.

Maku Kuki village is about 100 km away from Imphal. This village has about 39 people and 6 households (6.5 persons per household) of Kuki tribes. All the 6 households (100%) were sampled. All of them are practising jhum and all their land (9.58 acres) is under jhum.

Loss of Forest Cover

There is indiscriminate cutting and destruction of large quantity of forest trees and vegetation in the process of jhum in Ukhrul district. Though the exact information on this aspect is lacking in this region, yet people are aware of the degradation of forest. An attempt has been made to assess by field survey the cutting down of forest for the purpose of jhum in the sampled villages of Lungphu, Yeasom, Nungbi Khullen, Mongkot Chepu and Maku Kuki (Table 2).

A perusal of Table 2 shows that of the total sampled households in the five selected villages, 51 households were practising jhum and 62.58 acres of land

Table 2

Loss of Forest Cover due to Jhum in the Sampled Villages of Ukhrul District (2002)

S.No.	Name of the sampled villages	Total No. of households	Total No. of households practising jhum	Sampled households		Total land under cultivation in the sampled household (in acres)	Loss of forest cover in jhum in the sampled households		Total loss of forset cover in jhum in the village (in acres)
				Total No.	No practising jhum		Total loss (in acres)	Loss per household (in acres)	
1.	Lungphu	73	67	8	7	18.48	14.86	2.12	142.04
2.	Yeasom	46	40	8	7	15.20	10.81	1.54	61.60
3.	Nungbi Khullen	186	176	21	20	35.92	16.31	0.82	144.32
4.	Mongkot Chepu	104	95	12	11	11.90	11.02	1.00	95.00
5.	Maku Kuki	6	6	6	6	9.58	9.58	1.60	9.60
Total	5 villages	415	384	55	51	98.15	62.58	1.18	452.56

Source: Based on Field Survey (2002).

(63.75% of the total cultivated area) was under jhum. This shows that 62. 58 acres of land previously under forest has been cleared for jhum by the 51 households. On an average 1.18 acres of land under forest per household has been lost. If this calculation is extended to the individual villages and to the total number of households practising jhum, the loss of forest cover works out to be 144.32 acres in Nungbi Khullen village, 142.04 acres in Lungphu village, 95 acres in Mongkot Chepu village, 61.60 acres in Yeasom village and 9.60 acres in Maku Kuki village. Thus, in the five sampled villages, 384 jhum practising households have cut down 452.56 acres of forest for jhuming in a year. Hence, average area of forest lost in these five villages works out to be 90.51 acre per village. On this basis, if we calculate the total loss of forest cover per year for the purpose of jhum in the 222 villages of Ukhrul district, it works out to be about 20,093 acres or 8,135 hectares (Shah, 2003). Whatever area presently under jhum was previously under forest.

This assessment of loss of forest cover in Ukhrul district is more or less accurate if we compare it with the figures given by Manipur Remote Sensing Application Centre, based on satellite imagery (TM/IRS, LISS II of 1986–87 and 1993–94 on 1: 50,000 scale). In 1986–87, an area of 14,231 hectares was under current jhum while in 1993–94, the area under current jhum decreased to 8,460 hectares (MRSAC, 1995). The figure of 2002 assessment, based on field survey is 8,135 hectares under current jhum. This shows a further decrease in the area under current jhum from the previous years but the area under abandoned jhum or fallow land has increased (Table 3). This shows the continuation of jhum and the cutting of forests in Ukhrul district.

A perusal of Table 3 shows that an increase of 13.95 per cent of total area under jhum has been recorded. About the same per cent of decrease in area under forest has been recorded. This table again shows that jhum is the main cause of loss of forest cover in Ukhrul district.

Land Degradation

On the deforested hill slopes, soil erosion goes on and this is another integral process with jhum. Absence of soil conservation measures augmented with high rainfall results in increased run-off, erosion of top soil, decline of soil fertility and low crop yields.

Soil Erosion

The whole of Ukhrul district is mountainous with steep slopes and wet weather conditions. Results of slope analysis showed that 81.58 per cent of the area falls in very steep slope category and 16.28 per cent in moderately steep to steep slope category (Table 4). In such areas the intensity of soil erosion is very high.

Table 3

*Land Utilization in Ukhrul district (1986–87 to 1993–94)**

Sl. No.	Category	1986–97		1993–94		Percentage change (from 1986 to 1994)
		Area (in ha.)	Percentage to the total area	Area (in ha.)	Percentage to the total area	
1.	Built-up land	1,709	0.38	1,709	0.38	
2.	Fallow (Plain)	9,340	2.05	9,340	2.05	
3.	Current Jhum	14,231	3.13	8,460	1.86	–1.27
	(a) Abandoned Jhum					
	(b) 1–5 years A. jhum	20,781	4.57	29,082	6.40	+1.83
	(c) 5 and above years A. jhum	64,150	14.12	125,005	27.51	+13.39
	Total Jhum	99,162	21.82	162,547	35.77	+13.95
4.	Forest					
	(a) Primary Forest					
	(b) Secondary Forest	340,830	75.00	277,445	61.06	–13.94
5.	Water Bodies	1,785	0.39	1,785	0.39	
6.	Others	1,574	0.35	1,574	0.35	
	Total Ukhrul	454,000	100.00	454,000	100.00	

*Note: Latest data available

Source: Manipur Remote Sensing Application Centre, Imphal (1995).

Table 4
Slope Categories and Area Under Different Categories in Ukhrul District

S.No. Category	Slope (in %)	Area (in ha)	Percentage to the total area
1. Nearly level	< 1	00.00	00.00
2. Very gentle slope	1–3	1,174.00	0.25
3. Gentle slope	3–5	1,726.00	0.38
4. Moderatly sloping	5–10	346.00	0.08
5. Strongly sloping slope	10–15	6,465.00	1.43
6. Moderately steep to steep slope	15–35	73,975.00	16.28
7. Very steep slope	> 35	370,714.00	81.58
Total area of Ukhrul	—	454,400.00	100.00

Source: Manipur Remote Sensing Application Centre, Imphal (1995).

Field surveys revealed that the process of soil erosion begins when farmers enter the plots (hill slope under forest) either for selection of site or for cutting the forest vegetation. Their movement on slopes causes loose soil to disintegrate, forest litter and earthworm casting to slide down the hill. Jungle-cutting, burning, clearing and dibbling of seeds account for a considerable amount of loose soil material, ashes, earthworm casting and detached soil clods/stones to roll down the foot-hills. Through this process 3.7 tonnes of soil materials per hectare was reported to slide down the foothills (Singh, 1978).

With the onset of monsoon, soil erosion by water begins. Soil erosion from hill slopes (60–70%) under first year, second year, abandoned jhum (first year fallow) and bamboo forest was estimated to be 146.6, 170.2, 30.2 and 8.2 tonnes per hectare per year respectively (Singh and Singh, 1978). These observations clearly indicate that the second year of jhum cultivation is more hazardous than the first year. However, wide range of variation in soil erosion due to slope, crop canopy and agricultural operations had been recorded. Run-off and loss of organic carbon increased with increase in slope up to 21 per cent, but it decreased linearly with further increase in slope.

It is obviously true that high rainfall and steep hill topography is always associated with the problem of severe soil erosion, particularly when the land use system has biotic interference, The loss of soil through erosion is widely known in the sampled villages where survey was conducted. But the amount of soil lost in a particular year is not known due to non-availability of measuring facilities. Decreasing production of crops from first year to succeeding years shows the deterioration in the quality of soil through erosion. Moreover, high production rate of crops like paddy in one year cropping jhum fields mainly in Tangkhul Naga inhabited villages and low rate of production of paddy in 3 years continuous cropping fields in Kuki inhabited villages shows maximum loss of soil in the fields where intensive cropping is done.

Loss of Soil Fertility

The practice of jhuming affects the soil. Burning causes changes in the soil properties. Burning chemically alters a portion of the plant nutrient supply from an organic form to a mineral form in ash, which is often readily soluble

(Debyle and Packer, 1974). When water runs over or passes through this ash, the soluble components are flushed out and lost from the site in the run-off.

An attempt has been made to assess the soil fertility in the sampled villages of Lungphu, Yeasom, Nungbi Khullen, Mongkot Chepu and Maku Kuki. Soil samples were collected (before and after burning) from each of the village and were tested at Soil Testing Laboratory of Indian Council of Agricultural Research, Imphal, to assess the fertility status and changes in soil properties due to burning (Table 5).

Table 5

Changes in Soil Properties Before and After Burning in the Sampled Villages of Ukhrul District (2002)

Name of the sampled village	*Soil properties*	*Before burning*	*After burning*	*Changes*
Lungphu	pH	4.20	4.60	+0.40
	Organic Carbon	1.30	1.00	–0.30
	Available P_2O_5 (kg/ha)	4.20	4.25	+0.05
	Available K_2O (kg/ha)	250	550	+300
Yeasom	pH	4.00	4.40	+0.40
	Organic Carbon	1.15	1.00	–0.15
	Available P_2O_5 (kg/ha)	3.30	3.32	+0.02
	Available K_2O (kg/ha)	270	600	+330
Khullen Nungbi	pH	4.10	4.40	+0.30
	Organic Carbon	1.00	0.90	–0.10
	Available P_2O_5 (kg/ha)	3.70	3.71	+0.01
	Available K_2O (kg/ha)	285	570	+285
Mongkot Chepu	pH	4.50	4.80	+0.30
	Organic Carbon	0.94	0.85	–0.09
	Available P_2O_5 (kg/ha)	4.00	4.01	+0.01
	Available K_2O (kg/ha)	300	600	+300
Maku Kuki	pH	5.00	5.50	+0.50
	Organic Carbon	0.95	0.82	–0.13
	Available P_2O_5 (kg/ha)	4.96	4.97	+0.01
	Available K_2O (kg/ha)	300	650	+350

Source: 1. Base on Field Survey (2002).
2. Soil Testing Lab, Indian Council of Agricultural Research Imphal (2002).

More or less same results were observed in all the soil samples which were tested from different jhum sites. It was observed that the pH value of soil has increased slightly in the sampled soils after burning. It ranged between 4.00 to 5.00 before burning and 4.40 to 5.50 after burning. Percentage of organic carbon in the soils decreased after burning. It ranged between 0.94 to 1.30 before burning and 0.82 to 1.00 after burning. The amount of Potassium (K_2O) kg per hectare increased substantially after burning. It ranged between 250–300 before burning and 550 to 650 after burning. While the amount of phosphorus in kg per hectare is more or less same before and after burning. It ranged between 3.30 to 4.96 before burning and 3.32 to 4.97 after burning (Shah, 2003). The results show that the soils are weak acidic in nature. It is also known that if the pH value of soil is higher, the percentage of organic carbon is lower. Increase in pH is related to increase in the content of available potash.

Generally, these hill soils are rich in potash while inadequate in phosphorus content.

Conclusion

The study reveals that loss of forest cover and land degradation is an integral and continuous process in jhum. Large number of trees and vegetation have been lost in the preparation of jhum fields for cropping. Soil erosion starts when the jhumias enter the jhum site for cutting down the forest. Erosion is augmented with the onset of rain. Burning down of dried vegetation affects the uppermost layer of soil and brings changes in soil properties.

Deforestation in the north-eastern region of India has reached an alarming stage owing to jhuming. The tropical vegetation of the region including the evergreen, semi-evergreen and moist deciduous forest has been extensively fragmented. The primary cause of this fragmentation is jhuming. Also human pressure on land, encroachments on forest lands, illicit felling, topping for forest wood and sometimes even saw mills have led to deforestation. People living in lowland areas see the region as watersheds or forest resources rather than as agricultural reservoirs. The constant growth of hill population is widely viewed as a major threat to resources. This is one of the reason for poverty, political violence and growing criminal activity in South-East Asia, South America and African countries (Consultative Group on International Agricultural Research Report, 2001).

Several scientists feel that the pressure on these ecologically fragile areas will worsen with an increase in population. However, contrary to popular belief, population increase is not always the culprit for deforestation. A review of 70 different studies from around the world found that the impact of population increases on tropical hillside resources has been actually varied. In fact, aerial photographs and ground surveys from hilly areas of Kenya, Nepal and Rawanda show that tree cover increased with population growth. Bulging population and deforestation are definitely associated but once a certain density is reached, it is common to find farmers growing more and more trees in and around their plots. Communities in such, areas begin to value forests and thus regulate deforestation in ecologically critical niches like stream banks and water sources (Centre for Science and Environment, 2001).

The common belief is that poor people put pressure on land, thus degrading it. P.S. Ramakrishnan (Professor of Ecology at the School of Environmental Sciences, JNU, New Delhi) one of the few scientists who have studied shifting cultivation in the North-East blames external pressure, not the people (1979). In fact, there are cases where migration out of densely populated area led to land degradation as the cultivated terraces received less maintenance, or the farmers abandon the use of soil alternatives. Sara J. Scherr (Natural Resource Economist with extensive research experience in the tropics, visiting fellow at the Maryland University, US) mentions that this is observed through surveys in the hilly areas of Kenya, Nepal, Nigeria, Rwanda, Tanzania, Uganda and Indonesian Island of Java where population density is above 100 people per sq km and where the land requires intensive labour as the farming practices includes terracing, zero tillage, manuring composting or the use of inorganic fertilizers. In some higher rainfall areas, such as Mount Kalimanjaro (Tanzania), the Philippines and the Islands of Java, Roti and Savu, slash-and-burn agriculture developed into multi-storeyed home gardens and agro forests at high population densities. Scherr (1982) believes that these systems are both highly productive and sustainable.

The finding of this study also shows the increasing pressure on available land and consequently short jhum cycle. This is the only important difference in this age old agricultural practice. Earlier the farmers used to return to the same patch of land after 15–20 years but now they return after 3–4 years.

On the other hand it is not easy to stop jhum in the north-eastern region of India. The climate, the terrain, the food habits of the people, their, socio-economic and cultural needs, their self reliance and their desire to be close to nature all have a say in jhum. In other words their way of life, ceremonies, festivals and entire gamut of life are the product of jhuming system. This system is deeply embedded in the life of the tribals.

Efforts to substitute this system in India have been going on for almost a century. Alternatives like watershed based farming along with livestock rearing and planting of long duration timber crops has been suggested. If the farmer is employed elsewhere, his trees will give him good returns. Other suggestions include agro-forestry, horticulture and pisci-culture. Another suggestion is a three-tier system in which agricultural practices are carried out on the lower portion of the lower position of the slope with support of bench terraces as conservation measures. Horticulture takes place at the mid-portion of the slope with support of contour bunds and half moon terraces and finally, agro-forestry is carried out on the topmost portion of the slope with support of contour bunds.

Modifications are more applicable than alternatives. Emphasis should be given to modifications of jhum like alder-based jhum. Alder-based jhum is found to be most suitable modified jhum. Alder tree has root nodules, which improve soil fertility by fixing atmospheric nitrogen into properties. It also enhances crop yields and reduces soil erosion. This sustainable productive system has the potential to be adopted on the wider scale, Villages which are having very limited land area, alder-based jhum is found to be suitable to increase food production and to maintain the ecological balance. In addition, it also provides good qualities of timber and firewood from which money can be earned easily.

It is almost impossible to stop a practice that has been going on for centuries. To end a practice that is closely associated with their customs and traditions, the solutions has to be searched within the problem itself. Thus, it is the need of the hour to relook at the system again and find out some alternative or modification in jhum to minimize the loss of forest and soil erosion. Any modification or alternatives should be made keeping in mind that it is socially acceptable, economically viable and ecologically sustainable. Perhaps jhum needs to be studied more closely before it is condemned and replaced.

References

Centre for Science and Environment, *Down to Earth*, (2001), New Delhi.

Christanty L., (1986), "Shifting Cultivation and Tropical Soils: Patterns, Problems and Possible Improvements", G.G. Marten (ed.), *Traditional Agriculture in Southeast Asia*, Westview Press, Boulder, Co. pp. 226–240.

Consultative Group of International Agricultural Research Report, (2001).

Cramb, R.A., (1989), *Shifting Cultivation and Resource Degradation in Sarawak: Perception and Policies*, Borneo res. Bull., 20 (1), 22–49.

Debyle and Packer, (1974), *Journal of the North Eastern Council*, Shillong.

El Moursi, A.W.A., (1984), "The Role of Higher Agricultural Education in the Improvement of Shifting Cultivation Farming System in Africa", In, A.H. Bunting and E. Bunting (ed.), *The Future of Shifting Cultivation in Africa and the Task of Universities*, Proc. International Workshop on Shifting Cultivation: Teaching and Research at the University Level, lbadan, Nigeria, FAO, Roem, pp. 8–14.

Indian Council of Agricultural Research, Barapani, Shillong, (1999).

Indian Council of Agricultural Research, Imphal, (2002).

Manipur Remote Sensing Application Centre, Imphal, (1995).

Ramakrishnan, P.S., (1979), "Ecological Impact of Jhum on Forested Ecosystem of N-E India", Paper Presented in Annual General Body Meeting of Indian National Science Academy, Shillong.

Ranjan, R. and Upadhya V.P., (1999), "Ecological Problems Due to Shifting Cultivation", *Current Science*, Indian Academy of Sciences, Bangalore, Vol. 77, No. 10, Nov. 25.

Scherr, S.J., (1982), "Natural Resource Economist", Visiting Fellow at the Maryland University, US.

Shah, B., (2003), "Impact of Jhum Cultivation on the Ecosystem of Manipur: A case study of Ukhrul District", an unpublished Ph.D. Thesis, Department of Geography, Aligarh Muslim University, Aligarh.

Singh, A., (1978), "Shifting Cultivation and Soil Erosion Problems in North Eastern Hill Region", Paper Presented at XVI Annual Convention of ISAE, IIT, Kharagpur.

Singh, A. and Singh, M.D., (1978), "Effects of Various Stages of Shifting Cultivation on Soil Erosion from Steep Hill Slopes", *Indian Forester*, 106 (2), 115—21

Singh, A.L., Shah, B. and Khan, N., (2003), "Jhuming Cultivation: Root Cause of Land Use/Land Cover Changes in Ukhrul District", Manipur, *The Geographer*, Vol. 50, No. l., January, 2003, AMU. Aligarh.

Singh, A.L. and Shah, B., (2002), "Jhuming Cultivation: A Deep-Rooted Environmental Concern", *The Geographical Observer*, Vol. 34, 2002, Meerut College Geographical Society, Meerut, India.

Tobing, P.L., (1991), "Old as the Hills", *Down to Earth*, CSE, New Delhi, May, 31, 2001 April: 23–30.

WRI, (1990), *World Resources 1990–1991*, Oxford University Press, Oxford, pp. 383.

14 The Ecological Determinants of Seasonal, Altitudinally Directed Migration in Nomadic Livestock Keeping in Mongolia

Bat-Ochir Bold

In recent times interest in nomadism in Mongolia has increased amongst researchers into nomadic lifestyles. This is explained by the fact that in the period from the end of the 1920s until 1990, Mongolia was hardly investigated by scholars from western countries on account of the isolation caused by the political system. Mongolian nomadism is an example of mobile livestock keeping which even to the present time has little changed. Since 1990, mobile livestock keeping in Mongolia has been investigated under the most differing aspects. Research has been conducted by scholars from Japan, England, Canada, Germany and other countries. In all this research I see in general two main methods of approach: detailed case studies and generalized global research.

In this paper I would like to concentrate more on the generalized aspect. The main factors which give rise to seasonal migration in livestock keeping in Mongolia will in the main be treated from a geographic point of view. Although a good deal of the determinants of seasonal migration can in this way be understood, it should be borne in mind that there are also significant factors of influence which can be considered from the viewpoints of politics, economics, social anthropology and ethnography.

Seasonal migration within a region which is ecologically optimal for the yearly rotation of pasture land use is characterized from a geographic aspect by the following factors:

- climatic conditions, i.e., fluctuations in temperature, wind direction and precipitation,
- distribution of forage plants,
- distribution of salty soils,
- supply of water, and
- geographic borders.

By means of a description of all these factors and their close correlation it is possible to explain not only the ecologically determined migrations but also the nature of the nomadic lifestyle in Mongolia. In this way also the particularities and the generalities of Mongolian livestock keeping and of other forms of mobile livestock keeping in the world can be understood.

Climatic Conditions as a Determining Factor of Seasonal, Altitudinally Directed Migration

The general basis of nomadic livestock keeping in Mongolia, in the high mountains, the forest highlands, the steppes or the Gobi region, is a change of pasture to ecological areas which are climatically favourable in the respective seasons. This means that livestock keepers with their stock must avoid extreme climatic conditions caused by, for example, temperature, storms and precipitation. The effects of temperature, storms and precipitation on the choice of pasture land are different for each of the four seasons. It is useful to describe the climatically determined choice of pasture for each of the seasons separately.

Winter

Livestock keeping in Mongolia is still completely dependent upon climatic conditions, particularly in the hard winter. Migration at the start of winter into a region where temperature, wind and precipitation are relatively favourable for livestock keeping in the cold is inevitable. It is obvious that the average temperatures in the natural zones of Mongolia, in the high mountains, the forest highlands, the steppes and the Gobi region, differ. But even within these zones there are general differences in temperature depending upon orographic factors i.e., altitude, slope, exposure, and moisture, light and warmth conditions.

One of the most important reasons for migration to the winter pasturage is the search for the warmest mountain belt. The winter site is thus located at a point in this belt which is most favourable for humans and livestock. Many years of experience of Mongolian nomadic livestock keeping as well as the results of geographic and meteorological research show that the most favourable locations in winter as far as temperature is concerned—relatively independently of the nature zones of Mongolia—lie between the mountain crest and the middle altitudes of the respective regions. These favourable locations in the differing nature zones stand in close connection with the wind conditions i.e., the wind direction and the wind strength. The most frequent wind directions in Mongolia are westerly and north-westerly. The annual average wind strength is 4–8m/sec. This means that the south side or the south slopes of the mountains are always suitable for the winter as well as the spring pasturage. Livestock keepers as well take into account in which location less snow, as a result of the wind direction, is blown into the winter pasturage. Moreover, the wind strength must be considered in the choice of location, whether it is sufficient to blow away a possible snow covering. According to experience, livestock keepers choose their winter pasturage at a comparatively windy location if snowfall is high, or at a location most sheltered from wind otherwise. This shows that the evaluation of the wind in the choice of the winter pasturage depends a good deal on precipitation. The average annual precipitation for the entire country amounts to 200–250 mm. As well the yearly precipitation decreases from north to south from 400 mm to 100 mm.

The winter pasturage should in no case be carried out in low lying regions, where snowfalls are higher or snowdrifts can mount. The best pasture sites in winter are therefore those regions which provide a favourable combination of temperature, wind and precipitation for the livestock. Such sites lie in principle in all three nature zones in the belt between the higher and middle mountain ranges. The winter site is therefore located on a higher south slope of a mountain or in a mountain cleft away from the wind direction.

Mongolian livestock keepers not only seek regions protected from the wind according to natural principles but also make use of long-standing wind shelters (Mong.: Halha, Horoo). These shelters are constructed mostly from local building materials, from wood and stone in the highlands and forest lands, from compacted manure and straw in the steppes, and from stone in the Gobi region.

Winter shelters have been used in Mongolian livestock keeping for a long time to a greater or lesser degree. Autochthonous sources report that there were frequently in the 13th/14th centuries temporary additional shelters. The remains or traces of winter shelters from the 18th/19th centuries can even today be seen in rural areas. Since the end of the fifties, as a consequence of the policy of forced settlement, permanent shelters for winter and spring were intensively built. In 1993 there were in Mongolia about 76,100 shelters in all. Approximately half of these were winter shelters (*Mongolian Economy and Society in 1993*: 40). Logically, this should have had a positive effect upon the keeping and breeding of the livestock. However, the results were not as expected. It is worthy of mention that before this building of shelters on a massive scale the winter and spring sites were much simpler. Many livestock keepers do not consider from their experience the stabling of stock as a correct method. In traditional livestock keeping the opinion prevails that with stabling the stock should not be sheltered from the cold but rather from the wind, that is the stables should serve only as a wind shelter. Warm stabling makes the animals perspire, and the effect of a sudden change of temperature on a perspiring animal emerging from a stable accelerates the winter process of emaciation and enervation. In this respect the design of the shelters is of great significance (Sagdarsuren, 1984: 68). It would make an interesting comparison to represent loss of stock in connection with the construction of permanent winter shelters.

Spring

Approximately from the end of February until the start of May the day temperatures increase in the Central Asian highlands. With the collision of the northerly and north-westerly cold air and the southwesterly dry and warm air, air currents are activated and the wind direction is therefore relatively changeable. This has an effect upon the type of precipitation which frequently changes between snow and rain. The conditions which should be considered for the spring pasturage or site can be described in the following manner:

- A wind sheltered, relatively warm location. This is more important in spring than in the other seasons. From time immemorial Mongolian livestock keepers constructed only temporary wind shelters, provisional stables or mere fencing. In the last three decades however, as mentioned above, many permanent spring stables were built. At the end of the eighties there were in all more than 30,000 spring stables in the country. For the spring pasturage mainly those regions are used which lie in the belts of the lower mountain ranges. The spring pasturage therefore is located somewhat below the winter pasturage on the south face of a mountain between a middle altitude and the mountain foot. Although in general the lower the location the warmer it is, snow often remains at the mountain foot and in the lowlands until May, and nights can be cold and damp.
- Places situated away from lakes, morasses, gorges etc., in the windy spring the animals move frequently along the wind direction. Therefore the pasture must be chosen far away from the above sources of danger into which the animals could be driven by wind. This is another reason why livestock keepers in principle never conduct the spring pasturage in low lying places, or at least not in the first half of the season.
- At a sunlit mountain foot, where the first fresh grass grows and the pasture is as soon as possible free of snow.

Spring is the period in which calves are raised. This makes the pasture, which is located hard by the spring shelter, an important resource for both mothers and calves. A nearby mountain foot with fresh grass is from the middle of April advantageous for both. In order to preserve such a pasture for mothers and calves and to make use of distant pastures, livestock keepers travel often far away with the rest of the stock.

Summer

As far as livestock keeping is concerned, the summer months comprise the second half of May, June, July and the first half of August, during which the livestock put on weight. Average temperatures and the length of the summer period differ in the mountains, the forest highlands and the steppes. There is however a general climatic reason for the migration to the summer pasturage: to avoid the heat. While in winter and particularly in spring the livestock is protected from the cold wind, a windy and cool region is suitable for the summer pasturage. The climate of Mongolia is especially dry; 80–90 per cent of the annual precipitation falls in the short period of the humid months from May to September. For these reasons the suitable regions for the summer pasturage are in principle higher locations than the winter pastures, for example on mountain crests, where the winds blow more strongly. There are of course differences in connection with the differing nature zones. Since the inversion line of the Hangai and Hentii forest highlands lies lower, for example, the mountain crest on this inversion line should be avoided for the choice of summer pasture and accordingly the summer pasturage is chosen in the foothills below the winter and spring pasturages. In this case livestock keepers move with their stock often to a region with a plentiful water supply, where the inadequate coolth of the wind can be compensated with water i.e., to a site which lies not directly at a water source but amongst hills nearby to water and which is ecologically appropriate. If mountain crests and declivities are useable for pasturage and there is a water supply to satisfy the large need of the animals, then the most favourable conditions for a summer pasturage are met. Although in the Altai Mountains there are no plentiful water sources, this is all the same characteristic for a summer pasturage in this region.

Autumn

From the second half of August the variation in temperatures increases, nights are cooler and days often comparatively warm and free of wind. In this period the animals become restless on account of the increasing fluctuations of temperature. In order to ensure that the livestock remains well nourished from summer into autumn, migrations are carried out to a warmer location where the animals can feed more peacefully and can be fattened in autumn in a regular way. Such sites, which remain even until late autumn warm and free of snow, are located characteristically in low lying regions, as a rule in all nature zones. The migration from the highlands of the summer pasture, or from the hills in forest highlands lying near to water, or from the steppes, into lowlands is carried out often very slowly in order that the summer weight increase of the animals is not lost in the course of long migration treks. This migration from the heights into the lowlands is in principle in all regions the longest of all the seasonal migrations. In the zones where the migration treks from the summer to the autumn pasture are quite long—not infrequently more than 300 km—the migrations often last several weeks. As far as wind and precipitation are concerned, they play a comparatively lesser role in the pasturage of autumn than of the other seasons, since mild winds and low precipitation are characteristic for autumn, particularly in the lowlands.

Thus, the seasonal climatic differences, in particular the fluctuations in temperature, are important determinants in the frequency and distances of migration in traditional Mongolian livestock keeping, as we have attempted to describe.

The State of Vegetation as a Determining Factor of Seasonal Migration Viewed as Altitudinally Directed

All forms of pastural agriculture develop predominantly where, on account of mountain ranges, highlands and valleys, the annual cycle of forage plants exhibits large differentiations over small regions: for thus is the yearly balance of food attainable by means of relatively short migration treks. This means that migration occurs not only for climatic reasons but also for reasons connected with the seasonal food balance of the animals. Therefore, one of the most important factors of seasonal migration is the state of the vegetation: the seasonal division of forage plant resources and the distribution of forage plants. The four seasonal migration routes in traditional Mongolian nomadic livestock keeping have always been influenced by these two aspects of the vegetation.

We shall briefly consider their effects upon migration:

State of Vegetation

In the pastural areas about fifty million tonnes of forage plant material for livestock grow annually. This corresponds to the food requirements of about sixty million sheep each year (calculated in terms of sheep units, a horse, a cow and a yak are each the equivalent of seven sheep, a camel ten and a half sheep, and a goat seven-tenths of a sheep). The forage plant resources are certainly differently distributed throughout the seasons. Thus, for example, in the case of the food requirements in summer and autumn for the twenty-five million existing animals, there are 2.3 million tonnes of feed in excess. On the other hand there is a deficit in winter and in spring of about 5.5 million tonnes. This means that the food supply in summer and autumn is about 28.9 million tonnes and in winter and spring about 21.1 million tonnes (Cadraabal, 1982: 24).

The capacity of pasture land is, in the investigations which form the basis of economic policy and projects, in the rule evaluated by the annual average yield of forage plant material. Since the periods of vegetation in winter/spring and summer/autumn differ due to seasonal length (200 days in the former and 165 days in the latter) and to reserves of forage plants, it would be more accurate to evaluate the actual capacity of the pasture land taking into account the periods of vegetation when the reserves of forage plants are lowest. Thus, Mongolia has a reserve of pasture in winter/spring for about forty-eight to fifty million sheep. Moreover, it is a fact that in spring compared to high summer the protein content of forage plants is 3.5 times less, the juiciness is 2.5 times less and the time available to pasturing on account of the lesser daylight hours and of the cold is twice as short (Mojbuu and Bakej, 1996).

The state of vegetation in Mongolia is especially characterized by the fact that climatic conditions allow only a single short period of growth. In the middle of September the grasses and plants begin to die off. This is a sign of the coming dangerous phase of the year for the herds. The new vegetation does not appear before the end of March or the start of April of the following year and even then it is so sparse that it plays no great role in the upkeeping of the animals for yet another month. Thus, the animals must live for six or seven months on the dead plants which have remained after the period of growth. The decisive factor is the quantity of vegetation at the end of summer, which must be sufficient to supply the stock until the beginning of the next growth period. As a result of this, the need arises to exploit the yearly resources of forage plants by means of seasonal migrations with an optimal dividing of the pastural regions.

From the middle of winter to the first half of spring is the period of the smallest supply of forage plants, during which the capacity of the pasture land reduces by 60–75 per cent. In order to maintain a warm pasturage in this especially difficult period, the regions above the mountain slopes are made use of, in principle in all nature zones. By means of this the remains of forage plants, which have been preserved through the entire winter, are available in late spring.

In the second half of spring the period of vegetation begins. At this time too begins the active feeding period of the animals on the natural pasture. This season is characterized by the fact that the animals become especially debilitated. Therefore, they must be driven into a region where forage plants that are richer in calories grow. Such sorts of forage plants are plentiful amongst the vegetation lying beneath the middle altitude of a mountain.

In order to make use of pasture areas distant from the winter site, livestock keepers normally leave this site with the entire stock excluding the mothers and the calves. Such treks are frequently quite a distance. The point of the migration is to reach a pasture in which fresh grass can grow. Such pastures are as a rule located at sunlit mountain feet which are watered by the snow.

Regarded biologically, the livestock must put on weight in summer in order that a fat layer might form in autumn to act as protection against frost. Therefore, those types of plants are taken into consideration by livestock keepers so that weight increase and fattening can occur in the appropriate order. As far as vegetation is concerned, this is the actual reason for the summer pasturage. A negative consequence of a summer pasturage on a wrong pasture or amongst the wrong vegetation could be that fattening occurs too early, before the livestock has put on sufficient weight. In this case the livestock cannot survive the hard winter and the feed deficient spring.

In autumn the ripeness of the various plants gradually deteriorates. Although all forage plants in Mongolia complete their ripening in general during the summer, the period of ripening of individual forage plants is quite different. The period of ripening of vegetation is connected with altitude: from the mountain crest to the mountain foot. Later ripening occurs in the lowlands and in the valleys. Moreover, as a result of the ripening process and of low precipitation the plant concentrate is comparatively high and the water content always low. There is nonetheless a climatically determined difference in the preservation of water in the plants in connection with the altitude. The point of the autumn pasturage is therefore to seek still juicy green grass and a vegetation appropriate for the fattening of the livestock. Such sites are located in the lowlands.

Distribution of Forge Plants

Also the distribution of forage plants plays a large role in the direction of seasonal migration. Regarded generally, the vegetation density in Mongolia continually decreases in the direction from highlands and forest highlands to southerly or south-westerly steppes and lowlands. Moreover, they are in the highlands in both the Altai Mountains and the Hangai-Hentii Mountains separate vegetation zones exist which are dependent on the moistness, the amount of sunlight etc., of the respective microecology. However, in all nature zones it is possible to establish altitudinal belts of distribution of definite sorts of forage plants, corresponding to the respective attitudes, climatic conditions and soil types. In the steppes and the Gobi region there are few altitudinal vegetation and nature belts since the difference between highlands and hollows as far as pasture is concerned is too small.

On the 131.1 million hectares (84% of the country's total area and 98 per cent of the agriculturally used area) of pasture area of Mongolia about 2,300 types of plants grow, which belong to about 580 plant families. Of these about 600 types of plants are usable for the food requirements of the five most important sorts of animals (cp. Cerendulam, 1994: 3).

Of the above mentioned 2,300 plant types about 550 widespread types of forage plants have been quite well researched. According to the results of this research 47 per cent of these are frequently or willingly eaten and nutritious plants, 4 per cent are of very good food quality, 44 per cent are rarely or unwillingly eaten plants and 5 per cent are poisonous plants for animals (*Mallagaany*, 1966: 4). Of course the forage plants have different significance for the various animal types. From available research material concerning forage plants it can be determined which plant types are preferred by which animal types. This then makes it possible to ascertain, according to the distribution of the forage plants, in which direction the animals should migrate in the course of the seasons.

In general the seasonal food balance for all types of animals can be divided into two main periods: first, the active feeding in the period of vegetation, and second, the period of maintenance of weight gain and fattening in the inactive months of the vegetation. The time of vegetation is so short that it must be exploited to the full for an effective pasturage. Therefore, the attaining of a well-nourished condition of the livestock in the active vegetation period is divided into the following feeding periods:

1. the return of strength (Mong.: usan badailga) from April until May, the second half of spring,
2. the putting on of weight (Mong.: machan targa) from June until the middle of August,
3. the fattening (Mong.: ööchön targa) from the second half of August until the start of November.

Thus, the distribution of the forage plants is one of the most important factors for pastural migration, upon which the direction and distance of the migration depends.

Salty Soils as a Determining Factor of Seasonal, Altitudinally Directed Migration

The significance of salty soils for livestock and for seasonal migration has been neglected in investigations concerning nomadic livestock keeping in Mongolia. The reason for this is that the original seasonal migration to salty soils has become impossible or is overlooked, in particular in the last decades, as a result of newly established territorial-administrative borders. I consider it important that salty soils be considered as a basic motive for seasonal migration.

The salty soils of the pasture land of Mongolia are very rich in sodium (Na) and potassium (K) which are the most important elements for the growth and the strengthening of the bone tissue of the animals as well as the most important chemical constituents of bone marrow, magnesium (Mg) which is one of the most important elements of the muscle tissue, and copper (Cu) which is necessary for blood formation and for the functioning of the respiratory organs.

Thus, salts play a very important role in the putting on of weight and in particular in the fattening as well as the maintenance of a well-nourished condition of the livestock in autumn. Since the livestock do not obtain an adequate amount of the above mentioned elements from plants, salty soils are the only supplementary source of minerals. Salty soils are in no way used occasionally or seasonally but quite regularly. For example sheep must pasture on salty soils in average once every five to seven days in winter, every three to four days in spring, every four to seven days in summer and autumn (Gonchigzhav et al., 1980: 141). The regularity of course depends upon the mineral content of the soil. If the summer and autumn pasturage takes place on soils low in salt, horses, for example, must once every seven to ten days pasture on salty soils (*Mallagaany*, 1996: 167).

Thus, there arises the need to allow livestock to graze in regions rich in salts. Livestock keepers take into consideration when choosing a pasture not only favourable climatic conditions and the state of the vegetation but also the salt content of the soils. In the rule every seasonal pasturage must have salty soils available.

A physical-geographic investigation of the pattern of the distribution of salty soils in connection with the pasture land in Mongolia has only to a limited extent been carried out. In the forest highlands the proportion of salty soils is the smallest, while in the steppes and the Gobi and desert regions very high. This tendency is valid not only generally on a large scale but also locally: in the mountains almost nothing, at the mountain foot and in the valleys more frequent.

That the density of vegetation, as mentioned above, decreases gradually from the highlands and forest highlands to the steppes and the Gobi and the nutritional value of a food unit thereby increases, stands in close connection with the distribution of salty soils. Salty soils are divided in livestock economic practice according to the concentration of minerals into sparse and rich soils. In the lowlands the mineral content of salty soils is as a rule higher than in the highlands. According to experience as well as to the results of research the following regions in the steppe zones have been identified as having salty soils of good quality: Bor ovoo (in the Übröhangai Aimag, Bajanöndör Sum), Olon nuuryn gol (in the Bajanhongor Aimag, Bajanburd Sum), Erdene uul, Gutlyn ulaan uul, Rashaant (in the east steppe region) etc., (*Mallagaany*, 1996: 52).

Mongolian historical sources report that livestock keepers migrated in autumn with their stock often far into the south or the south-east, to the Gobi, to regions with soil rich in salts. This is still more or less the case today. But in this there are considerable difficulties in connection with the current of the administrative Aimag and Sum borders. In the Negdel period, 1959–1991, it became habitual to satisfy the needs of the animals by transporting salt. However, there is still today amongst the herdsmen in those Sums, within the borders of which there exists a region of salty soils, an attempt particularly in autumn to move with their stock into low lying areas. Here the direction of migration to salty soils corresponds with the other goals of autumn migration, namely the attainment of favourable climatic conditions and of a suitable vegetation.

Access to Drinking Water as a Determining Factor of Migration

Water plays differing roles in mobile livestock keeping. Above all, water is one of the most important constituents of nourishment. For example, a sheep requires three to four litres of water in order to digest 1 kg of dry food. The required amount of water of course depends upon the respective climatic conditions as well as the juice content of the plants. Sheep drink each time about two to two and a half litres of water. In summer sheep must receive water two to three times daily under the most favourable climatic and vegetative conditions, in spring and autumn once or twice daily and in a winter with heavy snowfalls once every second day. This is valid also for the other types of animals (Gonchigzhav et al., 1980: 144).

On hot summer days the need for water is even higher since the livestock are cooled down by means of water i.e., by perspiring. Therefore, livestock keepers in summer as well as in autumn migrate to the river valleys and lakes. Since on summer days there are at the water sites in the lowlands many mosquitoes, midges and horseflies, the summer pasturage is carried out on windy hills lying near to water. When the insect pests disappear in autumn on account of the decrease in temperature, it is possible to move downwards direct to the water source.

In winter the livestock require relatively little water on account of the sparse vegetation and of their physiological adaptation. The need for water in winter is mostly satisfied by the disintegration of acquired fat and by snow. A superfluous amount of water through drinking would decrease the resistance of the livestock to cold. Therefore, a pasture distant from a water source is most favourable in winter. The fact that regions lying nearby to water are cold in winter is another reason to seek out a pasture away from water sources and rivers. Moreover, the source of water is of great significance in the division and the divided use of the pastural regions. Watersheds delimit climatic- geographic belts and the gradings of vegetation both altitudinally and planarly, and thereby form comparatively isolated pasture strips or areas.

In historical sources and on ancient maps of Mongolia the locations of the ancient tribes were always represented in the vicinity of sources of drinking water—larger rivers or permanent fresh water lakes or abundant springs. Also in the later establishment of territorial-administrative borders the entirety of a pastural area and the availability of water were especially considered. For example larger rivers, lower river valleys or fresh water lakes in the lowlands were used as border lines of pastural areas. However, in the Negdel period this was of lesser significance. Owing to the organization of livestock keeping, the territorial-administrative divisions were made smaller and the number of such regions was massively increased. In 1990 there were in all 310 Sums in Mongolia.

The supply of drinking water is often ensured by the building of wells. In 1991 there were about 37,600 treadwheel wells (in the Gobi and steppe regions the wells must be dug deeper) and 3,600 built-up springs (*Mongolian Economy*, 1995: 40). This brought both advantages and disadvantages, If in the case of a scant network of natural water sites the pasture lands about these sites do not overlap, then the pastures remain unused for the greater part of the year or even over the years. They offer in the alternative planning of the nomads a natural food reserve, which is opened up only in those years when natural water sites are formed owing to a greater degree of moisture, since then sufficient water is available. These mostly unusable pasture zones are now opened up to the nomadic economy by means of the construction of the tread wheel wells. The nomads now make use of these wells even during periods of low rainfall, and as a consequence the natural food reserve is used up. The damage is even greater since the water potential of the wells is in most cases higher than the grazing potential in their vicinity. The mobility of the nomads, which was once forced by the exhaustion of the water sites, is being limited. A certain devastation of the vegetation of the pasture about the wells is also a consequence. The wells therefore are more or less depriving the nomads of their natural pasture reserves.

Geographic Borders as an Influencing Factor of The Direction of Migration

Not only climatic conditions, the state of the vegetation and the distribution of salty soils and water sources are of significance for the distance and frequency of migration, but also geographic borders. In general they are divided into ecologically and economically determined borders. The ecologically determined borders include mountain crests, larger rivers, cliff regions, deserts, valleys, etc. Besides these, which are determined by altitude and natural markings, there are also economic margins of the pasture land. Obvious borders are formed by settlements and industrial areas, by paths, roads and railways.

From the above the wrong impression might be gained that the borders of the pasture land are once and for all fixed. This is however in no way the case. The ecologically determined borders are relatively stable, but the economic margins are certainly not. They are formed in flexible adaptation as a consequence of technological and social-economic development. Thus there is, for example, in the lower river valleys of Orhon, Selenge and other rivers, which have their sources in the Hangai and Hentii Mountains and which today are located in the Selenge Aimag, hardly any pastural area. The reason for this is that the pasture land has been superseded as a result of the development of farming and industry in this region. The rate of industrialization in the Selenge Aimag is in average three times higher than in the other Aimags.

On the basis of what has been said about the ecological factors of seasonal migration, the ordering of the seasonal pasture areas in Mongolia i.e., how and in what relation to the ecological factors the four seasonal pasture regions lie, can be described. There are however two variations: first, the locations of the actual or the original pasture areas, in which case the economic or socially determined borders have no part, and second, the locations of the present pasture areas, in which case all economic and ecologically determined borders are considered.

Concluding Remarks

On the basis of the above discussed ecological determinants we can distinguish basic forms of seasonal migration.

In winter the best pasture regions lie, on account of the favourable combination of temperature, wind and precipitation, in principle in all three nature zones in the belt between the higher and middle mountain ranges. The winter site is therefore located on a higher south slope of a mountain or in a mountain cleft away from the wind direction.

The spring pasturage on the other hand takes place somewhat below the winter pasturage, on the south face of a mountain between a middle altitude and the mountain foot. Although in general the lower the location the warmer it is, snow often remains at the mountain foot and in the lowlands until May, and nights can be cold and damp.

The suitable regions for the summer pasturage are for climatic reasons located on the higher parts of a mountain, higher than the winter pastures, for example on mountain crests, where the winds blow more strongly. There are of course differences in connection with the differing nature zones. Since the inversion line of the Hangai and Hentii forest highlands lies lower, for example, the mountain crest on this inversion line should be avoided for the choice of summer pasture and accordingly the summer pasturage is chosen in the foothills.

In order to ensure that the livestock remains well nourished from summer into autumn, migrations are carried out to a warmer location where the animals can feed more peacefully and can be fattened in autumn in a regular way. Such sites, which in comparison to other areas remain even until late autumn warm and free of snow, are located characteristically in low lying regions, as a rule in all nature zones.

If we compare the forage plant types according to seasonal divisions and the altitudinal distribution of forage plants it becomes clear that the seasonal pasture regions, as far as the distribution of forage plants is concerned, stand in a belt-like relation with the differences in altitude.

Moreover, salty soils play a very important role in the putting on of weight in summer and in the fattening as well as the maintenance of a well nourished condition of the livestock in autumn. Since in the distribution of salty soils there exists the phenomenon that the mineral content is lowest in the highlands and highest in the lowlands, the seasonal migration should in the main be determined altitudinally.

Since on summer days there are at the water sites in the lowlands many mosquitoes, midges and horseflies, the summer pasturage is carried out on windy hills lying near to water. When the insect pests disappear in autumn on account of the decrease in temperature, it is possible to move downwards direct to the water source. The fact that regions lying nearby to water are cold in winter is a reason why it is necessary to seek out a pasture away from water sources and rivers during this season. Moreover, the source of water is of great significance in the division and the divided use of the pastural regions.

Geographic borders influence migration routes i.e., they determine in which direction and where the migrations should be carried out.

The effects of all these factors on the seasonal migrations are connected with differences in altitude. From this it follows that the seasonal migrations in principle in all nature zones are determined altitudinally. However, this view should not serve as an argument that in Mongolia there are absolutely no planar migrations. On the contrary livestock keepers undertake quite frequently planar migrations in the course of a season in order to extend the pasturage and to satisfy food requirements.

There are in respect to pasture and ecologically speaking two main regions of Mongolia: mountainous land and the steppes/Gobi, whereby mountainous land is divided according to the respective ecosystems into two sub-regions, the highlands in the Altai Mountains and in the Hangai-Hentii Mountains (Bazargür et al., 1989: 104–144). As far as pastural migration is concerned, there are thus three different types of migration corresponding to the defined nature zones: migrations in the Altai Mountains, in the Hangai-Hentii Mountains and in the steppes/Gobi region. In addition, livestock keepers conduct their seasonal migrations in the steppes in Central and East Mongolia and in the south-east Gobi region in quite different manners. This has probably been caused by the too closely drawn administrative borders, in which the actual pastural region has not been considered. In the first two forms the seasonal migrations quite evidently correspond to differences in relief and are altitudinally directed. There is however, a clear difference in connection with the order of the change of pasture, depending upon orographic conditions.

Seasonal Migration in The Altai Mountains

This mountain chain, which stretches from the north-west of Mongolia to the southern Gobi, is almost devoid of forests. The mountain system is surrounded by valleys and hollows. It borders onto the Altai valley of the northern Gobi which is covered with numerous saltpans and sand plains and which extends from the basin and over the broad valley of the Gobi lakes, as well as onto the Altai valley of the southern Gobi where there are substantially less lakes and sand plains but more gravel plains. This valley is comprised of three parts: in the west the Atschit valley, in the south the Alaschan Gobi and between the two the Transaltai Gobi. These orographic conditions enable livestock keepers to exploit to the full by means of migration the natural pasture from the mountain crests to the valleys and foothills. This permits a seasonal migration from the peaks to the lowlands in the order summer, winter, spring, autumn. There are however, in some regions other variations of the order e.g., winter, spring, summer, autumn.

Seasonal Migration in The Hangai-Hentii Mountains

The massive Hangai-Hentii Mountain system also borders on the great lake basin in the west and on the southern Gobi in the south, and is bordered in the east and south-east by the undulating eroded plains of the steppes and Steppe-Gobi which are covered with numerous saltpans. Between these two mountainous lands, the broad river valleys of the Orhon, the Selenge and the Tuul are situated. It is here characteristic that the south faces of the slopes and partly of the mountain crests (e.g., in Hangai) are cliffy and steep rising. The northern slopes and in part also the northern crests are on the other hand forested and levelled out. Mongolian forests, which cover in all only 9.6 per cent of the country, are comprised almost entirely of the forest regions of the Hangai-Hentii Mountain system and the Höbösgöl Mountains.

As a consequence the herdsmen are not able to make use of the mountain crests for pasturage, and in particular for the pasturing of sheep and goats, and therefore allow their stock as a rule to graze only as far as the inversion line.

Thus, it happens that the altitudinally directed seasonal migration from the peaks to the lowlands are ordered winter, spring, summer, autumn. However in a region where either there are crests devoid of forest or gravel and accessible to the livestock or the inversion line is higher than about 2,300 m, the first variation of migration as in the Altai Mountains is also possible.

Seasonal Migration in the Steppes

The high steppes of eastern Mongolia, which in the west are bordered by the Hangai-Hentii Mountains, are above all in the north covered with undulating eroded plains rich in salts. They are divided approximately through their centre by the voluminous Herlen River which flows into the Dalai Nuur. The east lying Lake Buir is situated on a huge plate-like high steppe. Here there is a relatively small difference in altitude between the highlands and the lowlands. For this reason seasonal migrations from the heights into the lowlands are ordered winter/spring, summer/autumn, whereby the pasturages for winter and spring are conducted at about the same altitude, as are those for summer and autumn. One sees here that this form is not only a planar migration determined by thermal and humidity conditions but also clearly altitudinally directed.

The extremely extensive and flat eastern and south-eastern Gobi valley is a steppe sparse in vegetation. Stretching valleys and flat hollows, partly covered with salt lakes and sand marshes, characterize the landscape. Desert steppe and desert alternate, ponds and small flat lakes are formed temporarily after heavy rainfalls. Although some scholars conjecture that migrations specially suitable for the region are carried out here results from research show no particular form of migration comparable to that in the steppes. I am of the view that seasonal migration within the steppes is incomplete on account of a delimitation of the original pastural strips by territorial-administrative borders. Through this the traditional, ecologically adapted migration routes and directions have in essence been simplified.

The climatic and vegetative conditions and the numerous saltpans and water reserves in these regions are exactly the requirements for an autumn pasturage, and in part too for a summer pasturage. From Mongolian historical sources it can be learnt that nomadic herdsmen for just this reason moved in particular in autumn into the steppes and desert steppe, whereby they covered distances of up to 750 km. Regarded geomorphologically, the lowland corridor from the eastern end of Mongolia, the Menen Steppes, to the southern end, the Galbin Gobi, is a geographic borderline of the flatlands of the north-western and south-western highlands and of the pastural regions of both sides.

As already mentioned, these are only the basic forms of migration. There are in addition numerous variations of and experiments in seasonal migrations connected with the respective conditions and geographic borders of the pasture land.

References

Bazargür D., Chinbat, B., Siirev-Ad'ya, S., (1989), *Bügd Nayramdach Mongol Ard Ulsyn Malchdyn Nüüdel* (Migration of Livestock Keepers in the Mongolian People's Republic), Ulan Bator.

Bügd Nairamdakh Mongol Ard Ulsyn Undesnii Atlas (National Atlas of the Mongolian People's Republic), 1990, Moscow.

Bügd Nairandach Mongol Ard Ulsyn Uur Am'sgal Gadragyn Usny Nööciin Atlas (Atlas of the Climatology and Hygrology of the Mongolian People's Republic), 1985, Ulan Bator.

Cadraabal, G. (ed.), (1982), *BNMA U-yn Mal aj Akhujg Khötlökh Sistem* (The Management of the Livestock Economy of the Mongolian People's Republic), Ulan Bator.

Cerendulam, R., (1994), Bilcheer Ashgilach Arga (Ways of Use of the Pasture Land), in: *Gov' Khangain Sonin Bichig,* No. 2.

Gonchigzhav, Ds., Zavdsmaa, R., Unenbüren, C. (eds.), (1980), *Mal aj Akhui* (Livestock keeping), Ulan Bator.

Gungaadas, B., (1984), *Ekonomicheskaya Geographia Mongolii* (Economic Geography of Mongolia), Moscow.

Map of Mongolia, Ulan Bator, (1993), 1: 1,000,000.

Mojbuu, D. and Bakej, A., (1996), Mal aj akhuig khögjüülekh bodlogyn zarim asuudal (Aspects of the Development Policy for the Livestock Economy), in: *Zasgiin Gazryn Medee,* 17/1/1996, No. 12 (452), S.2.

Mongol Orny Bilcheeriin Mal Mallagaany Arga Turshlaga (Practical Experience in the Pastural Livestock Economy of Mongolia), (1966), Ulan Bator.

Mongolian Economy and Society in 1993, Statistical Office of Mongolia.

Olzijchutag, N., (1989), *Mongol Orny Urgamalyn Aimgiin Toim* (Overview of the Distribution of Vegetation in Mongolia), Ulan Bator.

Sagdarsuren, O., (1984), Mongolcuudyn Bilceeryn Mal Mallagaany Tuuchen Arga Azillagaanuud (The Traditional Livestock Breeding Method of the Mongols), in: *Sinjlekh Ukhaan Am'dral,* No. 3.

Yunatov, A.A., (1950), *Osnovie Cherty Rastitelnovo Pokrova Mongolskoi Narodnoi Respubliki* (Fundamental Features of the Vegetation Cover of the Mongolian People's Republic), Leningrad.

15 Promoting Local Health Tradition Using Herbal Plants in the Socio-cultural Space of Kerala Using Geoinformatics

Rubeena T.A. and T. Vasantha Kumaran

Traditional medicine still meets the requirements of the people in the developing world. Following the Alma Ata declaration of 1978, efficient strategies have been adopted by many countries for integrating the traditional health practices into the national health systems. India is no exception to this. The country has by now developed a parallel stream of Indian medicine alongside the modern medical system as constituted by allopathy. However, the policies adopted for promotion of the Indian systems of medicine do not cover the traditional health practices of the local communities, which fall outside the realm of the classical medical systems like Ayurveda, Siddha, Unani and naturopathy. The Local Health Traditions (henceforth referred as LHT) which are nothing but the traditional health practices of the local communities making use of locally available herbs, have from time immemorial provided innumerable health care options as preventives and curatives and has thus played an important role in enhancing the health of the local populations.

> LHT are the oral traditions practiced by the local people, who by trial and error selected many plants from their habitat to treat the ailments confronted by them.
>
> (Gauniyal et al., 2005; Kumaran, 2002).

LHT as referred to in the study is devoid of any personalistic or supernatural beliefs and confines to the natural beliefs.

According to Aregbeyen (1996) there is growing evidence that

- herbal medicine is effective and acceptable and that:
- There is every reason to promote knowledge and understanding of which type of herb must be used for treating which ailment in the various communities.
- He suggests that, if properly developed, herbal medicine could be used to supplement PHC, and in promoting a spirit of self-reliance.

However, the LHT have moved into oblivion, with increasing access to modern medicine especially in Kerala, the state lying along the south-west coast of India. It might seem paradoxical that Kerala blessed with rich herbal wealth, which has stepped into sectors like health care tourism, has experienced a tremendous decline in the population employing LHT to meet the health requirements. The excessive commercialization of the health sector is one major reason for the current health seeking pattern. The commercialization has also made health care unaffordable to the majority of the people. Ayurveda, the classical medicine and allopathy, the modern medicine are increasingly becoming unaffordable with increasing price of medicines (Ekbal, 2000). This coupled with the excessive privatization of the health sector and the deterioration state of the public health facilities, is pushing the rural population further into the vicious poverty trap. The current research paper, therefore, looks into the reason for the popularity of particular medical systems and the viability of promoting LHT alongside the dominant systems of medicine. Geoinformatics has by now been used for mapping biodiversity and efficient management of the various natural resources. The herbal wealth which forms the valuable resource base for the LHT can be mapped in their habitats with the help of the local knowledge.

The objectives of the study are therefore:

- To assess and evaluate the relationship existing between the choice of a medical system and the physical and socio-cultural space.
- To evaluate the significance of ignorance of the local health traditions and non-availability of the resources and associated practitioners as influencing the limited use of the local health traditions.
- To assess the possibilities of employing geoinformatics in the promoting the use of the local health traditions at the local level and thereby providing affordable health care options at the finger tips of the lay people.

The Study Area

The study is carried out in the Kanjoor panchayat of the Ernakulam district of Kerala (Map 1). The village is bound on the east and south by the Periyar river. Lying along the flood plains of the larges river in Kerala, the region is blessed with fertile alluvium supporting all forms of agriculture. The village also has a well divined canal system belonging to the Kodanad-Edanad canal system. Toward the north the region has vast areas under paddy cultivation. However, the region is covered with nutmeg plantations and other crops making it a rich agricultural zone. However, tremendous land use change are taking place with more agricultural and virgin lands getting converted into residential areas, threatening the herbal population in the area.

The Kanjoor panchayat occupies an area of 14.32 sq km and has 4,885 households. The total population of the panchayat as per the 2001 census is 21,651 of which 10,624 are males and 11,027 are females. Of the total

Map 1

Location Map of Ernakulam District

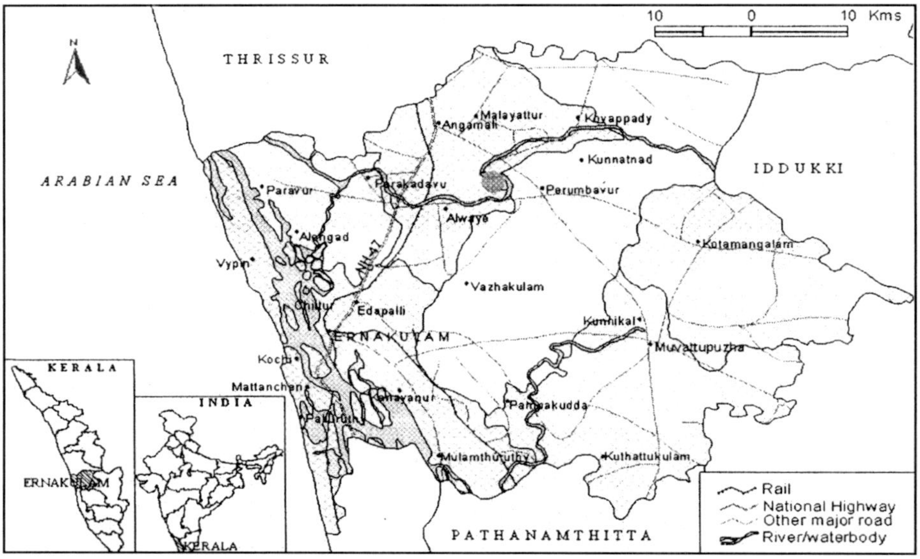

population of 21,651 the working population is 7,793. Of this 5,858 are men and 1,935 are women. The total work participation rate therefore is 35.9 per cent. The male work participation rate is 55.1 per cent and the female work participation rate is 17.5 per cent.

The panchayat is well served by educational, financial and health institutions (Map 2). There is one government PHC (Primary Health Center) (modern medicine) having one doctor and one private hospital with 40 beds and four doctors. The doctor at the PHC in Puthiyedam locality of Kanjoor, visits the health centre at Thuravankara, a remote locality in the panchayat, once every month. The remoteness of Thuravankara is attributed to the absence of public transport facilities. Though, the people of Thuravankara want better access to the PHC, the non-availability of transport facilities has limited their access. There are two private clinics having one doctor each, functioning in the panchayat.

There is a Government Ayurvedic dispensary with an Ayurvedic doctor and an Ayurvedic Research Centre functioning in Kanjoor, apart from a number of Ayurvedic pharmacies. There are three homeopathic clinics besides four traditional health providers who practise Ayurveda in the panchayat, though the exact number is not known.

The Methodology

The methodology employed consists of a few participatory, qualitative and quantitative tools apart from the cartographic tools employed. The participatory techniques were employed to enhance the quality of information and to have a wider perspective on the key issues addressed. The participatory

Map 2

Infrastructure Kanjoor

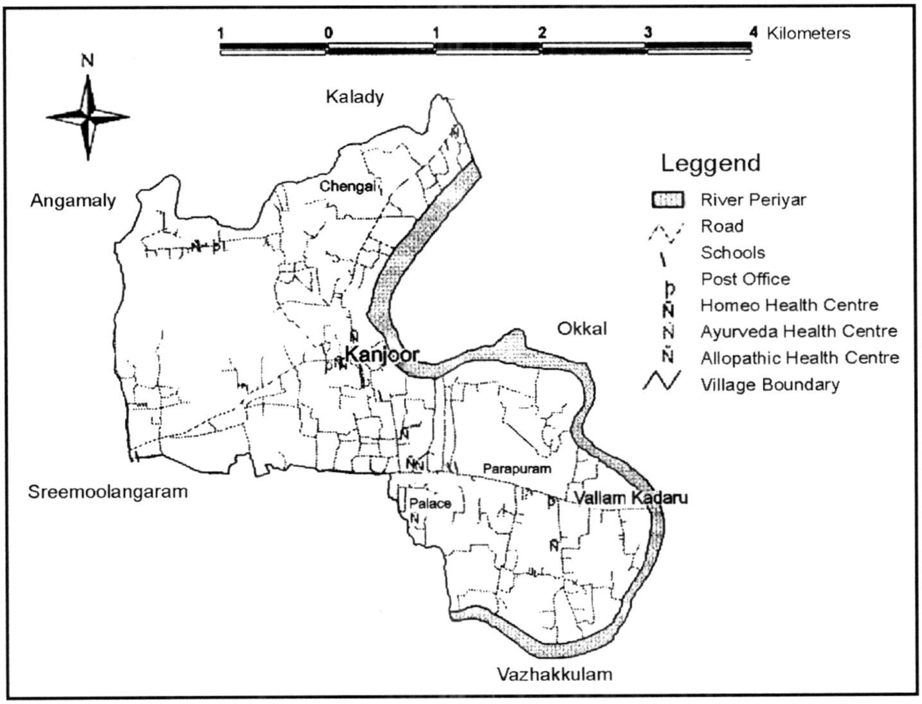

techniques helped to ensure the participation of the stakeholders who directly influence or are influenced by the use and access to local health traditions.

> PRA has been described as family of approaches, behaviours and methods for enabling people to do their own appraisal, analysis and planning, take their own action, and do their own monitoring and evaluation.
>
> Chambers (2002)

For the current research paper, data collected using historical profiling has been widely employed. Historical profiling is a participatory tool which helps to delve deep into the historical basis of the current socio-environmental set-up of the study area and ensures a better understanding of the major social and environmental changes and their effects on the local economy, environment and society (Bolt and Fonseca, 2001).

Qualitative techniques like key-informant interviews and focus group were used to collect information from the people. Quantitative methods, in particular experimental designs concentrate on the control of extraneous variables while qualitative methods are holistic (Morse and Chung, 2003). However, the information derived from qualitative and participatory techniques are not quantifiable and are just a wealth of ideas and information which would have never crossed the minds of the researcher employing quantitative tools alone. A questionnaire survey (for 245 households) was used to collect quantitative information which helped to quantify the information gathered during the participatory and qualitative appraisals. This was important to perform analysis and to obtain a clear picture of the relationship existing between the ignorance of LHT, non availability of traditional practitioners and natural resources to the use and disuse of LHT.

The Results and Discussions

The study reaffirmed that modern medicine remains popular among other systems of medicine (Kunhikannan and Arvindam 2000). Of the total 245 households, majority preferred modern medical system when compared to other systems. And the preference for local health traditions were limited (Figure 1).

Figure 1
Preferences-Medical Systems

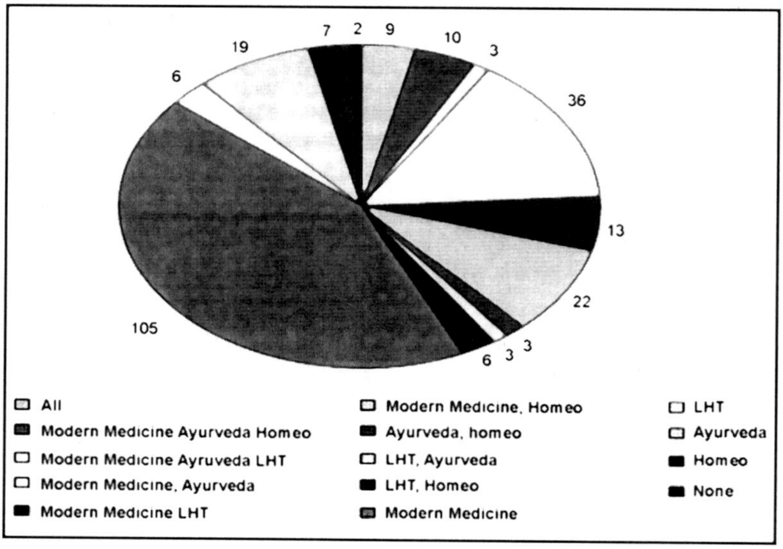

One hundred and five households preferred modern medicine solely for health care; while in 36 households Ayurveda was the most preferred medical system. In a significant number of households, more than one system of medicine was preferred. The preference for modern medicine increased during a time frame of fifty to hundred years and cannot be treated lightly. The time lines created by the villagers during the participatory workshops shows that there was a high scale dependence on LHT in the past (before 50 years) and modern medicine was rarely employed since it was inaccessible to majority of the population at that time and people were more knowledgeable about the LHT.

But the situation changed tremendously over the years, with modern medicine becoming accessible to the laymen and also popular due to the quick relief it offers to the symptoms of illnesses. The local health traditions gradually moved into oblivion with very few opting for them due to dwindling natural resources and reduced interest in the traditional knowledge. This coupled with the government strategy which gave more importance to modern medicine and sidelined all forms of traditional medicine led to the gradual decline of LHT. The choice of the medical system is a function of the socio-cultural and physical realities as existing in the neighbourhood. The findings of the historical profiling exercise showed that:

- there is tremendous change in the economic, socio-cultural and environmental situation as existing in Kanjoor, like other parts of the state;
- the land reforms helped to erase the caste differences though the class differences still remain;
- infrastructure has improved and modern health care facilities have become more accessible in terms of distance, but still remain inaccessible to the lay people in terms of cost;
- environmental changes are evident from the increasing pressure on land and the resultant land use changes from primary land uses to secondary land uses;
- the changed socio-climate has influenced the mindset of the people seeking efficient and quality health care, thus leading to a boom of private health care sector;
- the socio-cultural changes have led to a situation where the local health traditions are fast eroding, due to its unscientific nature and ignorance of these traditions; and
- the resource base also is threatened by the higher pressure and a consequent loss of herbal wealth.

These facts about the changed physical and socio-cultural space explain the current choice of modern medicine as the most preferred medical system by the people.

The realities which have led to the diminishing use of the LHT are manifold, many of which were highlighted in the discussions above. Of the manifold problems, leading to disuse of LHT, ignorance of LHT, non-availability of qualified traditional practitioners and resource scarcity were identified to be the key issues which need immediate reprisal.

Ignorance of the LHT

"The rate of knowledge erosion is faster than the rate of resource erosion" (Unnikrishan, 2004). Most people are unaware of the uses of even the most commonly available herbs in their neighbourhood. Being oral traditions, the knowledge of local health traditions cannot be retrieved once it is lost. The current situation points towards the need for an initiative to codify this knowledge base at the earliest. Creation of inventories is possible only through the active participation of the community and the traditional medicine men. Also care needs to be taken to serve the needs of the community and the traditional medicine men.

Ignorance as a factor controlling the use and access to the LHT is brought out clearly in the survey. Nearly 51 per cent of the households felt that ignorance of the LHT was the major constraint in limiting their use. Another 9.1 per cent felt that it is moderately important and 12.3 per cent felt it is important. The remaining 37.6 per cent felt that it is not a significant reason for the decline of the LHT (Figure 2).

Figure 2

Ignorances of Local Health Traditions

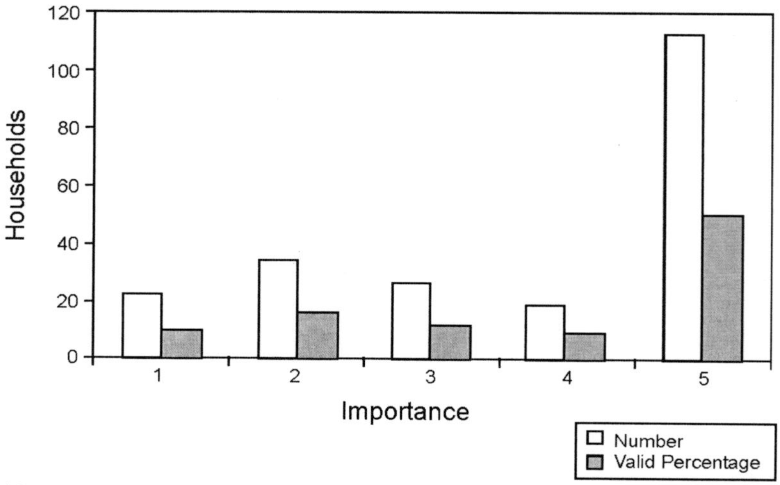

Adapted from Rubeena, 2006.

Non-Availability of Qualified Practitioners

The survey pointed out that non-availability of qualified traditional practitioners has led to a tremendous decline in the use of local health traditions. Thirty-seven per cent of the people who responded found non-availability of traditional medical practitioners as a very important reason, 14.2 per cent as important and 13.3 per cent as moderately important reason for the decline of the LHT (Figure 3).

In the words of a woman who attended the Participatory Rapid Appraisal (PRA) at Aryankavu, a neighbourhood in Kanjoor:

> During early days, when I was a child, we knew many traditional practitioners, who used to provide specialized treatments for may illnesses including migraine. But now, it is difficult to locate many practitioners who used to provide effective cures for illnesses for which medicines are not available in modern medicine. Many of them have died without having passed the information to the future generations. and the knowledge which was passed down by their ancestors has died out with them.

The non-availability of practitioners has resulted from:

- the governmental policies which fails to give due recognition to traditional practitioners and often hindering their practice;

Figure 3

Non-Availability of Traditional Medical Practitioners

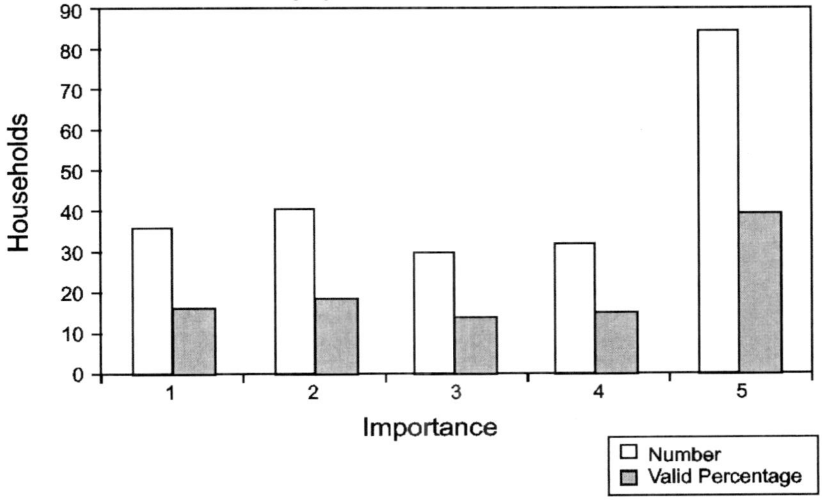

Adapted from Rubeena, 2006.

- the categorization of traditional medicine as unscientific; and
- reluctance to share the knowledge resources with others.

Though of late the government has shown interest in reviving traditional knowledge systems, there is a need for taking conservation measures on war footing (George et al., 2002). The declining number of qualified practitioners shows that there is an urgent need to conserve the knowledge bases, which might otherwise be lost with another passing generation. In Kanjoor, there are four families who still practice traditional medicine. But at the same time the members of the younger generation of these families are quite unaware and show little interest to these valuable traditional knowledge bases. A significant reason for this being the declining interest and the declining number of people, who approach a traditional medicine man. Also their contributions are seldom given due recognition by the community. During an interview with one of the traditional medicine men in the panchayat, he expressed his deepest fears on the existing situation where none of his children or grandchildren possess any amount of knowledge which his ancestors had passed down to him. Still he is optimistic that he could find someone who will be interested in this traditional knowledge in his own family.

He says that:

> Though I am worried that none of my children knows about this valuable knowledge, especially the panaceas, I feel that there will be someone who will be interested among my grandchildren to whom I can pass this on. However I have documented the traditional knowledge I possess and they may use it in future. But I do not know how useful it will be unless you know to diagnose diseases. Many people have come to me to learn but I haven't taught any of them due to time constraint.

Scarcity of Resources

Closely associated with the above two factors is the scarcity of resources in terms of herbs which are widely used in the practice of LHT. The tremendous changes in land use and land cover have led to tremendous loss of herbal wealth. The high scale development coupled with ignorance of these herbal resources has led to deterioration of the herbal wealth which once formed the basis of country medicine and home remedies.

Many households reported that scarcity of herbal resources and difficulty to locate them has led to reduced use of LHT.

Of the total respondent, 45.9 per cent rated resources scarcity as a very important reason for the decline in use of the LHT, whereas 14.2 per cent of the population rated it as moderately important and another 14.2 per cent as important factor influencing the reduced use pf LHT. A total of 24.7 per cent however rated it as slightly or not at all important (Figure 4). According to them the disuse of LHT is primarily due to ignorance and unwillingness to use the LHT.

Figure 4

Resource Scarcity

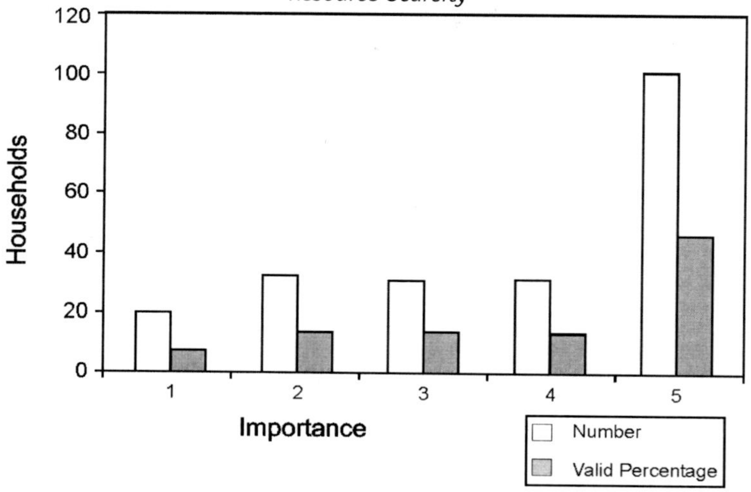

Adopted from Rubeena, 2006.

These results emphasize the need for generation awareness on the use of particular herbs found in their neighbourhood to enhance their use by community groups. This can possibly be achieved by actively involving the community in the process for reviving LHT. The steps which will help towards attaining this end include:

- preparation of inventories similar to the Peoples Biodiversity Registers[1], for LHT and resources supporting LHT and their spatial distribution within a village;
- making the community aware of the importance of using LHT and the need for conserving the resources used in LHT;

- providing options for affordable health care by complementing modern medicine with LHT; and
- calling for strict legislation to revive the LHT, as a step towards providing affordable and sustainable health care.

The Role of Geomatics (Figure 5)

The role of geomatics in the process of reviving the LHT starts with the basic mapping of the various herbal habitats and their spatial distribution within the village. Moving a step forward, the herbal resources can be mapped at ward levels and also the changes which have drastically affected the herbal population in the neighbourhood. Further the uses of the particular herbs found in the various localities in the panchayat can be integrated into the base map.

Figure 5

Promoting Local Health Traditions Using Geo-Informatics: A Framework

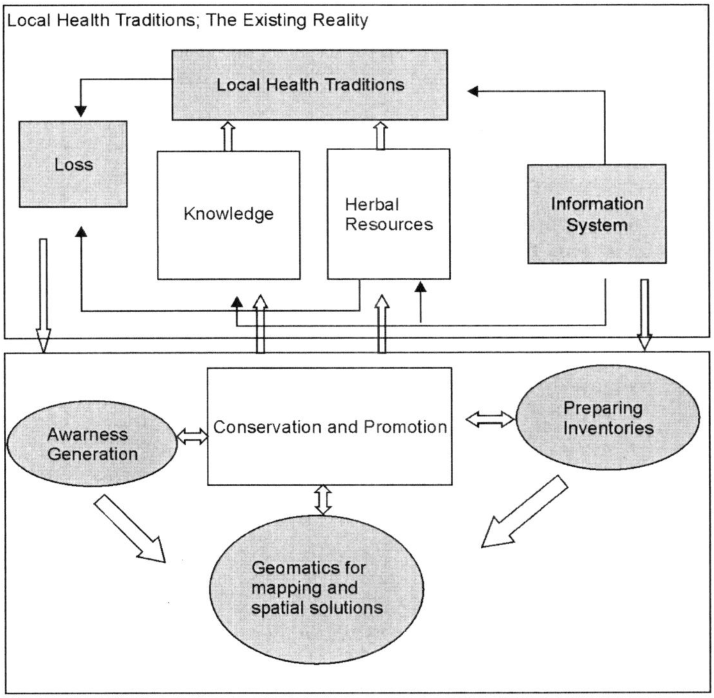

This procedure however, calls for active participation of the community members in the various localities, with support from the local and state governing bodies. Also the community needs to be trained in identifying the herbs which need technical assistance from members of the scientific community. Also the maps developed can be used by the community for a better understanding of the problems associated with the herbal wealth in the panchayat and to adopt planning strategies to best address the problems.

Thus geomatics will serve three ends:

- creation of an information system for the herbal habitats and the uses of the herbs in LHT;
- identification of vulnerable ecosystems and the reasons for vulnerability; and
- providing spatial solutions for the problems identified.

The creation of an information system at the local level will help the community to identify the herbs available in their neighbourhood and will help the people to access these resources when they need it. Also a periodic appraisal will help them to know about the resource erosion rate and the subsequent conservation measures to be adopted. The application of a herbal information system at the community level will help to mainstream the cause of LHT and their conservation.

Conclusions and Recommendations

From the foregoing discussions it has become evident that LHT are losing ground in Kerala, owing to ignorance of LHT, scarcity of resources and non-availability of qualified traditional practitioners. Since LHT are local cultural resources, the efforts to conserve them should start at the local level through preparation of inventories with the active involvement of the community.

The possibilities of employing geomatics in promoting LHT are manifold. But the success of employing geomatics lies in bringing together a number of key players ranging from the administrative elites, members of the scientific community, to the community members and the traditional medicine men. It needs to be emphasized that there is a digital divide and access to geomatic tools is not easy for the lay people unless supported under a governmental programme or scheme. The success of the initiative lies on the schemes adopted by the government and on the stringent measures adopted by the government in their implementation.

The support of the community is essential in bringing together the knowledge possessed by them and for efficient sharing of information among themselves. The roles of the traditional medicine men are significant in location herbs which are scarce and otherwise also in understanding their uses. The promotion of LHT using geomatics is possible only through the active participation of all these players.

Note

1. Peoples Biodiversity Registers: The Peoples' Biodiversity Register Project (PBRP) was taken up as a pilot project in the Ernakulam district of Kerala in 1997, under the direction of the State Planning Board for preparing inventories of the biodiversity resources and associated knowledge bases at the panchayat level, with the active involvement of the local community (Anuradha et al., 2001).

References

Anuradha R.N., Taneja B. and Kothari A., (2001), "Experiences with Biodiversity Policy Making and Community Registers in India," in, Swiderska K. (ed.), *Stakeholder Participation in Policy on Access to Genetic Resources, Traditional Knowledge and Benefit-Sharing: Case Studies and Recommendation*, Biodiversity and Livelihoods Issues No. 4, London, International Institute for Environment and Development.

Aregbeyen, (1996), "Traditional Herbal Medicine for Sustainable PHC", *IK and Development Monitor* 4: 2.
http://nuffic.nl/ciran/ikdm/4-2/articles/aregbeyen.html

Bolt E. and Fonseca, C., (2001), Keep It Working: A Field Manual to Support Community Management of Rural Water Supplies, IRC Technical Paper Series 36, IRC International Water and Sanitation Centre, Delft.

Chambers, R., (2002), *Participatory Workshops: A Sourcebook of 21 Sets of Ideas and Activities*, London: Earthscan.

Ekbal, B., (2000) "People's Campaign for Decentralized Planning and the Health Sector in Kerala: An Issue Paper", People's Health Assembly, People's Health Movement.

George, V., Rajasekharan, S., and Nair, G.M., (2002) "A Report on the Brain Storming Session held on 20.9.2002 and The Workshop held on October, 7 to 9, 2002", Botanic Garden and Research Institute Pacha-Palode, Thriruvananthapuram and Support by IK Programme The World Bank Washington D.C.

Gauniyal, Anil K., Rawat A.K.S. and Pushpangadan P., (2005), "Interactive Meeting for Evidenced-Based Complementary and Alternative Medicines: A Report", *Evidence-based, Complementary and Alternative Medicine*, 2: 2, London: Oxford University Press.
http://ecam.oxfordijournals.org/cgi/content/full/2/2/249.

Kumaran, T.V., (2002), "Local Health Traditions", a Paper presented at the Workshop on Ecosystem Approach to Human Health in Chennai (August), Organized by McMaster-York-Madras University Collaborative Research Programme at the CDMA, Chennai.

Kunhikannan, T.P. and Aravindan, K.P., (2000), "Changes in the Health Status of Kerala: 1987–1997", Discussion Paper No. 20, Centre for Development Studies: Thiruvananthapuram.

Morse, J.M. and Chung, S.E., (2003), "Toward Holism: The Significance of Methodological Pluralism", *International Journal of Qualitative Methods*, 2: 3. Article 2. Retrieved (April, 2006). http://www.ualberta.ca/~iiqm/backissues/2_3final/html/morsechung.html

Rubeena, T.A., (2006), "Local Health Traditions and Complementary Medicine in Kerala", A Doctoral Thesis, University of Madras, Chennai.

Unnikrishan, E., (2004), "Materia Medica of the Local Health Traditions of Payyannur", Discussion Paper No. 80 Kerala Research Programme on Local Level Development, Centre for Development Studies, Thiruvananthapuram.

16 Herbal Plants and Health Care Practices Among Tribal Population of Bastar Region, Chhattisgarh

A Study in Geography of Health Care

Kailash Choubey

Health care is defined as the active process by which an individual achieves physical and mental well-being. It means health care actively tries to avoid illness and it consists of ill-people and health care provider (Doctor). At present health care services are in a crucial condition, so the present 'crises' in health care delivery should be the concern of growing body of geographers. In no area geographical investigations and contributions are more useful than in problems associated with the delivery of health care services.

Health care practices covers all those services which protect and promote the health of community. Long before the development of modern medicine, indigenous health care practices had always been an integral part of Indian culture. With the advent of allopathic medicine this indigenous system was overshadowed but due to their rich heritage their importance was never reduced. The medicines of the indigenous system on the other hand have less side effects.

Tribal concept of the health has its basis in the holistic approach of the Indian system of medicine and of life itself. Man must be in harmony with his environment which is supposed to include supernatural power, nature and society. Indigenous health practices and blind faith on them have always been an integral part of Indian culture. Tribal people have enough knowledge of vegetative drugs used as both preventive and curative medicine. The tribes use nature and its products for treatment of various ailments of the body. They have their own specialists who serve them well and in whom they have complete faith and confidence.

Study Unit

Now Bastar region lies in newly formed state known as Chhattisgarh and extends from 17°46′ to 20°34′ N latitude and 80°15′ to 82°01′ E longitude with varying elevation. The study unit consists of five districts, with an area 39,114 km^2. The total population of Bastar is 26,72,651 (2001) out of which 13,29,248 were male and 13,43,403 female. The percentage of tribal population is about

70 per cent. The sex-ratio is 1,011 female per 1,000 male, while literacy rate varies from districts to districts and in study unit comparatively males are more literate than females. The main tribes of Bastar are Goud, Moria, Dhurva, Helba and Madia. It is on of the isolated area of the country where most of the tribes continue to live a primitive life below poverty line without proper shelter, food, drinking water and medical facilities. The tribals of the region in general constitute indigenous group sharing common language and culture, living under varying but difficult geo-ecological setting of hills and forests. Most of their economy revolves around food gathering, hunting and they are also good cultivators. The tribals without any exceptions are cheerful light hearted people. Anyone normally finds them laughing and joking among themselves: most of them bear a singular character for truthfulness and honesty.

A large segment of tribal population depends upon their traditional health care system supported by their own people known as Baiga, Ojha, (quacks) etc. Normally tribal people of Bastar use herbs and roots to treat sick persons.

Method of Approach and Objectives

Usually tribal people are habitual to use their traditional health care system for most of their health disorders. So it is the prime aim of the present study to find out the reality, use and effects of their own traditional health care system, particularly which is based on herbals/medicinal plants.

The present study is basically based on field-study, conducted by the author himself. During his project work most of the information was collected personally through oral-questionnaire method.

Various health care providers, Bagas, and local quacks in different tribal places of Bastar have been interviewed personally. All collected information have been analyzed and arranged scientifically and systematically. Local name of the herbs, plants and roots etc., which they used are also analyzed and arranged with the help of botanical and medical literature. It is very difficult to get the information regarding conventional ways of treatment from tribal people. Normally they do not want to disclose it. But in this study, author tried his best to get/collect the required information and method of treatment as well as herbs, roots, etc., which they used. This information was collected at least from one place in each tribal development block for each tribal group f the region.

The main objective of the present study is to find out the conventional health care practices mainly based on herbals/medicine plants, so its the main approach and aim is to find out the name of herbs, mode of treatment and their effect on the patient, as well as the tribal group which used them under their own traditional health care system. This type of medical care comes under the indigenous non-professional medical care system.

Discussion

Herbalists are widely accepted as health healers among the tribal societies whose knowledge of the properties and uses of herbs, roots and barks is supposed to be very extensive. Bastar has very extensive reserves of medicinal plants. The author himself conducted various field studies during his research activities, and some important claims of various tribes of Bastar are as follows.

Bastar tribes have their own method of family-welfare. The 'Marias' of 'Dantewara' claimed that a woman becomes infertile if three grains of 'Oria-Granulata' locally known as 'Devi-Dhan' (wild-rice) are taken at a time on the three consecutive days from the 4th day of menstruation. Another tribal health care provider (herbalist) claimed, another method to terminate pregnancy up to three months by the use of the root paste of the 'solanum-incanum' (Rani-Bhejri) applied around the naval of woman. One tribal group of nearby Bijapur of Bastar claimed that one again of Devi-Dhan is enough, if taken on five successive Sundays to attain sterility.

The 'Bhatras' of backwand and 'Halbas' of Bhanu-Pratpur use a root-paste of 'crepis-acanlis' locally known as 'Kaam-Roaj' to save the person from Botulism, in the same manner, the Halbas also collect the occphyllasmarag dina (chapoda) a kind of red tree branch and prepare a chutney (paste) to eat. It is a very effective medicine, which has wonderful therapeutic action on fever.

Beside above, the tribal of Bastar also frequently use some self-made herbal medicine for common and routine health disorder. For example during blood diarrhea, they take 'curd' with sugar while during vomiting they take mixture of lemon and sugar. Amla they used during heart problems and 'Jamun' (*Suzzgium-cwmin*) has been used during stomach pain. During diarrhea they also take cow's milk with lemon and they cure skin diseases by 'cherayata' (*Andrographispenigulata*). Some important claims obtained from field observation are as follows.

To avoid writing the detailed description of the use of herbal drugs and mode treatment etc., of the tribal people of the geo-medical unit, a new and more effective method has been used. The process has been written in short with the help of abbreviations.

The entire method of treatment has been divided into five major stages, i.e., (i) Disorder (ii) Botanical name of the herb/root etc. (iii) Local name of the same (iv) Place of observation (v) Mode/method of preparation and use of herb/root etc.

Abbreviations are as follows:

P = Place of observation
D = Health disorder
B = Botanical name of the material
L = Local name of the above
MPU = Mode of preparation and use of herb/root etc. for particular health disorder.

Observation I

P = Bijapur (South-Bastar)
D = Wound healing
B = *Bambuso-Arundinacca*
L = Baans

MPU = Green bamboo is scratched with a knife and powdered. Thus obtained is used externally to heal fresh wound.

Observation II

P = Naryanpur (Centre-Bastar)
D = Rheumatic Arthiritis
B = *Blepharis-Permum*
L = Raasnajadi
MPU = Juice extracted from root is given for three weeks daily.

Observation III

P = Bhopal-Pattanam (West-Bastar)
D = Anti-fertility
B = *Achyranthes Aspera*
L = Latkinna—Chidchidy
MPU = Root paste mixed with milk and given internally on empty stomach.

Observation IV

P = Sukma (South-Bastar)
D = Haemathuria
B = *Butea Superba*
L = Pharasbel
MPU = Paste of streamed bark given on empty stomach.

Observation V

P = Konta (South most Bastar)
D = Cough-cold
B = *Colebrookia*
L = Kudsaadi
MPU = Leaf is chewed with its juice swallowed to relieve cough.

Observation VI

P = Lohandiguda (Centre Bastar)
D = Oedema
B = *Flemingia*
L = Banrrahar
MPU = The juice extract from fresh root given internally in oedematic condition.

Observation VII

P = Pakanju (West Bastar)
D = Malaria
B = *Paederia Foetida*
L = Peeth-laha
MPU = The fresh stream is made into a ring and tied around the back of the patient.

Observation VIII

P = Barsus (Central Bastar)
D = Headache

B = *Chloroxylone swietaenia*
L = Birra
MPU = Paste of leaf and root is given internally. The paste is also used as balm.

Observation IX

P = Kondagawon (Eastern-Bastar)
D = Asthma
B = *Cassia-Auriculata.*
L = Anwl
MPU = The extract from the seeds is taken orally.

Observation X

P = Antagarh (North-Bastar)
D = Piles
B = *Psidium Gueyava*
L = Bihi Ki chal
MPU = The paste is used locally.

Observation XI

P = Keskal (North Bastar)
D = Abdominal Pain
B = *Ammania Baccifera*
L = Upka-Patti
MPU = Leaves boiled and after discarding the water the hot paste is applied around the abdomen.

Observation XII

P = Dantewara (Centre-Bastar)
D = Sore choked throat
B = *Mungifera Indica*
L = Ratti-Patti
MPU = Chew the leaves for relief

Source: Field Survey—Based on the author's personal interviews with respective health healers of the region.

Acknowledgment

Present paper is basically based on the part of the finding of the project on Tribal Health of Bastar sponsored by MAPCOST, in which author himself conducted the extensive fieldwork in the rural and remote areas of Bastar. The assistance of council has been thankfully acknowledged.

References

Choubey, Kailash, (1991,) "Life-style and Avenus among Tribal Population of Bastar", *M.P. Bulletin of the Tribal Research Inst. Bhopal.* Vol. XXVII. No. 2. Bhopal.

—, (1961), "Health Care Delivery System in India", *Geog. Review of India,* Vol. 58. No. 4 Calcutta.

17 Indigenous Knowledge and Biodiversity Conservation in Indian Desert

Durgadas Mukhopadhyay

One billion people live in dry regions of the world, which cover nearly 40 per cent of the earth's surface. Many live in rural areas and earn their living as farmers, pastoralists, herders, and woodcutters. What they all have in common is a reliance on natural resources—including biodiversity, which is declining at a rate unprecedented in recorded history.

"Desertification is hard to reverse, but in can be prevented", UN Secretary-General Kofi Annan said. "It will also help to preserve landscapes and cultures that date back to the dawn of civilization and is part of our cultural heritage", he added. "The world's deserts are facing dramatic changes as a result of global climate change, high water deprivation and salt contamination of irrigated soils".

According to an assessment by the United Nations Environment Programme (UNEP), not all of the changes created by desertification necessarily have to be harmful. Some have concrete benefits for the indigenous peoples and other inhabitants of the deserts, and even for the entire world. Deserts also promise new sources for medicines derived from their plants and animals, which are uniquely adapted to the severe and unpredictable habitat. Deserts are threatened by unsustainable human activities, over-extraction of water, and climate phenomena, all of which contribute to the extension of deserts into areas previously green with vegetation—a process known as desertification.

The Desert

In India about 107.43m ha, or 32.75 per cent of the total geographical area is affected by various forms and degree of desertification. Particularly the arid, semi-arid and sub-humid regions, commonly called dryland, represent fragile ecosystems that are susceptible to desertification. These regions are susceptible to frequent droughts that accelerate the process of desertification and increased its impact (Government of India, 2001).

The Thar Desert, also known as the Great Indian Desert, is a desert located in western India and south-eastern Pakistan. It lies mostly in the Indian state of Rajasthan, and extends into the southern portion of Haryana and Punjab states and into northern Gujarat state. In Pakistan, the desert covers eastern Sind province and the south-eastern portion of Pakistan's Punjab province. The area of the Thar Desert is 4,46,000 km^2 extending 805km long and about 485km wide, with 2,08,110 sq km in India. Of the Indian portion, 61 per cent falls in Rajasthan.

The soils of the arid zone are generally sandy to sandy-loam in texture. Stretches of sand in the desert are interspersed by hillocks and sandy and gravel plains. Some wildlife species, which are fast vanishing in other parts of India, are found in the desert such as the Great Indian Bustard (*Ardeotis nigriceps*), the Blackbuck (*Antilope cervicapra*), the Indian Gazelle (*Gazella bennettii*) and the Indian Wild Ass (*Equus hemionus khur*) in the Rann of Kutch (Gupta et al., 1975). Due to the velocity of winds in these areas the dust is blown off with it and deposited on fertile lands causing a problem of shifting sand dunes. The main occupation of the people in the desert is agriculture and animal husbandry. The Indian Desert is mainly inhabited by Hindus, Muslims, and Sikhs.

Traditional Environmental Knowledge

Traditional knowledge and practices have their own importance as they have stood the test of time and have proved to be efficacious to the local people. Some of these traditional practices are in the fields of crop production, mixed farming, water harvesting, conservation of forage, combined production system, biodiversity conservation, forestry, and domestic energy in the desert ecosystem.

The agricultural and natural resource management practices of indigenous people perfected over a period of time is termed as traditional ecological knowledge. Science should try to adopt traditional knowledge in contemporary situation, appropriately combined with modern scientific inputs wherever necessary. Recording and documentation of traditional knowledge requires close participatory research with communities, as they help in identifying and preserving traditional knowledge in various ways (Ramakrishnan et al., 1998). One should look at it critically, scientifically validate it by conducting research at grass-roots levels and then start analyzing this knowledge for formulating strategies for better land utilization. A broad-based study should also be conducted to find out how traditional knowledge can be applicable in the contemporary world. Through a variety of approaches, traditional knowledge, wisdom and technology, based on empirical knowledge accumulated over a long period of human evolution, traditional societies have learnt to conserve and enhance biodiversity. Sacred groves maintained by traditional societies demonstrate spiritual values attached to biodiversity (Berks, 1999). Ecological issues in sustainable development are tied up with social, economic, anthropological and cultural dimensions.

The Bishnoi Community

The Bishnoi community of the western Indian state of Rajasthan have, over centuries, made a unique blend of ecological sense and religious sensibility. The Thar Desert in India is full of ironies—one of them being the Bishnoi community of Rajasthan. Here, peace is maintained with aggression and robust health rubs shoulders with regular famine. Not to mention the fact that the Bishnois worship nature in all its manifestations. Here, women suckle motherless deer, die to save trees, go hungry to provide food for animals and live a strict life advocated by their guru Jambaji. Living amidst the barren wastelands interspersed with Khejri and babool trees, the Bishnois are a proud race. Johra Ram says "Any change in the world has to begin within the society. All this talk about nature and wildlife protection would be more effective if each individual was to believe in the earth as a living, breathing entity and fight for its survival the way we do". There is the story of Amrita Devi, a Bishnoi women who, along with more than 366 other Bishnois, died saving trees. In 1730 AD, Maharaja Abhay Singh of Jodhpur required wood for his palace. So he sent his soldiers to cut trees. Amrita Devi and other villagers hugged the branches while the soldiers chopped them down with the trees. This is still remembered as the great Khejarli sacrifice. The sect was founded by Guru Jambheshwar (b. 1451) after riots between Muslim invaders and the local Hindus. He had laid down 29 principles to be followed by the sect. Bish means 20 and noi means. 9. Killing animals and felling trees were banned. Before his death he has stated that the back was his manifestation after death and should be conserved. Bishnois are strong lovers of wild animals (Mukhopadhyay, 2004). It is because of their protection that in Bishnoi dominated areas deer, blue bulls and black bucks are seen grazing peacefully in their fields.

The origin of Chipko movement can be traced back to be related with Bishnois. Maharaja Abhay Singh, Ruler of Marwar (Jodhpur) state who wanted to fell green Khejri (*Prosopis cineraria*) trees at village Khejarli to burn lime for the construction of his new palace. Since, there was a lot of greenery in the Bishnoi villages even in the middle of Thar Desert, they protested to protect the trees and 363 Bishnois were massacred in the process. This incidence is considered to be the origin of Chipko movement. The brave Bishnoi lady Amrita Devi played a historical role in this incidence.

The community has a set of abiding laws including a ban on killing animals and on felling trees, especially their most sacred Khejri tree, which has numerous life-sustaining properties. Its dropped leaves can be used for fodder, its dropped branches for fuel and its fruits for food. Khejri trees stabilize sand dunes and they are said to increase yields of crops that grow close by. The *orans* also provide a protective habitat for the Indian gazelle and black buck. The Bishnois, a community in Rajasthan is an example of human beings living in harmony with nature (Gadgil and Folke, 1993). Among the 29 principles propounded by founder of the sect, three concern nature conservation. Cutting and lopping of green trees is strictly prohibited. They maintain groves, locally known as orans, for the animals to graze and birds to feed. Orans serve as important rechargers of rain water in the aquifers in the desert, where every single drop of water is precious. It is estimated that orans account for about 9 per cent of the desert area. In most orans, particularly in western Rajasthan, the dominant tree, Prosopis cineraria or Khejri, a tree species, is worshipped for its immense ecological value. They would not cut Khejri trees even from their agricultural fields. The tree enriches soil nitrogen, and during drought and famine, the bark of the tree is mixed with flour for consumption.

The community is also an apt manager of natural resources. In areas with sparse vegetation, scrubland and little water they manage the ecosystem with a fervour that refuses to diminish with time. Of late some tour operators have teamed up with local NGOs and started offering 'eco-cultural tours' that focus on providing cultural glimpses into the various tribal/non-tribal communities of Rajasthan. These tours come to stay amongst them in their houses in remote villages. Since Bishnoi settlements are known havens for all birds and animals due to the protection accorded to all wild animals and vegetation, these settlements are now attracting tourists in large numbers. Due to the steadily increasing number of eco-cultural tourists and the consequent rise in income, the village communities have been able to switch to LPG instead of fuel wood and have added incentives to conserve and sustainably manage their natural resources. This commitment to conservation gives the Bishnois a unique appeal as far as tourism is concerned.

Sacred Groves

Sacred groves or SGs in general, act as a nursery and storehouse of many of the ayurvedic, tribal and folk medicines. Species not under any immediate risk of extinction, if preserved in SGs, may have great potential of diverse uses in the future. The SGs may also preserve genotypes which may be useful in forest tree-breeding programmes It is evident that one of the important ecological roles of these groves is to provide a more dependable source of water for the organisms living in and around the sacred groves. All of these factors indicate that the conservation of scared groves is essential for maintaining local/regional biodiversity, the comprehensive health of a landscape, and preserving the socio-cultural integrity of local communities (Gadgil, 1985). There are several scared tree groves dedicated to temples in the Thar Desert. Communities zealously protect these groves against interference of any kind. These groves are excellent examples of biodiversity conservation. For example, it was religiously prohibited to cut any vegetation from the lands in the immediate vicinity of temples and religious places, known as oran (protected forest) lands. Collection of dry wood only is allowed for fuel and serious punishment is prescribed for using an axe in orans. In Barmer, Nagaur, Jodhpur, Pali, Sikar, Jhunjhunu, and Jalore districts of Rajasthan there are still 420 orans covering a total area of 100,140 ha.

In Jaisalmer, in the Bhadaria temple complex, the community, maintain an oran with a concentration of *Ziziphus nummularia* (Ber) trees known for

large fruits. Solar energy is used to extract underground water to irrigate the oran. Oran's resources are meant only for the animals and avifauna to ensure their survival in the harsh desert conditions.

Khejri—The Wonder Tree

Prosopis cineraria is a small to medium size tree, found mainly in the Thar Desert of Rajasthan in India. Common names include: khejri, jand, sangria (Rajasthan). *Prosopis cineraria* tree grows in dry and arid regions of India namely Rajasthan, Haryana, Punjab, Gujarat. The leaflets are dark green with thin casting of light shade. It coppices profusely.

Historically *Prosopis cineraria* has played a significant role in the rural economy in the north-west arid region of Indian subcontinent. This tree is a legume and it improves soil fertility. It is an important constituent of the vegetation system. It is well adapted to the arid conditions and stands well to the adverse vagaries of climate and browsing by animals. Camels and goats readily browse it. It is about the only indigenous tree species, which has withstood well the rigorous and exacting conditions of the Rajasthan desert. The root system of *Prosopis cineraria* is long and well developed. Growth above the ground is slow but below the ground the roots penetrate deeper and deeper for the sub soil layers. The tree is able to withstand the hottest winds and the driest season, and remains alive when other plants would succumb(Mann and Saxena, 1980). The tree is frost resistant. Because of its economic value, the tree is left standing in the arable land and the farmers regulate its population by adapting suitable agro-forestry management.

Prosopis cineraria provides wood of construction class. It is used for house-building, chiefly as rafters, posts scantlings, doors and windows, Persian wheels, agricultural implements and shafts. *Prosopis cineraria* is much valued as a fodder tree. The trees are heavily lopped particularly during winter months when no other green fodder is available in the dry tracts. There is a popular saying that death will not visit a man, even at the time of a famine, if he has a *Prosopis cineraria*, a goat and a camel, since the three together are said to sustain a man even under the most difficult condition. The leaves are of high nutritive value. Feeding of the leaves during winter when no other green fodder is generally available in rain-fed areas is thus profitable (Mukhopadhyay, 2004). *Prosopis cineraria* is most important top feed species providing nutritious and highly palatable green as well as dry fodder, which is readily eaten by camels, cattle, sheep and goats, constituting a major feed requirement of desert livestock.

Prosopis cineraria flower is pounded, mixed with sugar and used during pregnancy as safeguard against miscarriage. Water-soluble extract of the residue from methanal extract of the stem bark exhibits anti-inflammatory properties. *Prosopis cineraria* plant produces gum, which is obtained during May and June. The bark of the tree is dry, acrid, bitter with a sharp taste; cooling anthelmintic; cures leprosy, dysentery, bronchitis, asthma, leucoderma, piles.

The bark is used as a remedy for rheumatism, in cough colds, asthma. The plant is recommended for the treatment of snakebite. The bark is prescribed for scorpion sting (Kaul, 1976).

Prosopis cineraria has a very deep tap root system and hence it does not generally compete with the associated crops. Rural communities encourage, the growth of *Prosopis cineraria* in their agricultural fields, pastures and village community lands. Because of its extensive root system it stabilizes shifting sand dunes and is also useful as windbreak shelterbelt and in afforestation of dry areas (Mukhopadhyay, 2005). It fixes atmospheric nitrogen through microbial activities. It adds organic matter through leaf litter decomposition thus rejuvenating poor soils.

Gender Issues

Women in the desert are dependent on forest for food, fuel, fodder, medicines, and fibres. Moving in and out of forests, women have down the years accumulated knowledge in the sustainable use of natural resource as well as numerous insights of the value of biodiversity. Though women can play a significant role in protecting biodiversity, it has not been recognized (Mukhopadhyay, 1989). Their role in conservation has been overlooked. The loss of habitats and biodiversity mostly affects the underprivileged and many of them are women. A large number of women participated in mass movements like Chipko in the upper reaches of Himachal and Uttar Pradesh. They had a stake in protecting local biodiversity and natural resources like wood and water as additional burdens fell on them when these resources dried up.

Drylands pose different challenges for rural men and women because of their different roles, relations and responsibilities, opportunities and constraints, and uneven access and control of resources. By incorporating a gender perspective in policy, projects and programmes, innovative ways of combating dryland degradation and food insecurity can be promoted. The result is a more sustainable, relevant and equitable development based on women's and men's full and equal participation, on their respective local knowledge, and on ecological and socio-cultural factors (Inglis, 1993).

Sustainable Natural Resource Management in the Desert

In the Thar Desert, traditional systems of land and water use met environmental challenges in various ways. One group of indigenous cultivators (Paliwals) devised a rainwater-harvesting technique fully capable of growing winter season crops. As early as the 15th century, the Paliwal cultivators followed a unique practice of water harvesting and moisture conservation in suitable deep-soil plots. The bulk of natural resource base of the arid region is most suited to pasture based livestock farming. The traditional wisdom of the dryland farmer clearly manifest in the evolution of system of mixed farming—including crop and animal husbandry—which matched the potential and limitations of the natural resource base (Government of India, 2001). The pastoralist based on their centuries of experience, developed a unique method of water harvesting. With the commencement of rains, the population was divided into different caste groups and dispersed to their tobas (small dug out ponds) along with their livestock. The Jagirdar (feudal landlord) imposed animal grazing tax and periodic free gifts especially from owners of large flocks of sheep and goat, which acted as a strong deterrent against indiscriminate grazing. With the abolition of Jagirdari, the practice of realizing grazing tax was discontinued resulting in free ranging and consequent degradation of rangelands.

A useful indigenous technique of water conservation is called pitcher planting. Earthen pitchers with holes on one side are embedded near the root zone of newly planted seedlings to provide it with the required amount of water. This technique prevents loss of water either due to evaporation or seepage and helps in seedling establishment. This technique is still practised by melon cultivators in arid region of Rajasthan. Similarly, these farmers bury bushes in a chess board pattern to protect melon plants from getting buried by shifting sands (Kaul, 1976). Dryland farmers raised windbreaks (*matt*) around their fields and homesteads to protect crops and their livestock against hot desert winds. Lopping of trees such as *Prosopis cineraria, Azadirachta indica and Ailanthus excelsa* during winter season for leaf fodder is still a common traditional practice in arid and semi-arid regions of Rajasthan. This is scientifically sound, as it causes no damage to trees for they are dormant and by then all the food is translocated to the root.

Conclusion

Since many of these traditional systems are environment-friendly and sustainable, efforts should be made to restore them and support them with modern approaches to enable their effective mainstreaming in combating desertification. Some of the indigenous traditional technologies offer promising community-based dryland resource management. The participation of local communities would also lead to greater use of traditional practices. It is, therefore, necessary to document such knowledge base through a properly designed research programme and to analyze their economic, technological and socio-cultural sustainability.

References

Berkes, F., (1999), *Sacred Ecology: Traditional Ecological Knowledge and Resource Management,* Taylor and Francis, Philadelphia.

Borgerhoff-Mulder, (1988), "Behavioural Ecology in Traditional Societies", *Trends in Ecology and Evolution*, 3, pp. 260.

Brush, S.B., and Stabinsky, D., (1996), *Valuing Local Knowledge: Indigenous People and Intellectual Property Rights,* Island Press, Washington.

Burdak, L.R., (1982), "Recent Advances in Desert Afforestation", Dissertation submitted to F.R.I., Dehradun.

Gadgil, M., (1985), "Social Restraints on Resource Utilization: The Indian Experience" in J.A. McNeely and D. Pitt (eds.), *Culture and Conservation: The Human Dimension in Environmental Planning*, Croom Helm, Dublin.

Gadgil, M., F. Berkes and C. Folke, (1993), "Indigenous Knowledge for Biodiversity Conservation", *Ambio*, 22: 151–156.

Government of India, (2001), "National Action Programme to Combat Desertification", New Delhi.

Government of India, Ministry of Food & Agriculture, (1965), *Soil Conservation in the Rajasthan Desert*, Jodhpur.

Gupta, R.K. and Prakash, Ishwar, (1975), *Environmental Analysis of the Thar Desert*, English Book Depot., Dehradun.

Inglis, J.T. (1993), *Traditional Ecological Knowledge: Concepts and Cases,* IDRC.

Kaul, R.N., (1970), *Afforestation in Arid Zones* (ed.), Dr. W. JUNK N.V. Publishers, The Hague.

—, (1976), "Trees or Grass Lands in the Rajasthan—Old Problems and New Approaches", Indian Forester, 93: 434–435.

Mann H.S. and S.K. Saxena, (1980), *Khejri (Prosopis cineraria) in the Indian Desert, Its Role in Agroforestry*, Central Arid Zone Research Institute (CAZRI), Jodhpur.

McNeely, J.A and D. Pitt, (eds.), (1985), *Culture and Conservation: The Human Dimension in Environmental Planning*, Croom Helm, Dublin.

Mukhopadhyay, Durgadas, (1989), *Culture, Performance and Communication*, B.R. Publishing, New Delhi.

—, (2004), "Traditional Environment Knowledge in Indian Religion and Tradition", K.R.G. Nair (ed.), *Sustainable Development in India and Canada,* CIDA Sponsored Publication, Kaveri Publications, New Delhi.

—, (2005), "Indigenous Knowledge, Water Management and Sustainable Development", Paper and Documentary Film presented in UNEP sponsored.

—, (2006) "Folk Arts for Environment Education", Paper presented in the UNESCO World Conference on "Education in Arts", Lisbon, Portugal.

Ramakrishnan, P.S., (2002), *Coping with Uncertainties: Role of Traditional Ecological Knowledge in Socio-ecological Landscape Management*, School of Environmental Sciences, Jawaharlal Nehru University, New Delhi.

Ramakrishnan, P.S., K.G. Saxena and Chandrashekara (eds.), (1998), *Conserving the Sacred for Biodiversity Management*, Oxford and IBH Publishing Co., Pvt. Ltd. New Delhi.

Third World Environment Education Congress, Turin, Italy.

UNDP, (2004), *Sharing Innovative Experiences; Examples of Successful Conservation and Sustainable Use of Dryland Biodiversity*, New York.

18 Biodiversity and Conservation of Tropical Forest Fragments

A Case Study of Ghana, West Africa

Jennifer L. Hill and Paul J. Curran

Fragmentation of tropical forest has been described as the single greatest threat to global biological diversity (Laurance, 1999; Pimm and Raven, 2000). Fragmentation lowers species number and alters community composition as a result of both a reduction in forest area and a change in forest shape. The relationship between forest area and species diversity in the tropics is well documented (Bierregaard et al., 1992; Hill et al., 1994; Laurance and Bierregaard, 1997; Pimm, 1998) and this aspect of fragmentation continues to attract attention (May and Stumpf, 2000; Pimm and Raven, 2000 and, Hill and Curran, 2001). Fewer empirical studies, however, have examined the ecological effects of fragment shape, especially for fragments that have been isolated for decades or longer (Turner, 1996).

As fragmentation continues in many parts of the tropics, conservation efforts must focus on understanding the effects of forest spatial geometry on species diversity, as only then can biodiversity be influenced significantly through management. This is particularly relevant as synergies exist between forest fragments and their dynamic external environments (Cochrane, 2001; Laurance et al., 2001; Laurance, 2004). Positive feedbacks among deforestation, regional climate change and fire could pose a serious threat to some tropical forests.

Research presented in this chapter quantifies the influences of fragment area and shape on tree species diversity in Ghana, West Africa and it compares their significance with the influences of climate (average annual rainfall) and disturbance (fire burn, logging, agriculture). It begins by discussing the processes through which these variables influence species diversity, before using bivariate and multivariate regression to elucidate relationships between diversity (tree species number and composition) and the variables of forest spatial geometry, climate and disturbance. The chapter concludes by highlighting the benefits and limitations of landscape level biogeographical study for tropical forest conservation.

Methods

Study Site

Tropical forest in Ghana comprises fragments of differing area and shape, protected as extractive reserves and separated mostly by agricultural land (Hall and Swaine, 1981). The reserves were demarcated by the Ghana Forestry Department in the 1920s and 1930s (Varley and White, 1958) and the geographical distribution of 22 forest fragments was used in this study. Fragment spatial geometry has remained relatively stable for around 80 years, allowing considerable time for the ecosystem to respond.

A strong rainfall gradient exists across the forest zone. Average annual totals decrease from around 2,100 mm in the south-west (with 4–5 dry months per year) to around 1,000 mm in the north-east (with 8–9 dry months per year) (Hall and Swaine, 1981). This transition modifies forest cover from a closed canopy of predominantly evergreen trees in wetter areas to a more open and deciduous canopy in drier areas (Martin, 1991).

Disturbance has caused local species variation within the forests. Fire activity has spread from the DSFZ (Dry Semi-deciduous Fire Zone) south to both sub-types of MS (Moist Semi-deciduous) forest, where it is the cause of most degradation today (Hawthorne, 1994). Its effect does not penetrate into the WE (Wet Evergreen) and UE (Upland Evergreen) zones. There is a history of logging, but in semi-deciduous zones where it has been most prevalent, it occurs selectively according to a minimum girth system so that removals amount to only four to five trees per hectare (Hawthorne, 1993).

Measurement of Tree Species

Forest fragments were surveyed as part of a Forest Inventory and Management Project (FIMP) initiated in 1986 by the Overseas Development Administration and the Ghana Forestry Department. The total area surveyed was directly proportional to the reserve area within each forest zone (Wong, 1989). Sample plots of 1 ha were laid systematically across the forest reserves on a 2× 2 km grid. Each plot was subdivided into five 0.2 ha subplots, one of which was further subdivided to incorporate two 0.05 ha subplots. Trees counted within the sample plots were classified into two categories: ≥ 30 cm girth (mature) and < 30 cm girth (regenerating). Mature trees were identified and measured over the whole 1 ha sample plot, whereas regenerating trees were enumerated only within the two 0.05 ha subplots. The total sample size used in the study amounts to 384 of the roughly 700 Ghanaian tree species, recorded within 1,174 sample plots.

Tree species number within each forest fragment was determined using FIMP-specific programmes (Hill, 1994). Tree species composition was analyzed using four schemes of categorization: regeneration guild, rarity, phenology and dispersal. Guild types were based on qualitative abundance of seedlings in various extremes of the forest mosaic (Hawthorne, 1994). Species were allocated to guilds, according to their presumed germination requirements for sunlight, into pioneers, non-pioneer light demanders (NPLDs) and shade-tolerators. Pioneers regenerate only under canopy gaps in the forest canopy. NPLDs can survive in, or even require shade, as germinating seedlings, but they require canopy gaps to survive beyond the early parts of their life cycle. Shade-tolerant species are found as healthy saplings in forest shade. Rarity was based on frequency of occurrence of species. Based on a colour code (abbreviated from Hawthorne and Juam-Musah, 1993), rarity in Africa or Ghana merited gold or blue star ratings, whilst green star species were common both internationally and locally (Table 1). Information concerning phenology and dispersal was derived from ecological literature (Hill, 1994). Phenology was restricted to leaf shedding (evergreen or deciduous), whilst dispersal categorized the different means by which seeds were carried from the parent (animals or wind).

Table 1

Definition of Rarity Categories Applied to The Tree Species of Ghana

Category	*Descriptor*	*Average number of degree squares in Africa*
Gold	Fairly rare internationally and/or locally	7.8+/–3.8
Blue	Widespread internationally but rare locally or vice versa	24.5+/–12.6
Green	Common internationally and locally	69.2+/–49.8

Source: Hawthorne and Juam Musah (1993).

Measurement of Fragment Spatial Geometry

Area data for the forest fragments were derived from Hall and Swaine (1981). The areas of fragments used in this research ranged between 34.7 km^2 and 877.5 km^2. Fragment shape was quantified using the complex perimeter/area ratio. This compares the total external perimeter of a landscape fragment to the circumference of a circle having the same area and is therefore a measure of shape irregularity (Hill, 1994). As fragment shape becomes more irregular, the index increases without limit above a value of one. The complex perimeter/area ratios for the fragments were examined in this study. The measurements entered into the complex perimeter/area ratio were derived from a 1: 5,00,000 scale map of forest fragments (Gold Coast Survey Department, 1949a) digitized within an Arc/Info geographical information system. Area measurements obtained from the small-scale map did not differ significantly from those measured by ground-based survey ($p = 0.001$).

Measurement of Climate and Disturbance Variables

Average annual rainfall totals were derived from a 1: 5,00,000 scale rainfall map (Gold Coast Survey Department, 1949b). Forest fragments were located between isohyets and thus data were categorical. The measurements ranged from 1,200 mm to 1,850 mm. Figures for burn, logging and agriculture were taken from percentage visual estimates of their extent within FIMP sample plots (Wong, 1989). Average levels of disturbance across the sampled fragments were 15 per cent, 4 per cent and 10 per cent respectively for these variables.

Hypothesis Testing for Bivariate Regressions

Tree Species Diversity and Fragment Area

It was expected that tree species number would increase with fragment area. The positive relationship between species number and area has been attributed to ecological processes, explained by island biogeography theory, habitat diversity and successional development and to sampling effects, whereby larger fragments contain more samples that record more of the community (Hill et al., 1994; Lomolino, 2001). It was also expected that as the area of forest fragments increased, the relative proportions (percentages of sampled tree species numbers) of rare, shade-tolerant and evergreen species would increase with respect to common, deciduous and pioneer species (Laurance, 1991; Tabarelli et al., 1999). This is because as the area of forest fragments increases, there is a disproportionate increase in interior compared to edge forest (Turner, 1996). The model chosen to describe the species-area relationships was least squares linear regression.

Tree Species Diversity and Fragment Shape

Tree species were classified into 10 biological categories (Table 2) which, when partitioned into regenerating and mature trees, generated 20 shape-related hypotheses. The primary influence of fragment shape on species diversity

Table 2
Examples of Tree Species by Biological Category

Category	*Tree species*
Rarity	
Gold	*Anisophyllea meniaudii, Dactyladenia dinklagei*
Blue	*Balanites wilsoniana, Cylicodiscus gabunensis*
Green	*Celtis spp, Strombosia glaucescens*
Guild	
Pioneers	*Musanga cecropiodes, Trema orientalis*
NPLDs	*Parkia bicolor, Celtis zenkeri*
Shade tolerators	*Strombosia glaucescens, Celtis mildbraedii*
Dispersal	
Animal	*Celtis spp, Cola gigantea*
Wind	*Ceiba pentandra, Vernonia spp*
Phenology	
Evergreen	*Allanblackia floribunda, Celtis spp*
Deciduous	*Cylicodiscus gabunensis, Ceiba pentandra*

derives from the ratio of edge to interior environment (Ranta et al., 1998; Laurance, 2000). Fragmentation increases the proportion of edge zone in the landscape, altering microclimate and the competitive advantage afforded to any species (Kapos et al., 1997; Didham and Lawton, 1999; Zhu et al., 2004). Typically, fragment edges have large numbers of pioneer species (Laurance et al., 1998a and, Sizer and Tanner, 1999), whilst fragment interiors have large numbers of shade-tolerant species (Benitez-Malvido, 1998; Kellman et al., 1998). It was expected, therefore, that the proportion of pioneer and NPLD tree species would increase, and the proportion of shade-tolerators would decrease with increasing fragment irregularity because of the rising ratio of edge to interior environment.

It was expected that the proportion of rare tree species would decrease with increasing forest irregularity because rare species tend to be shade-tolerators which thrive in less disturbed interiors (Benitez-Malvido, 1998; Brokaw, 1998). Common species occur in both light-demanding and shade-tolerating guilds, which means that the balance between interior and edge environments does not dictate their proportions. It was hypothesized, therefore, that this category would be unrelated to fragment shape.

Increases in temperature and decreases in relative humidity along forest edges (Kapos et al., 1997) were expected to raise the number of deciduous tree species at the expense of evergreen ones resulting in 'induced seasonality' around forest peripheries.

Two opposing mechanisms allow either an increase or decrease in the expected proportion of animal-dispersed species with increasing irregularity of fragment shape. The first mechanism is interception; the movement of birds and mammals across a fragment boundary (Game, 1980). Interception was expected to increase with fragment irregularity because a longer boundary presents a larger target to mobile animals from most directions and an increase in interception means a larger proportion of animal-dispersed species. The second mechanism, circulation within a fragment, acts against interception. A more regular shape can provide sufficient undisturbed interior forest to allow animals to traverse around the area and minimize dispersal distances (Blouin and Connor, 1985). An irregular fragment will not provide such compact, undisturbed forest and so circulation was expected to decrease with increasing fragment irregularity.

A direct positive relationship with fragment shape was hypothesized for species with plumed seeds borne by wind. Plumed dispersal enables seeds covered with fine hairs to travel to around 5 km (Augsperger and Franson, 1987; Greene and Johnson, 1989) and forest fragments can intercept seeds over such distances according to the target-area hypothesis (Game, 1980). Consequently, an increase in plumed tree species was expected with increasing fragment irregularity. Relationships between tree species diversity and fragment shape were also tested using least squares linear regression.

Tree Species Diversity and Climate/Disturbance Variables

The numbers of tree species recorded in the forest fragments were expected to decline with increasing fire, logging and agricultural activity. This is due to direct species loss and altered ecosystem functioning (Hawthorne, 1993, 1994). Conversely, tree species numbers were expected to increase in forest fragments with high average annual rainfall totals as this reduces the stress of dry seasons (Terborgh, 1992).

Hypothesis Testing for Multivariate Regressions

Multiple and stepwise multiple regressions were undertaken to ascertain the combined association between tree species number and all environmental variables (spatial geometry, climate, disturbance) and to investigate the relative significance of the variables in accounting for tree species number. It was expected that a significant amount of the variation in tree species number would be accounted for by the environmental variables and that aspects of spatial geometry would feature significantly amongst these.

Results

Bivariate Regression Analyses

Regressions of tree species number and composition on forest fragments of differing area and shape identified the additional factors of climate and disturbance. Their influence was quantified by means of forest outliers in the regression plots (variance > 2). Forest fragments identified consistently as outliers were associated with extremes of rainfall and disturbance and were removed from further analyses. The number of forest fragments entered into regressions is indicated in the text.

Tree Species Diversity and Fragment Area

The logarithm of tree species number was related positively to the logarithm of fragment area for regenerating trees ($r=0.96$, $p=0.005$, $n=20$). Thus, fragment area accounted for 92 per cent of the variation in tree species number. This significant relationship was repeated for mature trees ($r=0.87$, $p=0.005$, $n=20$). The slope values of the regression lines were slightly higher for regenerating trees (0.31) compared to mature trees (0.25). Thus, the rate of increase of tree species with fragment area was higher for the regenerating group.

In terms of species composition, the relative proportions of rare (gold and blue star) tree species increased significantly with the logarithm of fragment area for the regenerating group ($r=0.69$, $p=0.005$ and $r=0.46$, $p=0.025$; $n=21$). This was not the case, however, with the common (green star) species, which provided between 60–72 per cent of the total species complement for fragments of all areas. There was a significant increase in the proportion of evergreen tree species with an increase in the logarithm of fragment area for mature trees ($r=0.68$, $p=0.005$, $n=19$). The result was reversed for deciduous trees, as they had a significant decrease in their proportion with an increase in the logarithm of fragment area (r= −0.72, p=0.005, n=19). The proportion of mature shade-tolerating tree species increased significantly with fragment area (r=0.49, p=0.025, n=20).

Tree Species Diversity and Fragment Shape

Selected relationships between tree species diversity and fragment shape are reported here with a full discussion in Hill and Curran (2003). With reference to guild, the proportion of regenerating pioneer trees increased with increasing irregularity of fragment shape (r=0.57, p=0.005, n=20) but the relationship between mature pioneers and fragment shape was insignificant. Mature shade-tolerators increased their proportions with increasing fragment irregularity (r=0.45, p=0.025, n=21), but the regenerating group recorded an insignificant relationship with fragment shape.

The proportions of common tree species were not related significantly to forest shape for regenerating or mature trees. The proportions of common species remained within a range of 63–72 per cent for regenerating trees and 58–66 per cent for mature trees. Tree species common in Ghana and Africa thereby formed the foundation of communities in fragments of all shapes.

The relationship between the proportion of evergreen tree species and the shape of forest fragments was significant only for mature trees. The correlation coefficient (r=0.44, p=0.025, n=20) translated into a coefficient of determination of 19 per cent. Fragment shape therefore accounted for around a fifth of the variation in the proportion of evergreen tree species, but no significant relationships were found for deciduous species.

There was no significant effect of fragment shape on the proportion of tree species dispersed by animals in the regenerating category, but a significant and positive relationship was found for the mature category (r=0.46, p=0.025, n=20). The coefficient of determination attained a value of 21 per cent.

Tree Species Diversity and Climate/Disturbance Variables

The number of tree species in forest fragments rose with increasing average annual rainfall for mature trees (r=0.57, p=0.05), but the relationship was insignificant for regenerating trees. Tree species numbers were also significantly negatively correlated with fire burn for both regenerating and mature trees (r= −0.8, p=0.001 and r= −0.69, p=0.01 respectively). As the area of fire burn increased within forest fragments, the number of tree species decreased. No significant relationships were recorded between tree species numbers and logging or agricultural activity.

Multivariate Regression Analyses

Significant relationships were recorded between the logarithms of both regenerating and mature tree species numbers and all six environmental variables (r=0.96 and r=0.95, p=0.001). For regenerating trees, 92 per cent of the variation in species numbers was accounted for by the variables and the value for mature trees was 90 per cent. Beta weights from multiple regression analyses indicated that out of the six environmental variables only three were significant when accounting for variability in the logarithm of tree species number. To elucidate their relative explanatory power, stepwise multiple regression was employed. Results indicated that the logarithm of fragment area accounted for 83 per cent of the variation in the logarithm of species number for regenerating trees and when fire activity was added, the coefficient of multiple determination rose to just over 94 per cent ($r=0.97$, $p=0.001$) (Table 3). The order of entry of the variables was identical for mature trees, but fragment area accounted for only 65 per cent of the variation in tree species number and burn added a further 23.5 per cent to bring the coefficient of multiple determination to 88.5 per cent. Average annual rainfall was also identified as a significant variable for mature trees, raising the coefficient of multiple determination to over 91 per cent ($r=0.96$, $p=0.001$) (Table 3).

Table 3

Significant Results from Stepwise Multiple Regression of the Logarithm of Tree Species Number with Seven Environmental Variables (values of r)

Variable	*Log number of tree species (regenerating)*	*Log number of tree species (mature)*
Log area	0.91	0.81
	83.0%	65.0%
Burn	0.97	0.94
	94.1%	88.5%
Rainfall		0.96
		91.4%
Significance	p=0.001	p=0.001

Discussion

Tree Species Diversity and Fragment Area

Large forest fragments in Ghana contained the greatest number of tree species, attributable to ecological processes and also to sampling effects (Hill et al., 1994; Lomolino, 2001). More regenerating than mature tree species were accumulated per unit increase in area, which bodes well for future persistence. Additionally, extra tree species found within larger forest fragments followed a predictable pattern rather than representing random species drawn from the community. The proportion of rare tree species increased with fragment area, whilst common species formed a foundation comprising around two-thirds of the total forest complement. As the area of forest fragments increased, raising the amount of interior compared to edge forest (Turner, 1996), the relative proportions of evergreen and shade-tolerant species increased with respect to deciduous pioneers. Forest fragmentation will thereby cause local extinction, particularly of rare, shade-tolerant, evergreen species.

Tree Species Diversity and Fragment Shape

At the landscape level, almost half of the relationships tested were significant and were used to support or refute the established hypotheses. A greater number of significant relationships were recorded for mature trees, but when regenerating trees did provide a significant relationship, the correlation was often higher. Thus, the shape of forest fragments influenced species composition to differing degrees depending on tree girth.

The guild category responded to edges clearly in the regenerating size group because it was defined by the amount of exposure that species require for healthy regeneration (Hawthorne, 1994). Level of exposure controls inter-specific competition between both seedlings and saplings, allowing a response to be measured quickly (Brokaw, 1985). Relationships may be weaker for mature trees because the present cohort of individuals is representative of a time when the forest was less fragmented. The forest fragments of Ghana have been isolated for about 80 years, which is shorter than the life span of many trees (> 100 years) (Martinez-Ramos and Alvarez-Buylla, 1998).

Pioneer species regenerate under canopy gaps, so their response to disturbance around forest edges is immediate (Laurance and Bierregaard, 1997; Nepstad et al., 1999). This explains the significant rise in their proportions with increasing fragment irregularity. Shade-tolerators thrive in low light conditions experienced under the forest canopy (Benitez-Malvido, 1998; Kellman et al., 1998), so it was expected that the proportion of regenerating species would decrease with increasing fragment irregularity. No significant relationship was found for this size group, however, which might be explained by the ability of shade tolerators to regenerate in high light environments. This demonstrates a 'fuzzy' demarcation of regeneration guilds (Hawthorne, 1996). With respect to mature trees, the rise in the proportion of shade-tolerators may refer to conditions prior to forest fragmentation.

The lack of significant association with shape displayed by the common tree species is as expected. Common species are often opportunist and consequently are found in both edge and interior environments, which prevents any significant influence of shape.

The phenology category displayed only one significant relationship with fragment shape. The lack of significant relationships in the regenerating group might be accounted for by the indeterminacy of phenology in young trees. Many rain forest species are evergreen in their early life before adopting a deciduous habit when mature (W.D. Hawthorne, *pers comm.*). This is partly due to the location of canopy trees in the forest interior environment when young and their transition to the exterior environment when mature. Thus, categorization according to phenology does not apply until trees are more mature when they may be (i) relics or (ii) actively shedding leaves due to water stress (Laurance et al., 1998b).

Relationships between the proportion of animal-dispersed tree species and fragment shape supported the established hypotheses. The lack of significant association in the regenerating group might be due to the operation of both interception and circulation (Blouin and Connor, 1985). These mechanisms oppose one another in the sense that increased interception eventually leads to a larger proportion of animals in irregularly shaped fragments, whereas increased circulation raises the proportion of animals within more regularly shaped fragments. A significantly greater proportion of animal-dispersed tree species occurred among mature individuals in irregularly shaped fragments. This suggests that the proportion of species dispersed by animals in irregular fragments has decreased over time. It is unlikely that this is the result of reduced animal dispersal into such fragments because of a change in their shape or proximity because these factors have remained stable over time. The reduction might be explained, however, by an overall loss of animals due to forest fragmentation, a process that has been documented in the tropics (Cordeiro and Howe, 2001; Harrington et al., 2001; Beier et al., 2002; Castelletta et al., 2005). A decline in animal seed dispersers is a process that requires a longer time to manifest than edge-driven changes (Turner et al., 1996). The significant results recorded here suggest such ecological change, but no data exist to test this theory.

Tree Species Diversity and Climate/Disturbance Variables

Fire activity was the second most important variable influencing tree species number across Ghanaian forest fragments. The association between regenerating trees and extent of fire damage was more significant than that for mature trees. This can be explained by the sensitivity of saplings and poles to ground fires (Hawthorne, 1994). Average annual rainfall proved to be a factor of tertiary significance for mature trees, despite its signature being dampened by the categorical data. This indicates a response of mature trees to large area factors, whilst regenerating trees are more sensitive to small area disturbance events. Logging and agriculture did not significantly influence the number of tree species across the sampled forest fragments, which is in accordance with reports of the Ghana Forestry Department (Ghana Timber Export Development Board, 1992). Over the last century, Ghana has made great efforts to reconcile biodiversity conservation with the production of timber, to progress towards sustainable management and to protect its tropical forest.

Application to Conservation

West Africa has the most fragmented tropical forests in the world (Rudel and Roper, 1997). Fragmented forests have lower species diversity and a decrease in species diversity is often associated with a decrease in ecological stability (Kiessling, 2005). It is important for conservation, therefore, to understand the effects of fragmentation on species diversity and to compare its impacts with those of other environmental variables.

Species-area curves from Ghana showed that large forest fragments contained the greatest number of tree species. The presence of outstanding residuals according to climate and disturbance, however, cautions against the predictive use of species-area curves when data are collected over a wide geographic area. Diversity figures are augmented by favourable climate and environmental heterogeneity and this renders prediction both location and scale specific (Rosenzweig, 1999). Species-area curves also need to be assessed with respect to the nature of species comprising them. The steep rising limbs of such curves in Ghana were composed of common tree species whilst the flatter sections represented increments of rare species. Rare and shade-tolerant species in Ghana will be safeguarded more effectively by the preservation of maximum species richness within larger forest fragments.

The research presented here supports the assertion that irregularly shaped forest fragments promote building phases along edges, which, in turn, raises the proportions of pioneer species (Laurance et al., 1998a; Sizer and Tanner, 1999). It is important to protect consolidated fragments in Ghana to ensure that species requiring interior habitat are maintained, especially in semi-deciduous areas, where fire and dry Harmattan winds can penetrate into the forest. Fragments with core areas provide safety for sensitive, shade-tolerant species and their genetic diversity (Brokaw, 1998; Kellman et al., 1998). They also provide sources of recolonization according to the rescue effect (Brown and Kodric-Brown, 1977). By maintaining a community of edge species, core species will be protected from altered micro-climates and disturbance regimes, especially when surrounded by harsh matrix vegetation (Didham and Lawton, 1999; Gascon et al., 1999, 2000; Laurance et al., 2000). A landscape-level conservation strategy is necessary, which maximizes area to perimeter relationships across fragments, protects fragment edges using buffer zones (managed ecotones) and minimizes matrix harshness. This might be aided by planting matrices with interior forest species (Matínez-Garza and Howe, 2003), accelerating succession far from fragment edges and promoting fragment immigration.

Tropical forest trees may persist as mature adults along fragment edges with little possibility of regenerating in situ (Turner and Corlett, 1996). They are effectively extinct locally, but are nevertheless recorded in the extant flora. This indicates that equilibrium for mature trees may only be achieved over long time scales, possibly centuries (Turner et al., 1996). This can lead to the build up of an 'extinction debt' where the true cost of habitat destruction is manifest only after a long time lag (Tilman et al., 1994). Forest fragments do not return rapidly to a microcosm of continuous forest but become organized by different processes. This requires a dynamic response by managers to work with new conditions and processes.

Four important management statements can be formulated from the results presented here. They are based on conserving maximum tree species number and rarity. Emphasis on other management aspects may modify the statements.

- Forest area is the most important consideration in reserve management. Large reserves contain the greatest number of tree species and include the highest proportion of rare tree species.

- Unless dealing with very small areas, shape is not a crucial consideration for the design of forest reserves. Maintenance of similar matrix vegetation surrounding fragmented reserves can act to buffer forests.
- Management of disturbance should be considered in preference to forest shape. Deforestation to produce a fragment may not be as important as internal degradation after it has been isolated.
- Locate a number of reserves across different forest types, according to climate, in order to capture the regional biota.

Ecological understanding at the landscape level, provided by the spatially explicit view of the biogeographer, can inform conservation management within the tropics. For biodiversity conservation to be successful, however, it must be comprised of a common vision among all forest stakeholders, set within the confines of sustainable livelihoods and development (Lugo, 2002; Dovie, 2003). Conserving tropical forest biodiversity remains a formidable challenge.

References

Augsperger, C.K. and Franson, S.E., (1987), "Wind Dispersal of Artificial Fruits Varying in Mass, Area and Morphology", *Ecology*, 68, 27–42.

Beier, P., Van Drielen, M. and Kankam, B.O., (2002), "Avifaunal Collapse in West African Forest Fragments", *Conservation Biology*, 16, 1097–1111.

Benitez-Malvido, J., (1998), "Impact of Forest Fragmentation on Seedling Abundance in a Tropical Rain Forest", *Conservation Biology*, 12, 380–389.

Bierregaard, R.O., Lovejoy, T.E., Kapos, V., dos Santos, A.A. and Hutchings, R.W., (1992), "The Biological Dynamics of Tropical Rain Forest Fragments", *Bioscience*, 42, 859–866.

Blouin, M.S. and Connor, E.F., (1985), "Is There a Best Shape for Nature Reserves?, *Biological Conservation*, 32, 277–288.

Brokaw, N., (1985), "Gap-phase Regeneration in a Tropical Forest", *Ecology*, 66, 682–687.

Brokaw, N., (1998), "Fragments Past, Present and Future", *Trends in Ecology and Evolution*, 13, 382–383.

Brown, J.H. and Kodric-Brown, A., (1977), "Turnover Rates in Insular Biogeography", *Journal of Biogeography*, 9, 339–344.

Castelletta, M., Thiollay, J.M. and Sodhi, N.S., (2005), "The Effects of Extreme Forest Fragmentation on the Bird Community of Singapore Island", *Biological Conservation*, 121, 135–155.

Cochrane, M.A., (2001), "Synergistic Interactions Between Habitat Fragmentation and Fire in Evergreen Tropical Forests", *Conservation Biology*, 15, 1515–1521.

Cordeiro, N.J. and Howe, H.F., (2001), "Low Recruitment of Trees Dispersed by Animals in African Forest Fragments", *Conservation Biology*, 15, 1733–1741.

Didham, R.K. and Lawton, J.H., (1999), "Edge Structure Determines the Magnitude of Changes in Microclimate and Vegetation Structure in Tropical Forest Fragments", *Biotropica*, 31, 17–30.

Dovie, D.B.K., (2003), "Detaining Livelihoods and Disputing Biodiversity: Whose dilemma?", *Ethics, Place and Environment*, 6, 27–41.

Game, M., (1980), "Best Shape for Nature Reserves", *Nature*, 287, 630–632.

Gascon, C., Lovejoy, T.E., Bierregaard, R.O., Malcolm, J.R., Stouffer, P.C., Vasconcelos, H.L., Laurance, W.F., Zimmerman, B., Tocher, M. and Borges, S., (1999), "Matrix Habitat and Species Richness in Tropical Forest Remnants", *Biological Conservation*, 91, 223–229.

Gascon, C., Williamson, G.B. and da Fonseca, G.A.B., (2000), "Receding Forest Edges and Vanishing Reserves", *Science*, 288, 1356–1358.

Ghana Timber Export Development Board, (1992), *Ghana: The Best Managed Forests in West Africa*. Takoradi, Ghana.

Gold Coast Survey Department, (1949a), *Road Map of the Gold Coast: Southern Section*, GCSD, Accra.

Gold Coast Survey Department, (1949b), *Atlas of the Gold Coast*, GCSD, Accra.

Greene, D.F. and Johnson, E.A., (1989), "A Model of Wind Dispersal of Winged or Plumed Seeds", *Ecology*, 70, 339–347.

Hall, J.B. and Swaine, M.D., (1981), *Distribution and Ecology of Vascular Plants in a Tropical Rain Forest: Forest Vegetation in Ghana*, Dr. W. Junk Publishers, The Hague.

Harrington, G.N., Freeman, A.N.D. and Crome, F.H.J., (2001), "The Effects of Fragmentation of an Australian Tropical Rain Forest on Populations and Assemblages of Small Mammals", *Journal of Tropical Ecology*, 17, 225–240.

Hawthorne, W.D., (1993), *Regeneration after Logging in Bia South Game Production Reserve, Ghana*. NRI, Chatham.

——, (1994), *Fire Damage and Forest Regeneration in Ghana*. NRI, Chatham.

——, (1996), "Holes and the Sums of Parts in Ghanaian Forest: Regeneration, Scale and Sustainable Use. *Proceedings of the Royal Society of Edinburgh*, 104B, 75–176.

Hawthorne, W.D. and Juam Musah, A., (1993), *Forest Protection in Ghana*. Forestry Department, Kumasi, Ghana.

Hill, J.L., (1994), "Landscape Spatial Geometry: Its Measurement and Importance". *Swansea Geographer*, 31, 25–41.

——, (1995) The Effect of Forest Spatial Geometry on Tree Species Diversity in Tropical Forest Patches, Unpublished Ph.D. Thesis, University of Wales, Swansea.

Hill, J.L. and Curran, P.J., (2001), "Species Composition in Fragmented Forests: Conservation Implications of Changing Forest Area", *Applied Geography*, 21, 157–174.

——, (2003), "Area, Shape and Isolation of Tropical Forest Fragments: Effects on Tree Species Diversity and Implications for Conservation", *Journal of Biogeography*, 30, 1391–1403.

Hill, J.L., Curran, P.J. and Foody, G.M., (1994), "The Effect of Sampling on the Species-area Curve", *Global Ecology and Biogeography Letters*, 4, 97–106.

Kapos, V., Wandelli, E., Camargo, J.L. and Ganade, G., (1997), "Edge-related Changes in Environment and Plant Responses Due to Forest Fragmentation in Central Amazonia", *Tropical Forest Remnants: Ecology, Management and Conservation of Fragmented Communities*, W.F. Laurance and R.O. Bierregaard, Jr. (eds.), pp. 33–44. University of Chicago Press, Chicago.

Kellman, M., Tackaberry, R. and Rigg, L., (1998), Structure and Function in Two Tropical Gallery Forest Communities: Implications for Forest Conservation in Fragmented Systems", *Journal of Applied Ecology*, 35, 195–206.

Kiessling, W., (2005), "Long-term Relationships Between Ecological Stability and Biodiversity in Phanerozoic Reefs", *Nature*, 433, 410–413.

Laurance, W.F., (1991), "Edge Effects in Tropical Forest Fragments: Application of a Model for the Design of Nature Reserves", *Biological Conservation*, 57, 205–219.

—, (1999), "Ecology and Management of Fragmented Tropical Landscapes—Introduction and Synthesis", *Biological Conservation*, 91, 101–107.

—, (2000), "Do Edge Effects Occur Over Large Spatial Scales?", *Trends in Ecology and Evolution*, 15, 134–135.

—, (2004), "Forest-climate Interactions in Fragmented Tropical Landscapes", *Philosophical Transactions of the Royal Society of London Series B Biological Sciences,* 359, 1443, 345–352.

Laurance, W.F. and Bierregaard, R.O., (eds.), (1997), *Tropical Forest Remnants: Ecology, Management and Conservation of Fragmented Communities*, University of Chicago Press, Chicago.

Laurance, W.F., Ferreira, L.V., Rankin-de Morena, J.M., Laurance, S.G., Hutchings, R. and Lovejoy, T., (1998a) "Effects of Forest Fragmentation on Recruitment Patterns in Amazonian Tree Communities", *Conservation Biology*, 12, 460–464.

Laurance, W.F., Ferreira, L.V., Rankin-de Morena, J.M. and Laurance, S.G., (1998b), "Rainforest Fragmentation and the Dynamics of Amazonian Tree Communities", *Ecology*, 79, 2032–2040.

Laurance, W.F., Delamônica, P., Laurance, S.G., Vasconcelos, H.L. and Lovejoy, T.E., (2000), "Rainforest Fragmentation Kills Big Trees", *Nature*, 404, 836.

Laurance, W.F., Lovejoy, T.E., Gascon, C., Pohl, L., Williamson, G.B., Delamônica, P. and Oliveira, A., (2001), "Effects of a Strong Drought on Amazonian Forest Fragments and Edges", *Journal of Tropical Ecology*, 17, 771–785.

Laurance, W.F., Lovejoy, T.E., Vasconcelos, H.L., Bruna, E.M., Didham, R.K., Stouffer, P.C., Gascon, C., Bierregaard, R.O., Laurance, S.G. and Sampio, E., (2002), "Ecosystem Decay of Amazonian Forest Fragments: A 22-year Investigation", *Conservation Biology*, 16, 605–618.

Lomolino, M.V., (2001), "The Species-area Relationship: New Challenges for an old Pattern", *Progress in Physical Geography*, 25, 1–21.

Lugo, A.E., (2002), "Can We Manage Tropical Landscapes? An Answer From the Caribbean Perspective", *Landscape Ecology*, 17, 601–615.

Martin, C., (1991), *The Rainforests of West Africa: Ecology Threats Conservation,* Birkhäuser Verlag, Basel.

Martinez-Ramos, M. and Alvarez-Buylla, E.R., (1998), "How Old are Tropical Rain Forest Trees?", *Trends in Plant Science*, 3, 400–405.

Matínez-Garza, C. and Howe, H.F., (2003), "Restoring Tropical Diversity: Beating the Time Tax on Species Loss", *Journal of Applied Ecology*, 40, 423–429.

May, R.M. and Stumpf, M.P.H., (2000), "Species-area Relations in Tropical Forests", *Science*, 290, 2084–2086.

Nepstad, D.C., Verissimo, A., Alencar, A., Nobre, C., Lima, E., Lefebre, P., Schlesinger, P., Potter, C., Motinho, P., Mendoza, E., Cochrane, M. and Brooks, V., (1999), "Large-scale Impoverishment of Amazonian Forests by Logging and Fire", *Nature*, 398, 505–508.

Pimm, S.L. and Raven, P., (2000), "Extinction by Numbers", *Nature*, 403, 843–845.

Ranta, P., Blom, T., Niemela, J., Joensu, E. and Siitonen, M., (1998), "The Fragmented Atlantic Rain Forest of Brazil: Size Shape and Distribution of Forest Fragments", *Biodiversity and Conservation*, 7, 385–403.

Rosenzweig, M.L., (1999), "Heeding the Warning in Biodiversity's Basic Law", *Science*, 284, 276–277.

Rudel, T. and Roper, J., (1997), "Forest Fragmentation in the Humid Tropics: A Cross-national Analysis", *Singapore Journal of Tropical Geography*, 18, 99–109.

Sizer, N. and Tanner, E.V.J., (1999), "Response of Woody Plant Seedlings to Edge Formation in a Lowland Tropical Rainforest, Amazonia", *Biological Conservation*, 91, 135–142.

Tabarelli, M., Mantovani, W. and Peres, C.A., (1999), "Effects of Habitat Fragmentation on Plant Guild Structure in the Montane Atlantic Forest of Southeastern Brazil", *Biological Conservation*, 91, 119–127.

Terborgh, J., (1992), *Diversity and the Tropical Rain Forest*, Scientific American Library, New York.

Tilman, D., May, R.M., Lehman, C.L. and Nowak, M.A., (1994), "Habitat Destruction and the Extinction Debt", *Nature*, 371, 65–66.

Turner, I.M., (1996), "Species Loss in Fragments of Tropical Rain Forest: A Review of the Evidence", *Journal of Applied Ecology*, 33, 200–209.

Turner, I.M. and Corlett, R.T., (1996), "The Conservation Value of Small, Isolated Fragments of Lowland Tropical Rain Forest", *Trends in Ecology and Evolution*, 11, 330–333.

Turner, I.M., Chua, K.S., Ong, J.S.Y., Soong, B.C. and Tan, H.T.W., (1996), "A Century of Plant Species Loss From an Isolated Fragment of Lowland Tropical Rain Forest", *Conservation Biology*, 10, 1229–1244.

Varley, W.J. and White, H.P., (1958), *The Geography of Ghana*, Longman, London.

Wong, J.L.G., (1989), *Ghana Forest Inventory Project: Seminar Proceedings*, 29–30 March, 1989, Forestry Department, Accra, Ghana.

Zhu, H., Xu, Z.F., Wang, H. and Li, B.G., (2004), "Tropical Rain Forest Fragmentation and its Ecological and Species Diversity Changes in Southern Yunnan", *Biodiversity and Conservation*, 13, 1355–1372.

19 Desert National Park

A Unique Biosphere Reserve for Conservation and Development of Biodiversity in India

Ved Prakash Meena

Severe environmental problems are waiting at the door of 21st century world humanity. In last few decades man has caused major destruction by his malactivities. Because of this the life on earth itself is endangered. Therefore, presently worldwide efforts are being made to protect environment and its factors. One of them is known as, "conservation of biodiversity". In order to conserve biodiversity many plans have been executed around the world. In India too many national parks have been established under Wildlife Protection Act, 1972. The Desert National Park of Rajasthan is one of them. It has also been made a biosphere reserve in the year 1986.

It is an attempt to explore the interrelationships between animals and plants of arid India and portray the plant communities and associations present here. In this respect the Desert National Park provides a rich feast to the ecologists to unfold the intricacies of a natural ecosystem.

Distinguished from a forest by being enclosed, and having no special laws or officers, national park is distinct from a public park. It is an extensive area of land, set apart as a national property to keep in its natural state for the public benefit.

National parks are areas usually of great natural beauty and interest, now established in many countries and serve specially for conservation of native flora and fauna, as well as, for recreation of human beings-specially those interested in wild life.

In addition to those in the United States and Britain, especially famous are the Canadian National Park in the Rockies, the Kruger Park in the Eastern Transvaal, Serengeti in Tanzania, Mt. Fuiji in Japan.

In Britain, national parks were set up in England by an Act of 1949. They are areas of natural beauty, but for the most part the land is farmed or otherwise used, and the parks include numerous villages and other settlements and the normal life of the countryside goes on in them. This is because the country is small and large chunk of land cannot be spared. They are, however, specially protected from building development. Lake District and North Park

Moors are the two of the national parks in Britain. The same Act set up Areas of Outstanding National Beauty (AONB), which are also protected.

In United States they are mainly extensive wild areas, often largely uninhabited and so serve both for recreation and for conservation of wildlife. The first to be set up was the Yellowstone Park in 1872, including parks of the States of Wyoming, Montana and Idaho, with an area of over 3,200 sq miles. It is said to have more geysers than all the rest of the world. Its creation really started the National Park movement all over the world. The United States has 350 other national parks, some of the famous being the Crater Lake, Grand Canyon, Mammoth Caves.

In order to retain the natural environment and to conserve natural habitat, no settlement has been allowed for scores of miles around the Crater Lake in the State of Oregon, not even a hotel or lodge. The tourists have to drive long distances to arrive at the Lake. They can go round it and learn about it only through a local guide. They are prohibited even to feed the birds or squirrels with a grain of food. There is not a single fish of any sort in the 2000 feet deep Lake, Thus is conserved the natural habitat.

National parks programme is today an extended activity of conservation of representative ecosystem in the form of biosphere reserves or nature reserves. Although initiated under the auspices of the UNESCO during 1973–74 the programme did not gain momentum for nearly a decade. Gradually 62 countries selected 223 areas to be brought under Biosphere Preservation Programme, but each country interpreted the philosophy of the programme as found convenient to it. After the promulgation of Wild Life Protection Act, 1972, Government of India, wildlife areas came to be reserved as national parks, sanctuaries, or closed areas, thus giving due importance to wildlife conservation. India so far has selected 13 sites as potential biosphere reserves. The number of such areas has kept on increasing year after year looking into their importance up to 2003 when India had a total of 93 national parks and 485 sanctuaries making a total of 578 wildlife protected areas, some parks and some sanctuaries specialised for protecting various forms of wildlife as birds, tigers, lions, rhinos, elephants, deer, Great Indian Bustard and others. In the present context some national parks are of international significance to the birds like Siberian cranes and the Great Indian Bustard. The former is called the Keoladev National Park while the latter–the Desert National Park, one each in the semi-arid and arid regions of Rajasthan respectively.

Undoubtedly, the incessant exponential growth of human race upon the earth, and more strictly in certain harsh areas, has brought about slow but subtle changes in the natural ecosystems, which are inimical to the humans to cause large-scale turbulence in nature and jeopardize the laws of nature. Hence, it is of paramount importance to study the nature's reserves against the backdrop of environmental degradation. It is this purpose that the present paper on Desert National Park has attempted.

The numerous life forms both floral and faunal that have adopted themselves to survive and propagate under harsh habitat conditions of the hot desert, constitute an invaluable stock of rare germ plasms, which are too valuable from the biological point of view. The desert being a fragile ecosystem, its further denudation and deterioration is likely to remove, for all times, one of the world's most engaging environments. It is imperative, therefore, that at least a part of the Great Indian Desert, in as much of its pristine conditions as possible, is kept further unmolested and restored to its original status to serve as a bench mark model of the desert as it once existed over a vast territory. This could be done only by establishing a Desert National Park.

The Great Indian Desert of the pre-partition period is the eastern section of the Saharan Arabian desert and was earlier known as "Tharparkar" desert (Tod, 1914). The name persisted for a long time as the name of a district in Sind province of Pakistan. It was later on known as the Thar Desert or the Great Indian Desert. Part of it is now in Pakistan and a larger part is in India. Though named Desert National Park it is indeed a sanctuary. The park area is situated between 25°47′ N and 26°46′ N latitude and between 70°15′ and 70°45′ E longitudes and extends into 3,162 sq km, partly in Jaisalmer and partly in Barmer districts of Rajasthan (Figure 1). Its northern part is under the influence of the Indira Gandhi Canal Command and the southern part under the influences of the Narmada Irrigation Project (under contemplation). The famous bulge of Jaisalmer is the core of the desert ecology and is under the defence establishment as well as under the command of the IGNP. The result is that there is hardly any sizeable area left for establishing a conservation unit even in the vast wasteland. The conservation unit is bounded by the Jaisalmer Sam area in the north, the Gadra Road–Barmer rail track in the south, the Gadra–Harsani–Satto defence road in the last and the international boundary in the west. This is the only available land. Its western boundary had to be shifted 20 km towards the east from the International boundary, on instructions from the defence department.

It is the solitary example of a desert park in the world. Its global reputation is due to its existing desert land ecosystem though developed in isolation, yet manifests a large variety of consequential changes in its habitat, seasonal variations in faunal and floral composition and diverse vegetation types. It is a unique ecosystem also because it is both man-made and well maintained

Conservation of Wildlife

Since the desert park is under the administrative control of wildlife wing, for planned working but no sacrosanct management plan has so far been prepared for the Desert National Park for wildlife preservation and its management.

Figure 1
Desert National Park

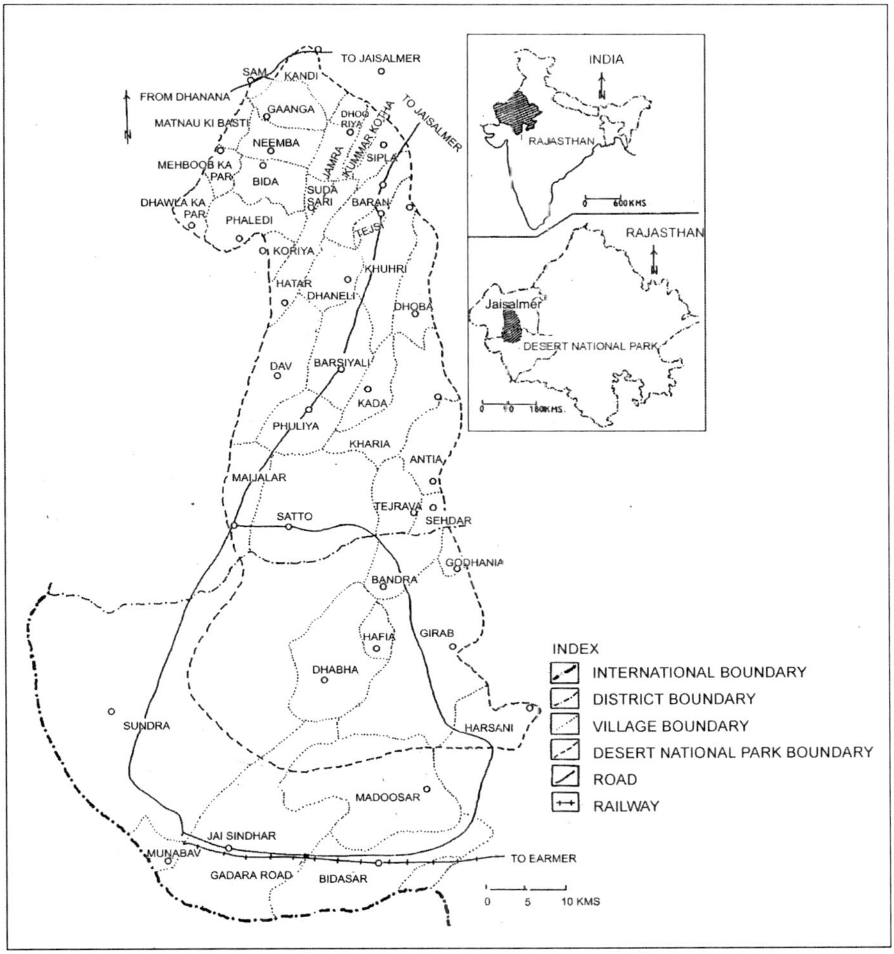

The Present Position of Desert National Park

The total area of the park being 3,162 sq km, it is the biggest sanctuary in the arid zone, all the rest being in the semi-arid and sub-humid zones. It has over 60 species of mammals and 6 species of reptiles.

The entire area was notified as a sanctuary under the Wildlife Protection Act (1972) and still has natural landscape, original vegetation, land forms, traditional land use, animal husbandry, fewer people and it is large enough to be an effective conservation unit.

Presently, the land of desert national park is being used in a diversified manner. Complete data of this is shown in the Table 1.

Since a single unit of 600 sq km representing all the landforms found within the desert was not available, two units of 300 sq km each have been selected, one in Jaisalmer and the other in Barmer district for the purpose of

Table 1

Land use in Desert National Park

District	Total No. of Village	Total area in ha	Forest in ha per cent	Unirrigated area in ha and per cent cultivated land	Cultural waste land in ha* per cent	Area not available for cultivation	Grazing and including oran, culturable waste and barren land	Grazing land as per cent of tatal area	Available grazing land/livestock	Total livestock population
Jaisalmer	28	2,75,272	–	10715 (3.89%)	250921 (91.15%)	13636 (4.95%)	2,64,513	96.09%	1.96 ha	1,35,241
Barmer	9	2,03,788	17.8 (0.84%)	54769 (26.88%)	108898 (53.44%)	38413 (18.84%)	1,65,375	81.15%	1.73 ha	95,812
Total	37	4,79,060	1708 (0.36%)	65484 (13.76%)	359819 (75.11%)	52049 (10.86%)	4,29,888	89.73%	1.86 ha	2,31,053

*Used as grazing land.

Source: Project Document–7. Ministry of Environment and Forest, Government of India, New Delhi.

total conservation. They serve the purpose of core areas of the national park. The two totally fenced and protected areas are not located at the centre of the national park nor are these insulated by buffer zones, as is generally the case with national parks and sanctuaries. Since the areas outside the Reserves, especially towards the international boundary in the west are heavily guarded and are likely to remain so indefinitely, the protected area will always be insulated from outside influence.

The Desert National Park has no ecological boundary. Agricultural fields, pastures, revenue lands and sanctuary land all are conjoint and at some places barbed-wire fencing is the only significant marking of boundaries, marking pillars to show its actual status.

It is a holistic and exploratory study unveiling the cause and effect relationships which are of great importance to unfold the cause of degeneration in a particular ecosystem and to suggest measures to infuse the basic components to revive it.

In order to study the ecology of biotic variations of this park a team under the leadership of K.S. Sankhala and another team under the Bombay Natural History Society had conducted research projects. K.A. Shankarmarayan (1987) had worked on eco-regeneration of Desert Biosphere Reserve in the Thar desert.

Geology and Landforms

It is surmised that the desert area was submerged under sea during the Jurassic, Cretaceous and Eocene ages. It was uplifted probably sometimes in the Upper Tertiary. It was during the late Pleistocene after the last Phase of Glaciation that the desertic conditions were set in. Small outcrops of Jurassic rocks comprising limestone, sandstone and shales are found in the midst of desert sands and alluvium. Marine deposition resumed in the Eocene period before a climatic accident created the desert.

Geomorphologically, it is largely sandy with shifting sand dunes, fixed dunes and interdunal flat lands, interdunal depressions, and saline flat lands. It slopes towards the south and north-west without any perennial river. The seasonal rivulets are mostly nallahs with limited catchment areas. The little flow of water is often locked by the inhabitants as 'Khadins' used both for their cultivation and for drinking purposes.

The soils are mostly sandy with low water holding capacity, 50 to 150 cm deep in the interdunal depressions, lacking in nitrogen, carbon and mineral salts, not suitable for cultivation. Its strong and gravelly areas are 'banjar' lands.

Climate

The climate is typically arid with 4 to 5 rainy days. Droughts occur quite frequently. January is the coldest month affected by the western disturbances when temperatures often go 3 to 4 degrees below the freezing point. Frosts damage the vegetation. May and June are the hottest months when the temperature may rise to 50°C and dunes travel long distances during the stormy summer days.

Settlement Pattern

The settlement pattern of the sedentary population in the area is of three types: (1) Entirely compact, (2) Entirely scattered, and (3) Villages having both compact settlements and scattered homesteads to which the majority of villages belong. The rural people mostly live in hamlets, small villages and 'dhanies' or homesteads. The dunal topographic conditions, the formation or dhani type settlements characterized by non-accessibility, scattered location of water sources, the socio-cultural heritage of certain castes, the lately prevalent 'jagirdari system' and the customs of bringing tenants for use of land and the practice of subsistence farming. There is a tendency among people belonging to the same caste to live together in the dhanies or in the village core. Historically the sequence of settlements in the villages has been first few families who settled there. Later people moved seasonally to their distant fields and then the households started settling in the fields to avoid long distant commuting. People living in dhanies take greater care of their crop land from close quarters while the people living in the core compact settlements enjoy a richer social life and make better use of the community facilities. There are 37 villages and 85 dhanies situated in Desert National Park (DNP). The occupation of the people residing in these are shown in the Table 2 and Figure 2.

According to landscape modification, the development of the park chronologically may be divided into three main periods:

1. The period of inception.
2. The period of development.
3. The period of maintenance.

The period of inception started in the 19th century and earlier when the occasional floods made this area of the desert park natural boggy and alkaline, with layers of saltpetre overlying the soil. After the retreat of the monsoon and drying up of the marshland, it was used as pasture for cattle, abandoned only during the famines and droughts. It was also a refuge area for the wild cattle of the Jaisalmer state. The state had been robbing this area of its vegetal cover by lopping trees, foliage cover, shrubs and hay.

It is interesting to find that the construction of 'khadins' which took place in gradual stages affected the development of the DNP. The work used to be done in years of famines when a large number of peasants and landless labourers used to be provided with relief work for their sustenance.

The management plans of the desert area were first prepared in the 1970s Though it is named Desert National Park, but in fact, it is a sanctuary as declared under the Wildlife (Protection) Act, 1972. Besides the main areas, other eight closed areas were selected for wildlife preservation and declared as Closed Areas under the Act. Today, these are called satellites of this biosphere reserve under the Director of Desert Afforestation and Pasture Development, Jodhpur till 1986, after which it came under the administrative control of the wildlife wing.

With the abolition of feudal kingship ushered an era of freedom from exploitation of vegetation and shootings of animals. But during the transition,

Table 2

Population and Occupational Patterns in Desert National Park.

District	*Number of villages*	*Area in ha*	*Population and density/ha*	*No. of workers and per cent or total population*	*Cultivators labourers (%) * to total workers*	*Agricultural hold (%)*	*Household industry (%)*	*Other workers +*	*Marginal +*
Jaisalmer	28	2,75,272 (0.04/ha)	11,938	3,785 (31.70%)	2,039 (54,00%)	191 (5.00%)	68 (1.80%)	1,487 (38.20%)	309 (7.76%)
Barmer	9	2,03,788	19,227 (0.09/ha)	6,948 (36.14%)	6,370 (91.68%)	19 (0.27%)	21 (0.30%)	538 (7.75%)	228 (3.18%)
Total	37	4,79,060	31165 (0.06/ha)	10,733 (34.43%)	8409 (78.35%)	210 (1.95%)	89 (0.83%)	2025 (18.87%)	537 (4.85%)

*Including graziers and pastoralists
+ Mostly graziers and pastoralists

Source: District Census Handbook, Jaisalmer and Barmer.

Figure 2
Deseert National Park Population Density

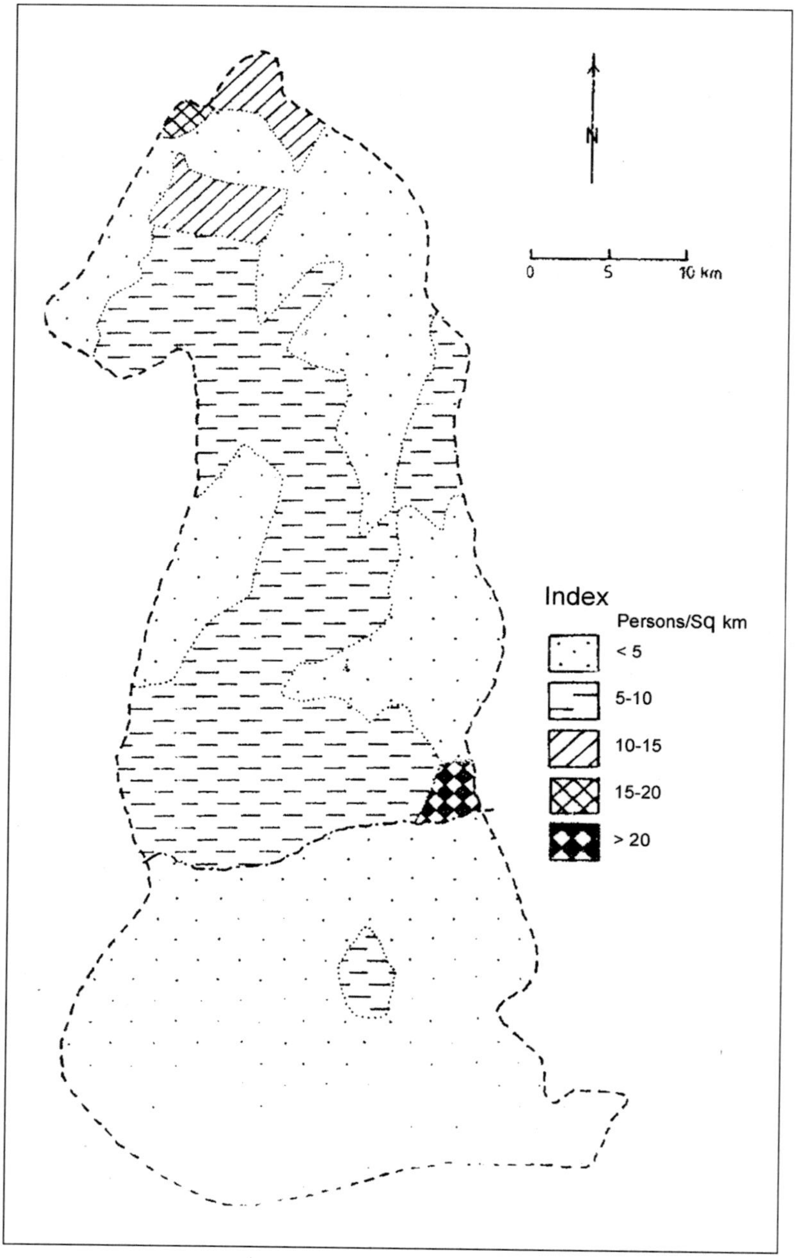

the neighbouring farmers encroached upon the peripheral lands of the park and its allotment to refugees by local administration. But these activities were prevailed upon by K.S. Sankhla with the concurrence of the then Prime Minister Mrs. Indira Gandhi.

Broadly speaking 16 types of land use are most significant. The practice of shooting the Great Indian Bustard, the State National Bird was banned in 1972, cutting of trees for fuel-wood and cattle grazing was banned in 1982 and also the temporary human settlements which existed for limestone making were also abolished. Grazing practices have been banned except in years when due to overload in of the Chelasari Nadi and khadin, there is immense growth of grasses. Establishment of human settlements inside this park and development of tourism has been continuing and thus the human population has been increasing being detrimental to the desert ecosystem.

The management of the desert park maintains wireless sets, water resources through tanks and tube wells for both the animals and staff posted at remote places, vehicles (including camels for mobility of the staff). It is good that there is no separate rest house of Desert National Park, nor their should be any if the ecosystem of the park is to be maintained.

Fossil Wood Park at Akal

About 15 km from Jaisalmer town on the NH–15 on the Jaisalmar-Barmer Road near the village Akal, is a scientific wonder protected in a small park named as 'Akal Fossil Wood Park'. There are several tree trunks lying prostrate on the ground in an area of 21 hectares. Twenty-five of the trunks are visible on the surface, the largest is 7 m. long and 1.25 m. in girth. About 18 crore years ago these tree trunks were real trees with leaves, roots, standing upright which grew in this part of Rajasthan and formed thick jungle. During submergence and emergence more than once, these trees were petrified. One can see large number of petrified logs and innumerable pieces of woods scattered on the surface in the area.

Water Resources

Water is the most precious natural resource of the region. The government has recently drilled tube well at Kuri, Miaji-alaar, Hasrani and Grab villages. Raw water is stored in tankas (covered tanks) in almost every house and is the only source of sweet water for drinking. The Indira Gandhi Canal ends up at Gadra village in Barmer district. Its Sagarmal Gopa branch would enter the DNP area right from 15 km from Sam village and traverse a distance of 125 km in the notified DNP area.

Environmental Adaptation of Animals

Several animals in this part acquire necessary water from their food. The gazelle and the hare can live long without water. They generally migrate to water points in the area. The gazelle, blackbuck, nilgai and the wolves, etc., congregate around large perennial rain pools. The hare, gerbil and gazelle lick dew on grasses. Dove, common vulture, sandgrouse, bulbuls, partridge, peafowl and the jungle babbler, etc., migrate in the vicinity of perennial rain pools during the drought season. Many of the mammals and birds of the desert drink considerable saline water having above 3,000 ppm salinity. Some scientists mention nasal salt excreting glands is some birds whereas other animals excrete extra salt with their urine.

Socio-Economic Aspect

The park is a valuable asset for harvesting fodder. In contemporary times, the human needs weighed a bit more than those of animals. So the villages small and large, surrounding the park had an access to the park to graze their domestic animals and have fuel wood. In modern times tourism industry is flourishing and generating local employment.

A few houses of the ex-Jagirdars and some others in villages within the park are built of sandstone brought from the towns. Otherwise, mostly the villagers build mud houses with roofs of brushwood. They put 2 m high thorn fencing around an extensive compound to ward off animals, thieves and also as protection against strong winds. Inside the compound, two or three jhupas (huts) are built depending upon the size of household. Bullocks and camels are tied in one corner of the compound while rest of the livestock are confined in separate enclosures. Fodder and grains are also stored here in specially built conical structures in jhupas of khejri wood and shrubs, locally available. The entire floor within the huts and compound is *kuchcha,* coated with cow dung, water and earth. Water is stored in earthen pitchers. Some families store rain water in tanks or pits. The members of the extended families or disjointed families build separate hutments constructed in the compound of the original dwelling. Shops, community facilities and even government functionaries are housed in the jhupas in the village core. In general, the people here are vegetarian. The desert people have developed certain peculiar food habits. Seeds and pods of several trees are harvested, dried and used as delicacies. Almost all the fruits of *Zizyphus nummularia (ber: jharberi)* growing in accessible parts of the desert are harvested and converted into flour for human consumption specially during drought years. The grass seeds add nutrition to their foods. Cattle grazing is allowed free except inside the enclosures. Sewan grass is of high nutrition value. In case of severe drought conditions, people generally migrate to distant states till the rains start in this area.

Activities have been mainly concerned with providing absolute protection to the flora and fauna within a particular territory against degradation by people living within or outside the territory. It has been recognized that this approach antagonizes the local people. There have been conflicts, sometime resulting in loss of life. The latest thinking is to involve the people living within the reserve in its management while ensuring sustained betterment of their lot. This philosophy of man being the hub of the biosphere will be the guiding principle in the operation of the Thar Biosphere Reserve. Rotational grazing of livestock will be the pivot around which the whole scheme will revolve. Many kinds of animals are kept in Desert National Park, which are shown in the Table 3.

Table 3
Livestock Population in Desert National Park

District	Number of villages	Livestock population, total and percentage						Total
		Camel	Sheep	Goat	Cattle	Donkey	Buffaloes	
Jaisalmer	28	5,359 (3,96%)	55,631 (41.1 16%)	38,638 (28.56%)	30,432 (22.50%)	5,081 (3.75%)	100 (0.07%)	1,35,241
Barmer	9	-	30,157 (31.47%)	43,605 (45.51%)	17,357 (18.12%)	4626 (8.82%)	67 (0.07%)	95,812
Total	37	5,359 (2.34%)	85,788 (37.12%)	82,243 (35.59%)	47,789 (20.68%)	9,707 (4.20%)	167 (0.07%)	2,31,053

Source: Office of Deputy Director, Desert National Park, Jaisalmer and Barmer.

There is great scope for tree planting at suitable sites in the village and desert posts and on water points primarily to provide shade. Shady and fast growing species like Neem *(Azadirachta indica),* Siras *(Abizzia),* Rohida (*Tecomella undulata*), khejri (*Prosopis cineraria*) and Peepal (*Ficus religiosa*) may be recommended. Gas supply and solar energy can save trees from felling for fuel.

Vegetational History of the Park

Several studies have shown that the grasses, the shrubs and the trees in the DNP area are well adapted to the environment. These are deep rooted tenacious enough to survive expanded droughts and yet efficient in putting up a good biomass. These are quite palatable and rich in mineral matter. What is more they occur in great variety, there being as many as 700 species, amongst which grasses alone account for 107 species. All told, the arid zone of Rajasthan has a well defined vegetation which is described as Tropical Dry Thorny Forest Type. The DNP consists chiefly of scrubby and thorny forest, the main dominated species are as follows: kikar (*Acacia nilotica*). pilu (*Salvadora Persica*), jai (*Salvadora oleoides),* ber (*Zizyphus mauritiana*), khejri (*Prosopis cincraria),* chonkra (*Acacia leucophloca*), karil (*Capparis decidua*). The dominant species of grasses like sewan, bhurat, phog grew in parts where land became usually dry just after the monsoon. Today, the grassy patches of the park are the remnants of the extensive closures, which once existed in the park stretching up to Sudasari village. The effects of climatic variations, droughts and floods, etc., also brought about considerable change in the natural biomass composition of the DNP.

It is interesting to note that this ecosystem was biologically less diverse prior to 1900 AD as compared to the later restricted exploitation of biomass and that the ecosystem of area has matured and diversified over the years. This is in contrast to the general observations of other desert ecosystems which have degraded due to the onslaught of both human and animal population. About 2,000 sq km of area of the park was once densely furnished with tanks in which rain water used to collect for some months during the year and a splendid thick forest existed. This is the famous Great Indian Bustard (GIB) shooting area of Jaisalmer. But as there was no restriction on cutting trees and grazing cattle, that serene verdant has disappeared. However, after 1980 some restrictions were imposed and developmental processes began to take place for the conservation of the natural environment after the recommendation of K.S. Sankhla in 1985. A detailed classification of desert species based on their adaptive characteristics is as follows:

Drought Escaping Plants

These include monsoonal plants which appear through 'seeds' just after the rains, grow profusely, flower and produce seeds within a duration of 4–6 weeks in which situation plants do not have to face harsh conditions of the desert species like *Mollugo cervina, Gisekia pharaceoides, Indigofera cordifolia, Tributus terrestris* etc.

Drought Evading Plants

These evade drought by conserving scanty moisture, e.g., *Pulicaria crispa, Digera muricata, Celosia agrentea, Tephorosia purpurea.*

Drought Enduring Plants

They are small sized plants tolerant of drought for longer period. Examples are *Convolvulus, microphyllus, Launea sp* etc.

Drought Resistant Plants

They have adaptive features through which they are able to face extreme conditions of desert. These plants may be further classified as (i) Succulents: *Cassigonun polygonoides, Euphorbia caducifolia, Opuntia* sp., *Aloe barbedense, Agave sp.*, (ii) Non-succulents are those that are true *xerophytes, Acacia senegal Prosopis cineraria, Capparis decidua, Salvador persica, Caltrop procera.*

Grasslands (*Lasiurus Scindicus*)

The grasslands which vary from dense tall sewan grass to low grasses, occupy large parts of the park area, mostly in the extreme north and extreme south, chiefly confined to the closures 1, 2, 3, 4 and 5. This sewan grass also grows more profusely in sandy soil areas. In the last over ten years these grasses occupied an indomitable area from Jaisalmer to Barmer. The grasslands have been constantly shrinking over the years due to encroachment and agricultural expansion. Thus, most of the closures have disappeared.

Ten large areas of woodlands in the park also have a continuous cover of tall grasses ranging from 50 to 135 cm in height interspersed with continuous growth of trees like khejri *(Prosopis cincraria),* babul with occasional growth of shrubs like ber and kair. The tall sewan grass and phog grow profusely during the rainy season. Owing to the diversity of vegetation and a profuse land cover of grassy patches, these areas are the most favoured sites to chinkara (deer), land birds and the mammalian fauna of the park. These tall grasslands have a predominance of shrubs like *Salvadora spp,* vilayati babul, kari, ber amidst the array of grasses like dhaman and bhurat. Micro-variations in the characteristics of vegetational diversities are also noticed on varied landforms in the park region. Flora of Desert National Park which are shown in the Table 4.

Table 4

Flora of Desert National Park

	Local Name	*Botanical Name*
(A) Trees		
1.	Khejri	*Prosopis Cineraria*
2.	Kikar	*Acacia Nilotica*
3.	Neem	*Azadirachta Indica*
4.	Hingota	*Balanites Egyptican*
5.	Ker	*Caparis Decidua*
6.	Jal	*Salvadora oeloides*
7	Jal (Khara)	*Salvadora persica*
8.	Rohaida	*Tecomela Undulata*
9.	Julinra	*Prosopis juliflora*
(B) Shrubs		
1.	Khimp	*Leptadema prypechnica*
2.	Aak	*Kalotropis procera*
3.	Senia	*Crotoleria burilia*
4.	Thore	*Euphorbia royaleana*
5.	Ber	*Ziziphus mauritiana*
6	Murali	*Lycium aurooeum*
7.	Lana	*Heloxylon selicomicum*
(C) Grass		
1.	Bhurat	*Cenchurus callaticus*
2.	Siwan	*Lasicirus sindicus*
3.	Bura	*Cymbopogom jwarincosa*
4.	Lamra	*Aristado mutica*
5.	Kucha	*Sasccharus griffinir*
6.	Murat	*Panicum turgidum*
7.	Dhaman	*Samccharus cibarisis*
8.	Bul	*Areva aomentcsa*
9.	Phog	*Calligomun polygonoides*

Source: Meena, V.P, 2000, Desert Ecology, Printer Publishers, Jaipur

Rajasthan, 'State Tree': Khejri, the Lifeline of the Desert and 'State Flower': Rohaida

The khejri (*Prosopis Cineraria*) has been one of the evergreen and long lasting friends of the dwellers of arid zones even in the period of scarcity such as droughts and famines since time immemorial. According to the Central Zone Research Institute, Jodhpur, there are only five important trees that grow in the Indian desert and the khejri is the most useful of all of them. It has the greatest valid and widest ecological amplitude in the arid and desert conditions. Khejri is the state tree of Rajasthan. Rohida is the state flower of Rajasthan These are normal and protected trees in the park. Desert National Park is an important research centre for them.

Faunal Community and Diversity

The park lowland has remained a battleground of many animal species which on account of vagaries of climate, frequent floods, droughts and natural catastrophe coupled with human invasion on grasslands induced few migrations of terrestrial animals. It is therefore, significant to consider the following faunal aspects of this park:

1. Food habits and breeding biology of the main wildlife species.
2. Case studies of rare and vanishing species like Great Indian Bustard, Chinkara, desert fox and migratory Imperial sandgrouse.
3. Migratory phenomenon of birds and reptiles.
4. Behavioural study of all the wild species with special reference reproduction.
5. Monitoring the whole Desert National Park from interrelationship of agricultural field produces and existing wildlife.
6. Study and monitor the reintroduction programmes.
7. Evaluation of population dynamics.

In the 1920s, blackbuck were abundant in the Jodhpur region. Up to the thirties of this century, the wild boar was so prolific around Jodhpur that people used to shun walking in the outskirts of the town after dark. The Asiatic wolf is also facing extinction in the desert region. The wild ass, now restricted in the Little Rann of Kuchchh was well distributed in parts of Rajasthan desert. What remains now of the mammals wildlife of the desert is a dim reminder of the past glory.

Amongst birds, some 300 species were reported by Huge Whistler. The Great Indian Bustard, houbara, Imperial sandgrouse are still in fair numbers in the desert,

The developmental activities during the years 1972 to 1990 in DNP improved its ecological conditions. By all these improvements a few fenced areas came up with artificial water and pasturage even in summer season for herbivores. Gradual growth in new vegetation and change in climo-vegetational factors the neighbouring population of herbivores ushered in the park closures of the DNP.

This region is rich in fauna, including bird life. It is the habitat of common sandgrouse, Imperial sandgrouse and the Great Indian Bustard. The spiny tailed lizard *(Uromastix hardwickil)* are common. The Indian fringe-toed sand lizard *(Acanthoductylus cantories)* and bronze grass skinks *Maboya marcularis* are common, movement markings of the snakes are seen on bare sand dunes. Ring doves, *(Slreplopeloa decaocto)* are very common. So are the grainvorous pigeon. *(Columbia livia)* and the sparrow (*Passer domestics)*, white cheeked bulbul *(Pycnonotus leucogenys)*, scavaenger vultures *(Neophron percnopterus)*, pariah kite *(Milvus migrans)*, gray parridges *(Francolinus pondicerianus)* are conspicuous. Crested larks *(Galerida cristata)*, small Indian skylark *(Alauda gulgula)*, the black bellied finch lark *(Eremopetrix grisea)* and white throated munia *(Lonchura malabarica)* are found. Among predatory birds shikara *(Accipiter badius)*, sparrow hawk *(Accipiter virgatus)* tawny eagle *(Aquila raapx)* short toed eagle *(Circaetus gallicus)* are seen perched on trees. The common sandgrouse *(Plerocles exustus)* are conspicuous. (Figure 3)

The Indian Gazelle are all over in small family groups. At times 8–10 individuals may group, whereas at some places one or two buck hold as many as 8–10 females. Sometimes, only feminine herds are observed. The desert fox *(Vulpes pusilla)* becomes conspicuous in the late afternoon.

This is the land of desert rats. There are two species, the hairy footed gerbil *(Ggrbillus gleadowi)* and the desert gerbil *(Meriones hurrianae)* are seen in the winters. The Imperial sandgrouse comes in thousands and their number may exceed over a million birds.

The rich insect fauna, which is a characteristic of this desert plays a remarkable role as the micro-consumer in the existing ecosystem. Although most elements of this fauna are found to be present almost throughout the year, they exhibit a high population build up particularly during the rainy season (July to September). They cause much loss to the growing vegetation. The termites attack the roots of plants. Among the termites, the most destructive species are *psammotermes rajasthanicus, Anacanthoytermes macrocehalus, Microterms spp and Odontotermes sp.* In certain pockets heavy losses of kharif crops due to the white grub *(Holotrichia sp.)* have been observed. Locusts and grasshopper breed in the region especially in the rainy season when sufficient moisture is available in the soft, sandy soil. These pests do much loss to the growing vegetation. Swarms of the desert locust *(Schistocerca gregaria)* visit occasionally, whenever suitable environment conditions are formed. The most noxious species of grasshoppers are *Ochrilidia affinis pockiloceruspictius, Pygromorpha bispinosa deserti, Acrotylus himbertianus* and *Crytacantharacis tetarica.* Some coleopteran *(Myllocerus sp., Crytozemia sp.)* are also active in the rainy season. The predatory habit of some insects like mantids *(Schizocephala bicornis, Statilia* sp.) Coleopteran *(Monochilus sexmachlatus)* and *(Camonutus gigas)* and spiders are quite marked in the rainy season.

The amphibian fauna of this region is restricted to one species of toad and five species of frogs. The Anderson's toad (*Bufo andersonil)* is widely distributed along the ephemeral water courses and temporary ponds. The desert is quite rich in reptilian fauna. Fauna of the Desert National Park are shown in the Table 5.

Figure 3
Desert National Park Fauna

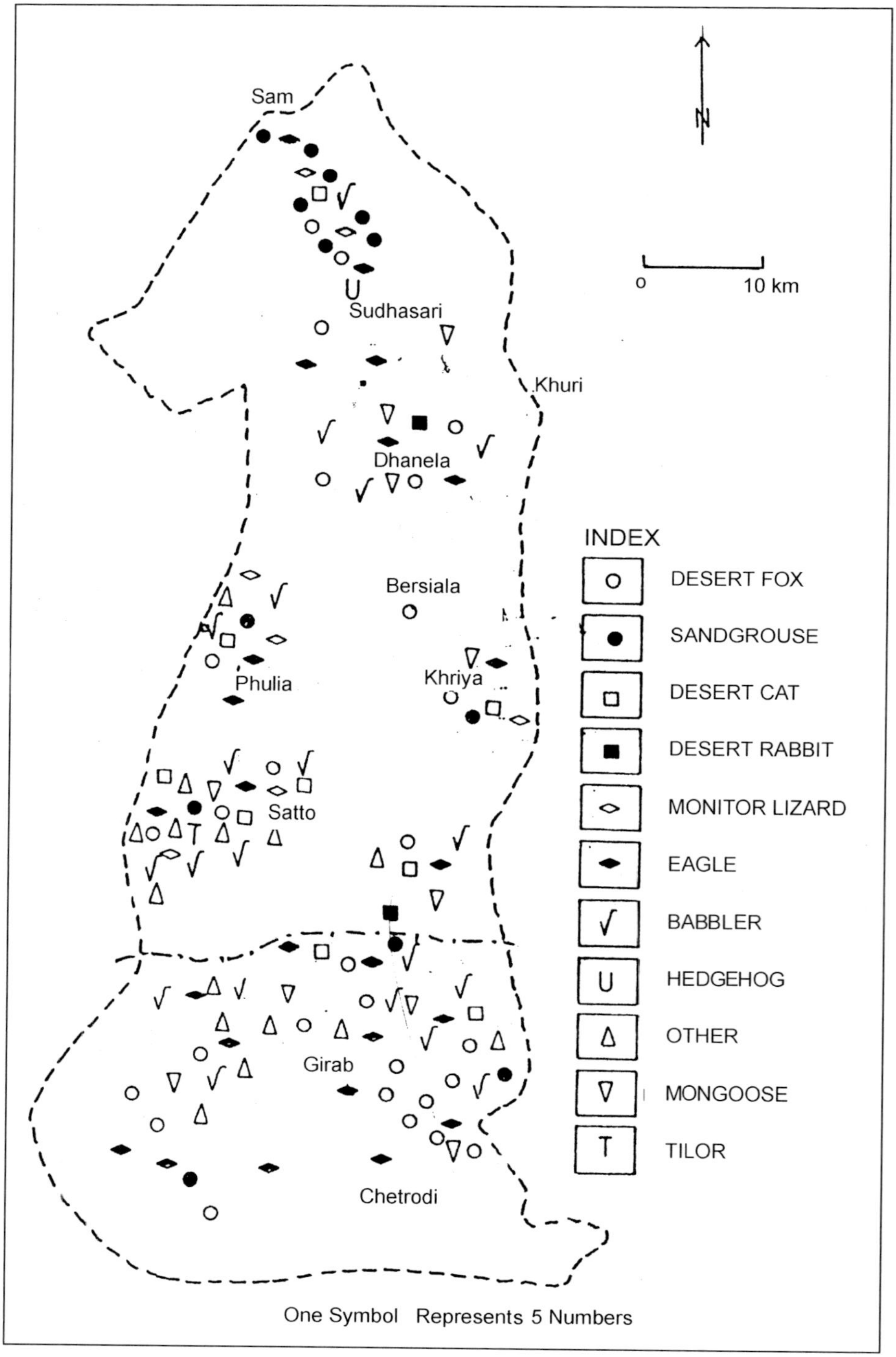

Table 5
Fauna of the Desert National Park

No.	Name	Zoological Name
A. Animals		
1.	Chinkara	Gazella gazella
2.	Jackal	Canis aurens
3.	Hare	Lul:s nigricoillis
4.	Hedgehog	Parachinus micropus
5.	Jungle Cat	Felis Chaus
6.	Fox	Vulpus bengalensis
7.	Mongoose	Herestes edwards
8.	Wolf	Canis lupus
B. Reptiles		
1.	Common lizard	Calotus varcidor
2.	Common lizard monitor	Varanus benalensis
3.	Dhaman	Ptyas mucosus
4	Krait	Bungarus caeruiens
5.	Russel's viper	Vipera russeli
6.	Spiny-tailed lizard	Uromaxtyx hardwicki
C. Birds		
1.	Babbler, common	Aggya caudata
2.	Babbler, jungle	Tudoieds somervillai
3.	Bee eater, blue-checked	Merops superciliosus
4.	Bee eater, small-green	Merops orientalis
5.	Bulbul, red-vented	
6.	Bulbul, red-whiskered	Otocompsa j ocosa
7.	Crow, house	Corvus splendense
8.	Crow, jungle	C. macrophychos
9.	Cuckoo	Euchynamis scolopaceus
10.	Dove, little brown	Stretopelia bengalensis
11.	Dove, red turtle	S. tranquibari ca
12.	Dove, ring Indian	S. deczotao
13.	Drongo	Diruus macrocerous
14.	Eagle. short-toed	C.rcatus gallicus
15.	Eagle, tawny	
16.	Great Indian Bustard	Choriotis nriceps
17.	Falcon, lager	Falco jugger
18.	Flycatcher	Rhepidura aureola
19.	Kite, common	Malvus mighans
20.	Lapuring, red wattled	Lobivanellus inducus
21.	Lark finch	Eremopterix griseo
22.	Myna, cornmon	Acridotheres tristes
23.	Owl, great horned Indian	Caprumulgus asiaticus
24.	Owlet, spotted	Athene brama

Cont'd...

Cont'd...

25.	Partridge, grey	Francolnus pondicerainus
26.	Robin. Indian	Saxecohdes fulicale
27.	Sandgrouse, common	Pteroceles exustus
28.	Sandgrouse, common	Tringa glareola
29.	Shrike	Asturbadius sp.
30.	Shrike, grey common	Tephrodornis pondiceranus
31.	Stilt, black-winged	Himantopus himantopus
32.	Tailor bird	Ortholunis sulorius
33.	Vulture, white-backed	Pseudogyps bengalensis
34.	Vulture, scavenger	Neophron peronoterus
35.	Woodpecker	Branchyternus bengalensis

Source: Meena, V.P. 2000, Desert Ecology, Pointer Publishers, Jaipur.

Rajasthan 'State Animal', Chinkara

The Indian gazelle or chinkara *(Gazella dorcus)* in different parts of the great Indian or Thar desert in western Rajasthan indicates that these animals are well-adapted for survival in arid and semi-arid regions.

The Indian gazelle is found in almost every type of ecosystem in western Rajasthan. Chinkara is the state animal of Rajasthan. Chinkaras are fully conserved in the park.

Rajasthan 'State Bird', The Great Indian Bustard

Godawan or the Great Indian Bustard *(Choriotis nigriceps)* is one of the rarest bird of India. It is half the size of saras crane with strong bare legs equipped with strong running *(Cursorial)* feet. It walks with confidence and grace. Godawan is the State Bird of Rajasthan. Godawan is the main attraction of the park.

Major Management Issues

The Thar is a desert area where the population pressure is not that high as to make it an unmanageable issue and, therefore there is every possibility of an eco-development based conservation programme becoming a success, provided it is backed by a strong political will and adequate administrative acumen. The park, with a legacy of the past for its multiple uses which were made from time to time, is at the moment a forbidden land for human intervention. Nonetheless, its former usage as a wild cattle enclosure and then as a great Indian bustard shooting reserve have expanded multiplicity of problems which have to be tackled for maintaining the biotic reserve. The park today has to galvanize together the habitat needs and food availability for majority of animals—the large mammalian fauna, reptiles, birds and other animals, as well as the human population in villages within the park area.

The isolated location of the park is both a boon and a disadvantage. Boon because there are not many human habitations in the immediate proximity of the park to disturb its natural ecosystem. Disadvantage because only a few managers and experts are available to take care of this national reserve far away from the urban metropolises. Every issue of this reserve or every component of this fragile ecosystem like wind, water, soil texture and biotic life on such land has to be studied thoroughly in scientific pattern, like the ten-year project at the Keoladeo National Park by the Bombay Natural History Society. Neglect of a single component may give irreparable loss to the nature of the park. The management objective of the park should aim at:

1. Preservation of a fragile and unique ecosystem as a natural heritage, and study the ecological behaviour of the desert life.
2. Provision of healthy outdoor recreation to the people of urban areas.
3. Involving the people living inside the park in the maintenance of the ecosystem and the safety and preservation of biotic life of the DNP.
4. Developing domestic and international tourism in desert ecosystems, restricted to the carrying capacity of the park.
5. Developing interesting and unique destinations for educational tourism.
6. Programmes of research projects for study of the Indian Desert ecosystem for their conservation.

The desert park management has to maintain a control on pollution from mining activities, grasses and weeds, management of wastelands, woodlands, grasslands, water resource points, and by ban on poaching.

Solution

The mystique of Desert National Park is its biodiversity within a big area, exhibited in a myriad forma of animals, birds, insects, aquatic plants and tree vegetation which has developed by relentless efforts, of man over the years. In their continued efforts the people living within the ecosystem should be educated and made aware of the great advantages of maintaining the quality of the desert ecosystems by dedication to the national park. They should be made to understand that the conservation of the natural system would also improve the quality of their life within the desert. They may be provided gas for their kitchens from the oil fields which have been explored in the vicinity in recent years. This would save the trees from being cut for fuel wood. Efforts may be made in scientific harvesting of the sewan grass for the upkeep and earning revenue to the park. The indigenous system of water harvesting and utilization system of water harvesting and utilization of the water of the intending Indira Gandhi Canal is a prime requisite as the biotic life may vanish with water famine.

The glory of this natural desert ecosystem can be enhanced if the aforesaid points are taken into consideration.

There is a need to save this great project of Indian planners. A final notification regarding declaring it a national park should be released without delay. A national desert museum should also be developed here. This project should be kept in natural ecosystem which is helpful in the prevention of international smuggling and infiltration experiences of this project tell us that tourism should be banned in wild animals reserve regions. In addition to this in this region permanent greenery should be developed in place of cultivating heavy grains. This Desert National Park needs to be saved from the dangers of Indira Gandhi Canal. The area of this Desert National Park should not be reduced. A research institute for desertic wildlife creatures should be opened here. On account of these grounds, Desert National Park will prove out to be a unique project and it will also become a multipurpose project.

Select References

Anonymous, (1977), "Desertification and its Control", ICAR, New Delhi, pp. 1–358.

—, (1983), "Progress in Arid Zone Research: 1952–1982", CAZRI, Jodhpur, pp. 1–65.

—, (1984), "Action Plan for Biosphere Reserves", *Nature and Resources*, Vol. XX (4), October-December, 1984, pp. 11–22, UNESCO, Paris.

Bhandari, M.M., (1978) *Flora of the Indian Desert*, Scientific Publishers, Jodhpur, 1–471.

Department of Science and Technology, Government of India, (1981), Report of the Working Group on Arid Zone Research, Mimeo, pp. 1–23 1.

Gupta, R.K. and Prakash, I., (1975), *An Environmental Analysis of the Thar Desert*, English Book Depot, Dehradun, pp. 1–484.

Maan, H.S. (ed.), (1980), *Arid Zone Research and Development*, Scientific Publishers, Jodhpur pp. 1–531.

Meena, V.P., (2000), *Desert Ecology*, Pointer Publishers, Jaipur. pp. 1–192.

Paroda, R.S., Mann, H.S. and Verma, C.M., (1980), CAZRI, Jodhpur, pp. 1–381.

Project Document, 7, (1988), Ministry of Environment and Forests, Government of India, pp. 1–100.

Rajasthan District Gazetteers: Jaisalmer, (1973), Directorate District Gazetters, Government of Rajasthan, Jaipur, pp. 1–381.

Rajasthan District Gazetteers: Jodhpur, (1979), Directorate, District Gazetteers, Government of Rajasthan, Jaipur. pp. 1–562.

Rajasthan District Gazetteers: Barmer, (1962), Directorate, District Gazetteers, Government of Rajasthan, Jaipur. pp. 1–227.

Sen, A.K., (1972), *Agricultural Atlas of Rajasthan*, ICAR, New Delhi, pp. 1–51 (Plates 34.)

Shankarnarayan, K.A., (1987), "Eco-regeneration and Sympathetic Resource Utilization in Desert Biosphere Reserve in the Thar Desert", *Biosphere Reserves: Proc. First National Symp*, pp. 1952–14.

Tod, James., (1914), *Annals and Antiquities of Rajasthan*, Vol. I, H. Routledge and Kegan Paul Ltd., London.

Whistler, H., (1938), "The Ornithological Survey of Jodhpur State", *J. Bombay Natural History* Society, 40: pp. 213–235.

20 Forest Productivity in Surplus and Deficit Water Resource Regions

A Comparative Study of Jhabua and Mandla Districts of Madhya Pradesh

S.P. Singh

Forest management is a universal and ancient human activity. However, it was only until UNCED—Earth Summit (Rio-de Janerio, Brazil, 1992) that a large number of participating commonly perceived the need for and the concept of Sustainable Forest Management (SFM).

Forest health and efforts taken either by the government or by the community is how to increase the forest productivity. The main emphasis and the major issue is concerned with forest cover and associated factors pertaining to enhancing the quality of forest products and productivity. For the ecological sustainability the forest health plays key role in sustaining the livelihood options and economic sustainability of the peoples living in the forest fringe areas or living in the adjoining regions. Tropical forests are vital for social, economic and ecological reasons. They play an important role in ecosystem processes (such as biogeochemical and hydrological cycles), they provide habitat for wildlife and serve as source of biodiversity, and they offer protection against soil erosion (Kermen et al., 2000; Sala et al., 2000; Pandey, 2001). In this era of global warming (Forest et al., 2002), tropical forest help mitigate the effects of climate change (Phillips et al., 1998; Schimel et al., 2001) and maintain biodiversity and ecosystem functioning.

Study conducted by Sankar et al. (2000) has tested and developed criteria and indicators for sustainable management of Tropical Plantation Forests in India. The Report finds out the relevance of eight set of criteria and 43 indicators in context of sustainable plantation. Singh and Singh (2006) found that there was an impact of joint forest management on ecological sustainability and suggested measures to tackle down the issues hindering ecological sustainability in the study area. The important issues were related to incidences of forest fire, grazing, encroachment etc. In another study, Singh and Singh (2006) found that participation of community member in controlling fire grazing, encroachment etc., is significant and the permanent pasture grazing land has increased.

Thus, it becomes essential to study and test the selected variables in the surplus water resource and deficit water resource region and its impact on the forest health and forest productivity in the study area. The following are the objectives of this paper.

Objectives

1. Identification of surplus and deficit water resource region.
2. To study the forest health, forest covers and forest productivity in the water surplus and deficit regions in the study area.
3. To assess the impact of water resources on permanent pastures, grazing land and status on land put to other uses in the study area.
4. To compare the collection of forest products and associated variables in the water surplus and water deficit region and its impact on economic condition of people.
5. To suggest on the basis of facts and findings.

Methodology

The present study is a field based one, therefore qualitative and quantitative assessment was done during the intensive village survey of Jhabua and Mandla district from Madhya Pradesh state. Both primary and secondary data was used in this paper.

Primary source of data is collected by using ecological quadrate method in which 10 × 10 m of plots were laid down in the selected Forest Protection Committees (FPCs) and Village Forest Committees (VFCs) in study area of Jhabua and Mandla district of Madhya Pradesh, to assess the tree density in the survey area.

Categorization of Water Surplus and Water Deficit Region

In the present study, based on the average annual precipitation received in the region are the criteria for the demarcation of water deficit and water surplus region.

Based on the average annual rainfall Jhabua district is selected as deficit water resource region and Mandla district as surplus water resource region.

The second criteria was based on the number of existing water bodies, availability of water resources in number of months and throughout the year in major/tributaries of Narmada River of Madhya Pradesh.

Comparative analysis was made for the selected attributes, to find out the changes in two districts in the condition of forest, Non-Timber Forest Production (NTFP) collection, income, population, cattle population and changes in the land used pattern in the study area.

Variables and Indicators Used in The Study

Eight variables and indicators were used in the study to meet with the objectives. These variables are related to changes in the forest area, population, rainfall, land use pattern, changes in level of peoples' participation in forest conservation, protection and development and changes in per capita forest area.

The Study Area

Jhabua district lies between 21°55′ to 23°15′ North latitudes and 74°2′ to 75°12′ East longitudes. The district is located on the south-western border of Madhya Pradesh and forms the core of western tribal belt of India. The total geographical area of the district is 6,782 sq km; which constitute 1.5 per cent of total area of the state.

The details of the selected forest ranges and VFCs from Jhabua and Mandla districts were given in Table 16.

Mandla district is situated on the east-central part of the state and lies between 22°12′ to 23°11′ North latitudes and 80°18′ to 81°51′ East longitudes. Its extreme length is about 133 km from north to south, and extreme breadth 182 kms from east to west. The tropic of cancer passes through the north of the district.

From the study area four VFCs (25 per cent of the total VFCs) were taken for in-depth analysis from Moti Nala forest range, and one FPC from Mawai Range, division of Mandla district of Madhya Pradesh. The study area, Moti Nala of Mandla district is 170 km from Jabalpur city. Mandla district is located near by the brim of Narmada river. Mandla district has good flora coverage, because of good moisture retainment in the soil, good water status. Moti Nala is 75 km from Mandla district. At the time of survey there were 95 VFCs in Mandla district and 16 VFCs in Moti Nala forest division.

Analysis

Table 1 shows the demarcation of surplus and deficit water resource regions in the study area based on average rainfall, location and catchments areas of rivers.

It is evident from the Table 2 that the geographical area of water surplus region of Mandla district is higher than the water deficit water region in Jhabua district. The average elevation from sea level was recorded highest in Mandla district while the physical appearance of Jhabua district is undulating plateau with steep hilly terrain, eroded slopes and degraded forest. In water surplus regions of Mandla district, it forms ridge system. It is also evident from the table that the forest cover in surplus water resource region of Mandla district is higher than the deficit water resource region, Jhabua district.

Table 3, indicates the changes in forest cover, in study area of Jhabua and Mandla district. The per capita forest area has slightly changed in Madhya Pradesh due to the formation of Chhattisgarh state, as the total population has now split up into two states, while per capita forest cover in Mandla district has registered a change by 0.003 sq km due to formation of other district from Mandla territory.

Table 1
Profile of the Study Area

Attributes	*Jhabua (water deficit region)*	*Mandla (water surplus region)*
	Criteria for demarcation of surplus and deficit water resources	
Average Rainfall (in mm.)	845 mm	1292.3 mm.
Location longitude and latitude	21°55′ to 23°15′ North latitudes and 74°2′ to 75°12′ East longitudes.	22°12′ to 23°11′ North latitudes and 80°18′ to 81°51′ East longitudes.
Rivers	The Narmada and the Mahi. None of the rivers flowing through the area are perennial, excepting the Narmada which has no practical significance for the area, while all the rivers convert into dry beds during summer with some ponds.	Catchments of river Narmada and its numerous tributaries. The important tributaries of Narmada are the Banjar, the Halon, the Burner, the Kharmer, the Balai and the Hingna. The basin of Banjar has rich cover of sal forests and the richest rice tract, a level country for about 18 km above Mandla where it joins the Narmada. Rivers Burner and Halon flow through deep rocky beds in wild and undulating jungles. Halon also waters plains of Raigarh-Bichhia. River Kharmer waters the plains in middle and south of Dindhori tahsils. The Gaur river which rises in the district forms for a short distance its boundary with Jabalpur district. Alone and Thawar rivers are tributaries of Wainganga drain the Pathar area on south west district.

Table 2
Status of Geographical and Forest Area in Water Surplus and Water Deficit Region

	Jhabua	*Mandla*
Geographical area	6,782 sq km; which constitute 1.5 per cent of total area of the State.	13,269 (thousand hectares)
Average elevation from sea level	450 m	450–950 m
Physical appearance of region	An undulating plateau with steep hilly terrain eroded slopes and degraded forests.	Ridge system that runs east-west from which spurs bifurcate, on the north creating valleys of varying extents. The large valleys commence at the base of the spurs and are encompassed by the main chain of hills. These hills and the major spurs tend to have flat and extensive tops locally called dadar.
Forest area (in thousand hectares) (Based on FSI Assessment 1997)	823 sq km	5,720
Dense forest	364 sq km	4,628
Open forest	459 sq km	1,092

Table 3

Status of Forest Cover and per capita Forest Area

State/ District	Geographical area (in 000' ha)	Forest cover (in sq km)		**Population		Per capita forest cover	
		1991	1997	1991	2001	1991	2001
Madhya Pradesh	4,43,42,000	1,55,414	*1,54,508.4	6,61,81,170	6,03,85,118	0.0023	0.0025
Jhabua	6,782	1,197	*1,882.24	11,30,405	13,96,677	0.0010	0.0013
Mandla	13,269	5,720	*6,666.9	12,91,263	8,93,908	0.004	0.007

Source: * Madhya Pradesh Forest Department 1995, and FSI 1997.
** Population Census 1991, 2001.

Table 4, reveals the information of forest resources and available tree species in surplus water and deficit water regions. It is evident from the table that in Mandla district most important species, other important species and common species are higher than the Jhabua district.

Table 4

Forest Resources and Available Tree Species

Type of forest	Tree species in Jhabua	Mandla district
	Tropical dry deciduous type,	
Most important species	Teak, Tendu,	Sal (*Sal robusta*), Teak (*Tectona grandis*), Saj (*Terminalia tomentosa*)
Other important species	Mahua, Salai, Anola, Babul, Dhawara, Palash, Khair, Anjan, Teak, Bamboo Achar Ber, Tendu etc.	Tinsa (*Ongeinia dalbergioides*), Dhaura (*Anogeissus latifolia*), Dhamin (*Grewia tiliaefolia*), Bija (*Pterocarpus marsupeum*), Lendia (*Lagerstromea parviflora*), haedu (*Adina cardifolia*), Koha (*terminalia arjuna*), Palas (*Butea frondosas*), Harra (*terminalia Chebula*), Mahua (*Bassia latifolia*), Aonla (*Phyllanthus emblica*) tendu (*Diospyros tomentosa*), Khamer (*Gmelina arohorea*), Jamun (*Ugenia jambolana*) and Achar (*Buchanania latifolia.*)
Common species	Bamboo, eucalyptus, Babul,(*Acacia arabica*) Neem (*Melia Azardirachta*) are very rare but Bar (*ficus bengalensis*) Pipal (*Ficus religiosa*) are found in open country.	Some other common species are Babul, (*Acacia arabica*) Neem (*Melia Azardirachta*) are very rare but Bar (*ficus bengalensis*), Bamboo, Pipal (*Ficus religiosa*) are found in open country.

Table 5, reveals the information of human resource in the Jhabua and Mandla district. In the water deficit area the population was recorded higher as compared to Mandla district. As per Census 2001, it is also clear from the table that 0.13 per cent of population of Madhya Pradesh lives in Jhabua district and 0.08 per cent of population lives in Mandla district. The decadal growth was lower 14.66 per cent in Mandla district, as compared to Jhabua district. The sex ratio Mandla district 1,002 per thousand is higher than national sex ratio.

Table 5

Human Resource

Population (as per census 2001)	*Jhabua*	*Mandla*
Code	*23*	*42*
Tribal concentration	Highest tribal concentration in Madhya Pradesh. The tribal ratio in the district is 83.5 per cent as compared to the provincial average of 23.0 per cent	About 70 per cent consists of tribal population
Dominated tribes	The area is dominated by Bhils, Bhilalas, Patliyas	Gond, Baiga and Parikar tribes
Males	701,742 (50.24)	446,487 (49.9)
Females	6,94,935 (49.76)	4,47,421 (50.1)
Total	1,396,677 (0.13)	8,93,908 (0.08)
	(1981–1991)	(1991–2001)
Jhabua	42.16	23.56
Mandla	24.17	14.66
	Male female ratio	
Jhabua	977	990
Mandla	990	1,002
	Density	
Jhabua	167	206
Mandla	134	154

Status of Pasture and Grazing Land

Table 6, shows the depleting permanent pastures and grazing lands. It is evident that the area of permanent pastures and grazing lands was declined during 1985 to 1998–99 in the study area by –1.99 per cent in Jhabua district while –1.78 per cent in Mandla district. In context of Mandla and Balaghat region of Madhya Pradesh the permanent pastures and grazing lands has shown a contracting and declining trends (Singh, 1996). In Madhya Pradesh, it was recorded –0.48 per cent as compared to 1990.

Table 6

Contracting Status of Permanent Pastures and Grazing Land ('000' ha)

District	*1985*	*1990*	*1991–92*	*1998–99*	*(% declined)*
Jhabua	47.1 (6.93%)	42.2 (6.24%)	41.0 (6.06%)	33.53 (4.94%)	*–1.99
Mandla	41.7 (3.14%)	40.1 (3.04%)	38.4 (2.8%)	18.09 (1.36%)	*–1.78
Madhya Pradesh	Not available	2772.6 (6.27%)	2738.1 (6.17%)	2568.02 (5.79%)	**–0.48

Source: Commissioner of Land Records, M.P. per cent to total geographical area.
* as compared to 1985, ** as compared to 1990.

Table 7, shows the status of barren land and unculturable land in the study area. It is evident that the area of barren land and unculturable land had declined during 1985 to 1991–92 in the study area by –4.55 per cent on contrary it has increased in Jhabua district by +0.93 per cent in 1998–99, with a decline of –1.31 per cent in Mandla district. In Madhya Pradesh it was recorded +4.52 per cent in 1998–99, as compared to 1990. It is evident from the table that there is a direct positive correlation between water resources. The status of barren land and unculturable land in water deficit region Jhabua is higher than the water surplus region of Mandla district. This proves that there is an impact of availability of surplus water and deficit water on the changing status of barren and unculturable land.

Table 7

Status of Barren and Unculturable Land (in '000' Ha)

District	*1985*	*1990*	*1991–92*	*1998–99*	*(% declined/ increased)*
Jhabua	105.8 (15.57%)	88.9 (13.15%)	74.2 (10.98%)	111.95 (16.50%)	*+ 0.93
Mandla	64.2 (4.84%)	60.2 (4.53%)	57.5 (4.33%)	46.89 (3.53%)	*–1.31
Madhya Pradesh	Not available	2210.2 (4.99%)	2020.3 (4.55%)	4217.209 (9.51%)	** +4.52

Source: Commissioner of Land Records, M.P. per cent to total geographical area.
* as compared to 1985, ** as compared to 1990.

Table 8, shows the status of culturable waste land in the study area. It is clear from the table that the area of culturable waste land has decreased during 1988 to 1998–99 in the study area by –0.24 per cent under the category of 'A' column and +2.69 per cent under the category of 'B' column in Jhabua district, while –1.68 per cent under the category of A column and –0.17 per cent under the category of B column in Mandla district. In Madhya Pradesh it was recorded –0.15 per cent under the category of A column and –0.88 per cent under the category of B column, in the year of 1998–99 as compared to 1988.

It could be concluded on the basis of Table 6, 7 and 8, that the positive and negative trend in the percentage changes of permanent pastures and grazing land; barren and unculturable land and culturable waste land has increased in per cent area in the land put to non agricultural uses.

The impact of surplus water resource region and deficit water resource region could be viewed in terms of the livestock and animal statistics. The changes occurred during the comparative period of 1985 with the year of 1991–92 indicates that the higher the number of bovine animals is confined to the availability of fodder resource in the water surplus region as compared to water deficit region of Jhabua district.

The livestock and animal statistics (As per 1985 and 1991–1992 year) in the study area shows:

Table 8

Status of Culturable Waste Land (in '000' Ha)

District	1988		1990		1991–92		1998–99		% Change	
	'A'	'B'	'A'	'B'	'A'	'B'	'A'	'B'	'A'	'B'
Jhabua	3.8 (0.55%)	4.6 (0.68%)	3.9 (0.57%)	4.4 (0.59%)	5.3 (0.78%)	7.1 (1.05%)	2.166 (0.31%)	25.50 (3.37%)	*(–0.24%)	*(+2.69 %)
Mandla	28.1 (2.11%)	7.2 (0.54%)	20.7 (1.56%)	8.3 (0.62%)	20.1 (1.51%)	7.8 (0.58%)	5.832 (0.43%)	5.009 (0.37%)	*(–1.68%)	*(–0.17%)
Madhya Pradesh	791.8 (1.79%)	427.3 (0.96%)	749.0 (1.68%)	425.0 (0.95%)	Not available		7293.61 (1.64%)	3936 (0.08%)	*(– 0.15)	*(–0.88%)

Source: Commissioner of Land Records, M.P. per cent to total geographical area: + increase; – decrease
'A': land that can be brought under cultivation immediately.
'B': land that can be brought under cultivation after some improvement.

* As compared to 1988

The population of bull and bullocks, cows and others has increased in Jhabua district by 5.06 per cent and decreased by 6.4 per cent in Mandla district. The population of livestock has increased in Madhya Pradesh by 0.89 per cent.

In case of he buffalo, she buffalo and others has increased in Jhabua district by 15.19 per cent and decreased by 0.33 per cent in Mandla district. The population of livestock has increased in Madhya Pradesh by 7.4 per cent.

There was increase in the population in the same year of sheep, goats, horse, pigs others in Jhabua district by 20.1 per cent and decrease by 5.59 per cent in Mandla district. The population of livestock has increased in Madhya Pradesh by 1.6.

The livestock population in Jhabua and Madhya Pradesh is increasing. This increasing livestock population has an impact on ecological sustainability in the study area. The per capita availability of permanent pastures and grazing lands in the study area of Jhabua and Mandla district has decreased.

Table 9, shows the per capita availability of permanent pastures and grazing land per thousand cattle. It is clear from the table that per capita area in hectare has decreased in Jhabua and Mandla district respectively, which also indicates increasing pressure on permanent pastures and grazing land. The per capita area for bull, bullocks cows and others has decreased by 0.0147 ha; for he/she buffalo and others by 0.1354 ha; and for sheep, goats, horse, pigs and other by 0.0639 ha in Jhabua district, while in Mandla district, the per capita area for bull, bullocks cows and others has decreased by 0.0012 ha; For he/she buffalo and others by 0.0262 ha; and for sheep, goats, horse, pigs and other by 0.008 ha. It is evident that reducing per capita area in hectares in future may become an alarming situation.

Table 9

Per capita Availability of Permanent Pastures and Grazing Land

Per capita availability for livestock	*Jhabua (water deficit region)*			*Mandla (surplus water region)*		
	1985	*1991*	*Changes*	*1985*	*1991*	*Changes*
Area of permanent pastures and grazing lands (in hectares)	47100	41000	–6100	41700	38400	–3300
Per capita area for bull, bullocks, cows and others (in hectares)	0.0887	0.0733	–0.0147	0.0711	0.0699	–0.0012
Per capita area for he buffalo/she buffalo and others (in hectares)	0.5175	0.3821	–0.1354	0.3446	0.3184	–0.0262
Per capita area for sheep, goat, horse, pigs and others (in hectares)	0.2091	0.1452	–0.0639	0.3237	0.3157	–0.008

Above Table also indicates that in the water surplus region of Mandla district the per capita area available to bovine animals is lower as compared to water deficit region of Jhabua district because of good growth and productivity of fodder grass resources per hectare. Whereas such growth and productivity of fodder grass even is not recorded 20 per cent per hectare area in Jhabua district.

Status of Forest Fire and Damages Caused by Forest Fire on Forest Health and Productivity

For the ecological sustainability the control on forest fires are very important. From the records of forest department there is not a recorded major forest fire. Some of information obtained from the selected VFCs/FPCs in the study area of Jhabua and Mandla district reveals that small fire incidents had taken place. The detail are given in Table 10.

Table 10

Number of Forest Fire Incidents Taken Place and its Impact

Impact on	*Jhabua*	*Mandla*	*changes*
Incidences of forest fires	15	7	–8
Destroys the new recruits and regeneration (area in hectares)	45	38	–7
Burns other trees (number of burnt trees)	120	150	30
Not obtaining fuel wood from the effected area (per cent of responses)	90	95	5
Non-availability of fodder and grasses till first rainfall (per cent of responses)	85	90	–5
Disappearance of wild animals (per cent of response)	75	85	–10
Clears the way in forest because of leaf litters in forest (per cent of response)	50	45	–5

Source: Survey and Discussion with VFCs/FPCs members.

It is clear from the table that incident of forest fire was recorded highest in water deficit region and lowest in Mandla district. Study conducted by Singh et al. (2006) suggested that the incidents of forest fires and damages caused by fire could be controlled through community participation and effective involvement of committee members. The major reasons of the forest fire is man made, which generally occurs in the collection period of Non-Timber Forest Products (NTFPs) and Non-Wood Forest Products (NWFPs) by throwing the lighted end tails of Indian rural cigarette (*Bidi*). Most of fire incidence take place in the summer season due to high temperature because of combustion (Negi, 2000) which results in low moisture containment.

Impact of Forest Fire

The impact of forest fire is given Table 10. It is clear from the table that forest fire has destroyed almost 95 per cent of new recruits and regeneration in the 45 and 38 hectares of land in the study area of Jhabua and Mandla district. Another impact of forest fire is seen on the number of trees burnt which were recorded 120 and 150 in the study area. The impact of forest fire is also seen on less quantity of NTFP collection in the study area of Betul division and how the less quantity of collection has resulted in the monetary loss of Rs 60,000 (Singh et al., 2006). The impact of forest fire is also seen on "not obtaining fuel wood from the effected area"; 92.5 per cent of people have supported this statement. Whereas 87.5 per cent of people responded in favour of that "fodder and grasses are not available till the first rainfall"; on the other hand 80 per cent of people viewed that because of fire the wild animals start disappearing from the effected site and shift to other dense forest patches. The 42.5 per cent lowest responses were recorded as clears the way in forest because of leaf litters It could be concluded that forest fire is one of the important hindrances in the ecological sustainability. The control of forest fire is essential for ecological sustainability (Artsybashev, 1983; Singh et al., 2006).

Employment Opportunities Provided by the Forest Department in the Study Area

Table 11, shows the condition of employment in the study area, after Joint Forest Management (JFM) programme. It is evident from the table that in Jhabua district all types of works continued up to 117 days and 90 people have obtained employment as labour in the selected VFCs/FPCs. People's have generated wages worth of Rs 1.83 lakh. While all types of the work continued for 170 in Mandla district and 144 people have obtained employment as labour

Table 11

Employment Opportunities Provided by the Forest Department in the Study Area after JFM

Description of work	*Jhabua (Water deficit region)*			*Mandla (Water Surplus region)*		
	No. of days work obtained	*No. of peoples*	*Wages earned (in Lakh Rs)*	*No. of days work obtained*	*No. of people*	*Wages earned (in Lakh Rs)*
Construction work	60	45	1.35000	60	48	1.44000
Plantation work	20	15	0.15000	20	19	0.19000
Nursery development work	7	8	0.02800	10	12	0.06000
Land development work	30	22	0.33000	30	20	0.30000
Works related to Khudai and removal of weed,				20	15	0.15000
Felling of *sal borer* affected trees	–	–	–	10	18	0.09000
Road repair work	–	–	–	20	12	0.12000
Average income of the people (in Rs)			2064			1631.94
Total	117	90	1.858	170	144	2.35

Source: Survey

in the selected VFCs/FPCs and people have generated wages worth of Rs 2.35 lakh. The wages earned by the people as labour is higher by Rs 0.492 lakh (20.9%) as compared to Jhabua district. The average income of the people earned from wages in Jhabua district is lower by Rs 433 (20.9%) as compared to Mandla district. It is also evident that after JFM programme, the role of forest department has emerged as key in providing employment opportunities thus keeping a check on the migrants from the district.

Economic Sustainability from Fuel Wood and Pattern of Rural Energy Consumption in the Study Area

Fuel wood is derived mainly from forest and the key to its production lies in forest policy and management. Earl describes fuel wood as 'forest energy' and refers to the 'forest capital stock' and 'forest productivity'. A section of his book *Forest Energy and Economic Development* is devoted to so-called natural forest types and uses a map of forest types to make an estimate of the renewal and removal of energy stored in the world's forest resource (Earl, 1975). The FAO report on food and agriculture for 1976 also refers to the cutting of wood from the world's 'Forests' and use of half of it as fuel wood (FAO, 1977a)

Although enough work has been done on rural energy, yet a detailed comprehensive analysis is required to understand the factors which influences the rural energy consumption, such as economic, social, and technical factors, but also agricultural and environmental conditions, which must all be taken into consideration, when supply and demand analysis is required.

Thus, assessing economic sustainability of fuel wood is very important in context of rural energy consumption in the study area of Jhabua and Mandla district of Madhya Pradesh.

Fuel Wood Collection by the Households

Table 12, reveals the attributes related to collection of wood fuel in the study area of Jhabua and Mandla districts of Madhya Pradesh.

It could be concluded on the basis of the table that less distance was travelled by the people in the surplus water resource region. The 2 per cent of the households has been reduced in context of wood fuel collection. The gender participation in collection of wood fuel was recorded highest as compared to Jhabua district both in case of male and females. The weekly and yearly trips were recorded highest in Mandla district. It is also evident from the table that 50 per cent of less time was recorded as average time spent per household in collection of wood fuel. The quantity collected and brought per trip by the household was higher in Mandla district. The average collection of wood fuel in a year by the sample households recorded was 1,820 quintals by 50 households in the study area. The increase in collection of wood fuel per household also indicates the level of extraction of wood fuel from the forest and if estimated at the market price the annual extraction of wood fuel per household could be assessed in money value.

Table 12

Fuel Wood Collection by the Households

Non-wood forest products	*Jhabua (water deficit region)*	*Mandla (water surplus region)*
Distance from forest (in km)Average Distance	1.3	0.800
Number of households collecting wood fuel	50	50
No. of family members going to collect wood fuel		
Male	37	40
Female	145	158
Number of weekly trips	3.0	3.5
Number of yearly trips	156	182
Average time spent per household in collection (in hrs)	4.0	2
Quantity of fuel wood brought per trip by household (in kgs)	18	20
Weekly collection of wood fuel by household (in quintals)	0.54	.70
Average collection of fuel wood per household annually (in quintals)	27.0	36.4
Average collection of fuel wood by sample households annually (in quintals)	1350.0	1820.0

Source: Survey.

Table 13, gives the information about the extracted wood fuel by the household in a year from the forest in Rupees, if estimated at the base year price of the local market. It could be concluded from the table that the wood fuel extracted before JFM was Rs 1.35 and 1.82 lakhs in Jhabua and Mandla district of Madhya Pradesh, while the per household earning if estimated was Rs 0.027 and 0.0364 lakh in Jhabua and Mandla district of Madhya Pradesh respectively.

Table 13

Estimated Value of Wood Fuel Extracted from Forest at Market Price (in Lakh)

Attributes	*Jhabua (Water deficit region)*	*Mandla (Water surplus region)*
Market value of fuel wood extracted from forest @ Re 1.00 kg annually the households (Base year price 1995).	1.35	1.82
Per household earning indirectly from fuel wood collection in lakh Rs (if valued)	0.027	0.0364

Source: Survey.

The households earning has increased by Rs 0.3405 and 0.3888 lakh in the study area of Jhabua and Mandla district of Madhya Pradesh. The overall estimation of income has increased by 17.5 per cent. While the per capita income estimated from the removal of wood fuel removal from forest has increased to Rs 0.0345 and Rs. 0.0450 lakh Rs in Jhabua and Mandla district of Madhya Pradesh. The per capita income has increased overall by 20.2 per cent. The estimation of income from the extraction of wood fuel is free to the people.

Fuel Wood Trade

The forest energy resources include all woody plants that provide fuel. Eight categories may be defined: fuel forests, timber forests, shelter forests, sparse woods, shrub areas, economic forests (excluding timber) homestead gardens and orchards.

Three parts of the tree can be used as fuel; the branches, top of the trunk, and bark of the main trunk. The quantity of fuel wood obtained from each tree depends on its variety. In general, broad leaf trees have more branches and therefore yield more fuel wood than coniferous trees. bark is 10 per cent of the trunk in weight.

Some of the study deals in the trade of wood fuel and its impact on the livelihood of fuel wood seller. Study concluded that alternative rural energy system need to be developed at low or cheap cost so that the withdrawal of wood fuel should be checked and depletion of the forest resources be controlled (Prasad et al., 2001; Bhattacharya et al., 2001).

It could be concluded from Table 14 that 45 per cent of the total earning from the sale of wood fuel comes to their hand in the month of rainy season (because of high prices of wood fuel as compared to other season.) while other 55 per cent of income earned from two other seasons. Average annual income, per household from the sale of wood fuel was recorded highest in the water surplus region of Mandla district as compared to Jhabua district

Table 14

Income Generated from the Sale of Wood Fuel per Households

Attributes	*Jhabua (water deficit region)*	*Mandla (water surplus region)*
Number of households selling wood fuel.	30	35
Number of trips per week to sell wood fuel	2.8	2.9
Yearly trips to sell wood fuel	145	150
Wood fuel selling per trip in a week (in quintals)	6.0	7.2
Average distance travelled by the fuel wood seller (in km)	7.2	6.8
Income earned by selling wood fuel to the households per week @ of Rs 1.50 per kg (in Rs) Base year price 1995.	900	1080
Annual income from the sale of wood fuel (in lakhs)	1.30	1.62
Net income from the sale of fuel wood per household per year (in Rs)	4,333.0	4,628.0

Source: Survey, Data is based on the information as stated by the households.

Fuel Wood Consumption

Table 15, shows the fuel wood consumption in the study area of Jhabua and Mandla district of Madhya Pradesh.

Table 15

Fuel Wood Consumption by the Households

Attributes	*Jhabua (Water deficit region)*	*Mandla (Water surplus region)*
Average fuel wood consumption per household daily(in kg)	6.5	7.5
Average fuel wood consumption per household daily during winter season 4 months (in kg)	8.5	12.5
Difference in consumption of wood fuel as compared to other seasons	+2.0	+2.5
Per capita per year collection (in quintals)	4.029	5.83
Per capita consumption of wood fuel in 8 months in quintals	2.32	2.88
Per capita consumption of wood fuel in 4 months during winter season (in quintals)	1.51	2.00
Per capita consumption per year (in quintals)	3.83	4.88
Balance wood fuel as stock with households after consumption (in quintals)	0.199	0.95
Per capita per day in 8 months in kg	0.970	1.20
Per capita per day for 4 months during winter season in kg	1.26	2.00

Source: Survey

It could be concluded that 7.5 kg of wood fuel was consumed by per household in the study area of Mandla district which was recorded higher than the Jhabua district.

The per capita per day consumption of wood fuel was 0.970 and 1.20 kg in other seasons or 8 months in Jhabua and Mandla district respectively. The per capita per day consumption of wood fuel was 1.26 and 2.00 kg in winter season or 4 months in Jhabua and Mandla district respectively.

The per year per capita collection and per capita consumption was recorded higher in winter season as compared to other seasons in the water surplus region of Mandla district. The balance wood fuel has been available as stock with households after consumption.

Non-Timber Forest Products or NWFP Non-Wood Forest Products

The role of Non-Timber Forest Products (NTFPs) is growing globally (Grimes et al., 1994; Perez and Arnold, 1996; Wickens, 1991). However, most of the 'developing countries', central policy still categorizes NTFPs as 'minor' forest products, resulting in less emphasis upon these products than upon timber within forest management programmes and policies. Yet NTFPs are frequently the primary motivating factors for local emphasis and underscores contrasting valuation of NTFPs between policy makers and local people. The government have undervalued forest, or focused only on timber, and this has led to clearance or degradation of forests.

NTFPs are seen to be important in three crucial aspects (Arnold and Perez 1996): (a) sustainable use of forests, (b) the livelihood systems of very large numbers of people, and (c) meeting commercial demands. Theoretically, the above three issues are important for any government, and first two should be more important; the practice, however, is contrary. The third has always been the major concern of many governments in developing countries (Falconer, 1990). Most of the world's forest is under direct control of governments and the rest is also heavily influenced by government policy (Repetto, 1988) and so government's value judgments for setting management priorities.

In this context the protection rights of green consumers and redressal of grievances were stressed by (Singh et al., 2006). The quantum and quality of collection of NTFPs varies in Jhabua and Mandla district which has a direct impact on the livelihood and income generation of the people. Taking some threads from the earlier work done by the author, attempts to find the collection, consumption and pricing of NTFP, distance travelled, time taken etc., variables in water surplus and water deficit regions (Singh, 2004)

Collection of NTFPs and Other Forest Products

Table 16, indicates the changes in collection of NTFPs and other forest products in the study area of Jhabua and Mandla district of Madhya Pradesh.

Table 16

Collection of NTFPs and Other Forest Products in the Study Area (in qt.)

Non-wood forest products		*Jhabua (water deficit region)*	*Mandla (water surplus region)*	*Difference of amount collected higher in the water surplus region*
Mahua flowers		124	158.0	+ 34
Mahua seeds		115	130	+15
Chironjee		2	2.8	+0.8
Aonla		6.0	15	+9
Sal seeds		—	140	+140
Harra		—	58	+58
Baheda		—	60	60
Bai bedang		—	72	72
Gums		0.25	5.0	+4.75
Bel patra		2	15	+13
Fodder/grasses		3516	45	+10
Chakoda		—	23	+23
Medicinal Plants	Karriya	—	0.257	+0.25
	Jal Peeri	—	0.30	+0.30
	Van Jeera	—	1.18	+1.18
	Safed Musali	0.05	0.20	+0.15

Source: Survey.

Valuation of NTFPs and other Forest Products

Table 17, shows the estimated income from the collection of NTFPs and other forest products from forest in the study area of Jhabua and Mandla district of Madhya Pradesh. The debate on valuation of nature and the environment, green national accounting and economic growth is a classic and important topic in environmental economics. (Ekko et al., 2001). Discussion on the definition and meaning of national income began at the beginning of the twentieth century with the work of (King, 1919) and Pigou's famous, *The Economics of Welfare* (Pigou, 1920).

Table 17

Estimated Income from NTFP and Other Forest Products (in lakh Rs)

Non-wood forest products.		*Jhabua*	*Mandla*
Mahua flowers		0.744	0.948
Mahua seeds		0.92	1.04
Chironjee		0.09	0.126
Aonla		0.003	0.0075
Sal seeds		—	1.12
Harra		—	0.464
Baheda		—	0.36
Bai Bidang		—	0.288
Gums		0.007	0.14
Bel patta/Mahul patta		0.006	0.045
Fodder/grasses		0.028	0.036
Chakoda		—	0.069
Medicinal Plants	Karriya	—	0.0033
	Jal Peeri	—	0.003
	Van Jeera	—	0.014
	Safed Musli	0.0015	0.80
	Total	1.7928	4.6518

Source: Survey—Per capita income is estimated based on the price in local market

Fabricant (1947) argued about the depletion of natural resources should be taken into account as a cost component in the system. The debate renewed at the end of the 1960s when serious concerns grew about pollution and the exhaustibility of resources. Especially, Mishan's classic contribution, *The Cost of Economic Growth* (Mishan, 1967) induced a lot of discussion. He also challenges the traditional way to measure welfare or utility through the concept of national income because its failure to include the deterioration of the environment. The concept of how to make correction in national income, in this regards (Hueting, 1991) argue that the valuation exercises should explicitly account for future generations and their rights on particular environmental services. This suggests that the concept of a sustainable or green national income is also appropriate from a green national product.

It could be concluded from the table that collection of NTFP and other products from the forests was recorded highest in water surplus region of Mandla district as compared to Jhabua district.

Table 18, shows the per capita collection of NTFPs and other forest products in the study area of Jhabua and Mandla district of Madhya Pradesh.

Table 18

Per capita Collection of NTFP and other Forest Products (in qt.)

Non-wood forest products.		*Jhabua (Water deficit)*	*Mandla (Water surplus)*
Mahua flowers		1.078	1.180
Mahua seeds		0.958	1.0
Chironjee		0.05	0.082
Aonla		0.16	0.32
Sal seeds		—	0.82
Harra		—	0.55
Baheda		—	0.53
Bai bedang		—	0.73
Gums		0.010	0.11
Bel patra		0.017	0.33
Fodder/grasses		0.32	0.42
Chakoda		—	0.37
Medicinal Plants	Karriya	—	0.031
	Jal Peeri	—	0.021
	Van Jeera	—	0.047
	Safed Musli	0.003	0.0061

Source: Survey.

The per capita collection of NTFP was recorded highest in all NTFPs. This increase in per capita collection has a impact on per capita earning of people in the study area of Jhabua and Mandla district of Madhya Pradesh.

According to Campbell (1988, cited in Tiwari and Campbell, 1995), small scale forest based enterprises, many of which rely on NWFPs, provide 50 per cent of the income for about 25 per cent of India's rural labour force.

At national level over 50 per cent of forest revenue and about 70 per cent of forest export revenue comes from NWFPs mostly from unprocessed and raw forms (Tiwari and Campbell, 1997; Prasad, Shukla and Bhatnagar, 1996).

Table 19, shows the per capita income if valued at base year market price from NTFP and other forest products. It is evident from the table that per capita income was recorded highest in NTFPs by more than 2.57 times in water surplus region of Mandla district as compared to water deficit regions.

Table 19

Per capita Income if Valued at Base Year Market Price from NTFP and other Forest Products (in Rs)

Non-wood forest products.	*Jhabua (water deficit)*	*Mandla (water surplus)*	*Difference*
Mahua flowers	646.0	708.0	+62
Mahua seeds	766.0	800.0	+34
Chironjee	225.0	369.0	+144
Aonla	8.0	16.0	+8
Sal seeds	—	656.0	+656
Harra	—	440.0	+440
Baheda	—	318.0	+318
Bai bedang	—	292.0	+292
Gums	28.0	297	+269
Bel patra	5.0	99.0	+94
Fodder/grasses	25.0	33.0	+12
Chakoda	—	111.0	+111
Medicinal Plants Karriya	—	40.0	+40
Medicinal Plants Jal Peeri	—	20.0	+20
Medicinal Plants Van Jeera	—	58.0	+58
Medicinal Plants Safed Musli	30.0	266.6	+236.6
	1733.0	4523.6	+2790

Source: Survey.

Per capita income is estimated based on the price in local market.

Pricing of NTFPs and other Forest Products

Table 20, shows the price of NTFP and other forest products in the study area of Jhabua and Mandla districts of Madhya Pradesh.

Table 20

Pricing of NTFP and other Forest Products. (Rate per Quintal)

Non-wood forest products	*Jhabua (water deficit)*	*Mandla (water surplus)*
Mahua flowers	600	600
Mahua seeds	800	800
Chironjee	4,500	4,500
Aonla	50	50
Sal seeds	—	800
Harra	—	800
Baheda	—	600
Bai bedang	—	400
Gums	2,800	2,700
Bel patra	300	300
Fodder/grasses	80	80
Chakoda	—	300
Medicinal Plants Karriya	—	1,300
Medicinal Plants Jal Peeri	—	1,000
Medicinal Plants Van Jeera	—	1,250
Medicinal Plants Safed Musli	3,000	4,000

Source: Survey.

Market Price in the nearest local market .

It could be concluded from the table that the prices have increased in the local market of NTFP and other forest products. There is no price variation in the study area before and after JFM programme, on the contrary the price has increased in the same proportion as seen in the districts. The increase in prices per quintal after JFM has economic impact on people's earning in the study area.

Time, Distance and Mode Factor in NTFPs Collection

Table 21, reveals the average time taken in hours and distance travelled by the people to collect NTFPs and other forest products.

Table 21

Average Time Taken and Distance Travelled by the People to Collect NTFP and other Forest Products Before JFM

Non-wood forest products (in Rs)		Jhabua		Mandla		Average	
		Time in hrs	Distance traveled (km)	Time in hrs	Distance travelled (km)	Time in hrs	Distance travelled (km)
Mahua flowers		4 to 5	7 to 8	4 to 5	5 to 6	4 to 5	6 to 7
Mahua seeds		4 to 5	7 to 8	4to 5	5 to 6	4 to 5	6 to7
Chironjee		5 to 6	6 to 7	5 to 6	6 to 7	4 to 5	6 to7
Aonla		4	5 to 6	5	4 to 5	4.5	5.5
Sal seeds		—	—	5 to 6	6 to 7	5 to 6	6 to 7
Harra		—	—	4 to 5	7 to 8	4 to 5	7 to 8
Baheda		—	—	4 to5	7 to 8	4 to 5	7 to 8
Bai bidang		—	—	4 to5	7 to 8	4 to5	7 to 8
Gums		6 to 7	10 to 12	6 to 7	9 to 12	6 to 7	9 to 12
Bel patra/mahul patta		3 to 4	5 to 6	3 to 4	4 to 5	3 to 4	4 to 5
Fodder/grasses		5 to 6	8 to9	5–6	5 to 6	5 to 6	8 to 9
Chakoda		—	—	4 to 5	5 to 6	4 to 5	5 to 6
Medicinal Plants	Karriya	—	—	3 to 4	6 to 8	3 to 4	6 to 8
	Jal Peeri	—	—	4 to 5	5 to 6	4 to 5	5 to 6
	Van Jeera	—	—	5 to 6	8 to 9	5 to 6	8 to 9
	Safed Musli	5 to 6	9 to 10	3 to4	4 to 6	5 to 6	9 to 10

Source: Survey.

Time Factor in Hours

It could be concluded from the table that people were taking less time in hours in collection of NTFPs in Mandla district as compared to Jhabua district. The declining trend of time frame is a good indicator of availability of NTFPs and other forest products in abundance in the water surplus area.

Distance Factor in Kilometres

It could be concluded from the table that people are travelling in kms to collect NTFPs and other forest products in the study area of Jhabua and Mandla district of Madhya Pradesh. The decline in the distance travelled in km indicates availability of NTFPs and other forest products in forest area in the study area.

Conclusions

It is evident from the above analysis that the water resources have direct impact on the health and forest productivity. The forest cover and tree density was found in very good condition in the surplus water resource region of Mandla district as compared to deficit water resource region of Jhabua district. Most important species and other important tree species were recorded higher in the water surplus region, while the quantity and quality of tree species varies from the deficit water resource region.

The conversion of cultivable soil into barren or waste land was seen very low in water surplus regions as compared to water deficit region.

The availability of water in rivers and stop-dams helps in maintaining ecological balance in the ecological degraded zone by maintaining the disturbed water cycle, which in turn depends upon the precipitation, temperature etc. Many studies related to impact of forest on rainfall and temperature reveal that the forests are very important in maintaining moderate ecology. Most of the rivers which flow in the rainy seasons, later dried up in December or January. The water availability in the rivers and stop-dams has improved in the study area but not up to the people's expectations, due to unpredictable rainfall and unsustainable water harvesting practices. Many conflicts have taken place over the use of water from river and stop-dams for irrigation purpose; problem of water will become intense for cattle population. Therefore, new techniques as suggested by the people should be incorporated for storing water in stop-dams, water availability in number of days should be assessed and use of water per hour by the farmers should be fixed as per power of engine/pump. It is also suggested that effective soil and water conservation works should be taken up as per people's priority and they should be trained in aspects of water conservation works. However, engineering measures would be helpful in soil and water conservation work.

The per hectare productivity of fodder and grasses could be increased in the available land area for pasture and grazing land by modern agricultural inputs, which may yield high productivity. The issues related to increasing barren and unculturable waste land could be minimized by maintaining fertility and moisture containment in soil and checking soil erosion through developing micro-watershed. However, the vegetative fencing or vegetative barrier would be helpful in controlling erosion at the slope of hills in the study area to slow down the water run-off.

The increasing population is the greatest issue among all issues as it shallows the fruits of development. As per the increasing demand elasticity by the increasing population has to be controlled down by following preventive measures, family planning etc. Population of the study area has an impact on reducing per capita forest area and density is increasing per sq km which results in spatial problem. The increased forest productivity results in declined per capita availability of forest products.

In the study area the issue related to increasing cattle population could be tackled down by selecting productive cattle population. This is essential because the per capita availability of fodder and grasses per cattle is reducing, and results in the health of livestock. From the ecological point the increasing cattle is a great menace to young recruits and regeneration.

The issue related to incidence and occurrence of pest attack on certain tree species like *Sal roubsta*, *Tectona grandis*, *Buchnaian lanzan*, need to tackled down (such as sal borer, disease in sal trees, and pest attack on *Tectona grandis*,) by adopting and new techniques in pest control methods. The low cost methods of pest control in the large areas are suggested.

The volume of production of identified NTFPs was found higher in surplus water resource region as compared to deficit water resource region of Jhabua district. The regenerative capacity of important tree species, less distance travelled by the villagers, time consumed in collection of NTFPs was recorded highest in Mandla district.

Acknowledgement

Author sincerely acknowledges Prof. D.K. Bandhyopadhyaya, Director, IIFM, Bhopal, Prof. Ganesh Kawadia, Prof. and Head, School of Economics, Devi Ahilya University, Indore, for their continuous inspiration and support to come up and develop a new concept of water resources and forest productivity. This paper was presented in the One Day Panel discussion on "Natural Resource Management in the Catchments area of Big Dams" organized by School of Economics, at Devi Ahilya University, Indore on 28 November, 2006.

References

Arnold, J.E.M. and Perez, M.R., (1996), "Farming the Issues Relating to Non-timber Forest Products Research", M.R. Perez and J.E.M. Arnolds (eds.), *Current Issues in Non-Timber Forest Products Research*, Centre for International Forestry Research, Bogor, Indonesia.

Artsybashev, E.S., (1983), *Forest Fires and their Control*, Oxonian Press Pvt. Ltd.

Bhattacharya, P., Bharati Joshi, Parag Dubey, S.P. Singh, and Rishi Kapoor, (2001), "Forests and Displaced People: Fuelwood Collection and Trade as a First Step Survival Strategy, in Wood Fuel Trade in India" Food and Agriculture Organization of the United Nations, Bangkok, July, 2001. pp. 113–132, RWEDP Report No. 57.

Earl, D.E., (1975), *Forest Energy and Economic Development*, Oxford.

Earth Summit, (1992) Rio-de Janerio at Brazil UNCED.

Ekko, C., V. Ierland, J.V. der Straaten and Herman R.J. Vollebergh (eds.), (2001), *Economic Growth and Valuation of the Environment*, Edward Elgar Publishing Limited, UK.

Falconer, J., (1990), "The Major Significance of 'Minor' Forest Products", Community Forestry, Note 6. Food and Agriculture Organization of the United Nations, Rome.

Food and Agriculture Organization (FAO), (1977a), "The State of Food and Agriculture, 1976", Rome.

Forest, C.E., P.H. Stone, A.P. Sokolov, M.R. Allen, and M.D. Webster, (2002), "Quantifying Uncertainties in Climate System Properties with the Use of Recent Climate Observations", *Science,* 295: 113–117.

FSI, (1998), The State of Forest Report (1997). Forest Survey of India, Ministry of Environment and Forest, Dehradun.

Grimes, A., Loomis, S., S. Jahnige, P., Burnham, M. Onthank, K., Alarcon, R., Cuenca, W.P., Martinez, C.C., Neill, D., Balick, M., Bennett, B. and Mendelsohn, R., (1994), "Valuing the Rain Forest: The Economic Value of Non-timber Forest Products in Ecuador", Ambio 23 (7), 405–410.

Hueting, R., (1991), "Correcting National Income for Environmental Losses: A Practical Solution for a Theoretical Dilemma", in R. Costanaza (ed.), *Ecological Economics, The Science and Management of Sustainability*, New York: Columbia University Press, pp. 94–213.

King, W.J., (1919), "The Wealth and Income of the People of the United States".

Kremen, C., J.O. Niles, M.G. Dalton, G.C. Daily, P.R. Ehrlich, J.P. Fay, D. Grewal, and R.R.P. Guillery, (2000), "Economic Incentives for Rain Forest Conservation Across Scales", *Science*, 288: 1828–1832.

Mishan, E.J., (1967), *The Cost of Economic Growth*, London: Penguin Harmondsworth.

Negi, S.S., (2000), *Forest Fires*, International Book Distributors, Dehradun.

Pandey, D.N., (2001), "A Bountiful Harvest of Rainwater", *Current Science,* 293: 1763–1763.

Perez, M.R. and Arnold, J.E.M., (1996), "Current Issues in Non-timber Forest Products Research", *Centre for International Forestry Research and Development, Bogor,* Indonesia.

Phillips, O.L., Y. Malhi, N. Higuchi, W.F. Laurance, P.V. Nunez, R.M. Vasquez, S.G. Laurance, L.V. Ferreira, M. Stern, S. Brown, and J. Grace., (1998), "Changes in the Carbon Balance of Tropical Forests: Evidence from Long-term Plots", *Science,* 282: 439–442.

Pigou, A.C., (1920), *The Economics of Welfare*, London McMillan and Co.

Prasad, R., Shukla, P.K. and Bhatnagar, P., (1996), "Leaves from the Forest: A Case Study of Tendu Leaves in Madhya Pradesh", Jabalpur, Madhya Pradesh, India, Centre for Environment and Sustainable Development.

Prasad, Ram, N.B. Majumdar, P. Bhattacharya, B.R. Phukan and S.P. Singh, (2001), "Wood fuel Trade in India", in *Wood Fuel Trade in India*, Food and Agriculture Organization of the United Nations, Bangkok, July, 2001. pp. 51–76, RWEDP Report, No. 57.

Repetto, R., (1988), "Overview", R. Repetto and M. Gills (eds.), *Public Policies and the Misuse of Forest Resources,* Cambridge University Press, Cambridge pp. 1–41.

Sala, O.E., F.S. Chapin III, J.J. Armestro, E. Berlow, J. Bloomfield, R. Dirzo, E. Huber-Sanwald, L.F. Huenneke, R.B. Jackson, A. Kinzig, R. Leemans, D.M. Lodge, H.A. Mooney, M. Oesterheld, N.L. Poff, M.T. Sykes, B.H. Walker, M. Walker, and D.H. Wall., (2000), "Global Biodiversity Scenarios for the Year 2100, *Science"*, 287: 1770–1774.

Sankar, S., P.C. Anil and Amurth in collaboration with Ram Prasad, S. Raghavan, Abey George, S.P. Singh, Jyothi Krishnan. D. Duggaya, M. Mishra, B. Sreekumar and S.C. Prasad, (2000), "Criteria and Indicators for Sustainable Plantation Forestry in India Centre for International Forestry Research", (CIFOR) Bogor, Indonesia with Support from Australian Centre for International Agricultural Research (ACIAR).

Schimel, D.S., J.I. House, K.A. Hibbard, P. Bousquet, P. Ciais, P. Peylin, B.H. Baraswell, M.J. Apps, D.Baker, A. Bondeau, J. Canadell, G. Churkina, W.Carmer, A.S. Denning, C.B. Field, P. Friedlingstein, C. Goodale, M. Heimann, R.A. Houghton, J.M. Melillo, B. Moore III, D. Murdiyarso, I.Noble, S.W. Pacala, I.C. Prentice, M.R. Raupach, P.J. Rayner, R.J. Scholes, W.L. Steffen, and C. Wirth,

(2001), "Recent Patterns and Mechanisms of Carbon Exchange by Terrestrial Ecosystems", *Nature*, 414: 169–172.

Singh, R.D. and Singh S.P., (2006), "Impact of Joint Forest Management on Ecological Sustainability and Related Issues: A Comparative Study of Jhabua and Mandla Districts of Madhya Pradesh", pp. 145–152, *Indian Journal of Millennium Studies: An International Journal,* Vol. 1(1), March 2006, Serials Publications, New Delhi.

—, (2006) "Involving and Evolving: Impact on Ecological Sustainability and Related Issues", pp. 34–37, Vol. XXI No. 2., *Wastelands News*, November, 2005–January, 2006, SPWD, New Delhi, India.

Singh, S.P., Phukan, B.R, Mishra, Manish, Mishra, Apurva, (2006), "Regional Planning for Fire Prone Zones of Betul Forest Division through Community Participation", Paper presented in the 37th International Regional Science Symposium held at Indian Institute of Management, Bangalore during 6–8 January, 2006.

—, (2006), "Impact of Forest Fire on NTFPs: "A Case Study of Betul Forest Division", M.P. Souvenir and Abstract of XII Silviculture Conference (February 1–3, 2006) Organized by Forest Research Institute, Dehradun.

—, (2006), "Mapping of Forest Fire Damages and Fire Induced Disease Management Through Cartographic Techniques: A Study of Betul District, Madhya Pradesh", Souvenir & Abstract of National Seminar on Environmental Conservation and Mapping held at National Atlas and Thematic Mapping, Kolkata, organized during 2–4 March 2006.

Singh, S.P., (1996), "Why Permanent Pastures and Grazing Lands are Contracting? A Search of Responsible Factors in Balaghat and Mandla Region", Proceeding of National Workshop Environmental Impact on Human Health, (2–3 February, 1996) Organized by Maharaja Autonomous College Chattarpur. M.P.).

Singh. S.P., (2004), "Socio-Economic Issues in Sustainability of Joint Forest Management in Madhya Pradesh", Ph.D. Thesis, Department of Regional Planning and Economic Growth, Barkatullah University, Bhopal.

Singh, S.P., (2006), "A Quest for Sustainable Development and Management of Natural Resources: A Future Road Map", *Indian Journal of Development, Research and Social Action,* An International Journal, January–June, 2006 GRP, Serials Publications. New Delhi: 2 (1) 31–52.

Singh, S.P. Mishra, M. and Phukan, B.R., (2006), "Role of JFMCs in Sustainable Fire and Fire Induced Disease Management: A Study of Betul District", M.P. Report Submitted NAEB, MoEF Sponsored Project Report IIFM Bhopal.

Singh. S.P., Mishra Manish, Mishra Apurva, (2006), "Protection Rights of Green Consumers: A Case Study of Jhabua and Mandla districts of Madhya Pradesh", Published in proceeding of National Seminar on "Consumer Protection and Welfare". Organized by Department of Economics, Dr. H.S. Gour University, Sagar. 13–14 Oct. 2006. pp. 99.

Tewari, D.D., and Campbell, J.Y., (1995), "Developing and Sustaining Non-timber Forest Products: Some Policy Issues and Concerns with Special Reference to India", *Journal of Sustainable Forestry*, 3 (1); 53–77.

Tewari, D.D., and Campbell, J.Y., (1997), "Economics of Non-timber Forest Products", J.M. Kerr, D.K. Marothia, K. Singh, C. Ramaswamy and W.B. Bentley (eds.), *Natural Resource Economics: Theory and Application*, New Delhi, India, Oxford and IBH.

Wickens, G.E., (1991), "Management Issues for Development of Non-timber Forest Products", *Unasylva*, 165 (42), 3–80.

21 Implications of an 8-Year Control Programme for European Buckthorn in Saskatoon, Canada

O.W. Archibold and L. Delanoy

European buckthorn (*Rhamnus cathartica* L.) has established and formed a sizeable disjunct population in the South Saskatchewan River valley in the vicinity of Saskatoon, Saskatchewan. Although the first evidence of European buckthorn in Saskatoon was in 1923, according to records from the W.P. Fraser Herbarium at the University of Saskatchewan, it was not until the 1930s that its abundance began to increase. At this time the species was tested as a potential shelterbelt species at the federal government Dominion Tree Nursery located in what is now a suburb of the city. Although European buckthorn was never officially released as a hedge plant, because it was discovered to be an alternate host for oat crown rust (*Puccinia coronata*) (Harder and Chong, 1983), it did escape, and has become locally naturalized. Since that time the shrub has spread along the South Saskatchewan River valley up to 56 km north and 21 km south of its source location.

Degradation of natural areas due to the establishment of dense, even-aged stands of European buckthorn is of primary concern to conservation authorities (USGS, 2000). European buckthorn readily invades native woodland and once established outcompetes local species. Conservation authorities throughout North America have initiated significant campaigns to eradicate the shrub. Although predominantly located in the river valley, it has successfully established in favourable adjacent upland sites such as aspen groves.

Conservation of cultural and natural heritage resources of the South Saskatchewan River valley in Saskatoon is the mandate of the Meewasin Valley Authority (MVA). MVA is engaged in various programmes to manage remnants of native vegetation with the goal of establishing historically authentic ecosystems with high biodiversity. Although the focus is on proactive management, such as reintroducing natural disturbances to rejuvenate native vegetation, reactive weed control measures are often warranted as the first step in restoration when a species is particularly troublesome.

MVA initiated a European buckthorn control programme in 1998 and has completed its eighth year of mapping and control work. Because the species is dioecious, and because European buckthorn spreads only by seed, eradication effort is more effective if directed against the female, fruiting plants. In an attempt to contain the future spread MVA has, to date, treated over 3,47,000 fruiting stems. A consequence of treatment disturbance is the development of new fruiting stems from immature seedlings and from seed reserves; an additional 2,50,000 fruiting stems are expected to mature as a result of this initial release effect. To date MVA has not targeted male stems. Male stems, in the short-term, can act as a deterrent to maturation of immature female stems. Over the long-term, control of male stems would be necessary to effectively restore damaged natural areas and increase biodiversity.

Impact of European Buckthorn on Community Diversity

Considerable information has been generated, mainly from the US Midwest, regarding the impact of European buckthorn. Reduced recruitment of native tree species has been noted in semi-open canopy forests, such as oak woodlands in Minnesota (Pearson and Gillette, 1999), as well as the crowding out of herbaceous understory species, such as meadow-rue (*Thalictrum dioicum*), large flowered trillium (*Trillium grandiflorum*), two-leaved Solomon's-seal (*Maianthemum canadense*) and northern starflower (*Trientalis borealis*) (Gale, 2001). In Michigan, European buckthorn displaces aspen by creating a competitive multi-layered understory (Miller, 2005).

Local vegetation studies indicate the preferred niche for European buckthorn is green ash (*Fraxinus pennsylvanica*), Manitoba maple (*Acer negundo*) and aspen (*Populus tremuloides*) woodland (Delanoy, 1996; Godwin and Thorpe, 1992; Lineman, 1997; Thorpe and Godwin, 1994). In the mid-1990s it was concluded that European buckthorn was one of the most serious of the invasive exotic species in the region. European buckthorn had invaded Saskatoon's riparian woodlands and so increased in numbers to frequently become the predominant understory species in all size classes from seedling to tall shrub stages. The ability of European buckthorn to regenerate and grow under the shade of other shrubs and trees, and to out-compete and replace native species in the understory, along with its sheer numbers, raised a bleak prospect for the riparian communities which are amongst the most diverse in the region (Thompson and Hansen, 2001).

Hall (1973), made no reference to European buckthorn in the Saskatoon region, whereas Lineman (1997), ranked European buckthorn thickets as one of 8 distinct tall shrub/tree communities. Other local vegetation inventories indicate a distinct distribution pattern for European buckthorn in relation to plant community and site conditions. For example, upland sites show very high invasion rates of European buckthorn into aspen groves (Thorpe and Godwin, 1994). Elsewhere, it is most prevalent in dry forest patches characterized by open Manitoba maple and green ash (Godwin and Thorpe, 1992).

European buckthorn also is prominent in drier upslope shrub communities comprised of species such as saskatoon (*Amelanchier alnifolia*), choke cherry (*Prunus virginiana*), snowberry (*Symphoricarpus occidentalis*) and rose (*Rosa* sp.). However, European buckthorn is less common in moist forest where balsam poplar (*Populus balsamifera*) and white and river birches (*Betula papyrifera, B. occidentalis*) dominate imperfectly drained seepage sites. Similarly, Delanoy (1996), reported that European buckthorn was absent from areas of obvious high groundwater conditions.

While floodplain sites remain relatively free of European buckthorn, Hall (1973) predicted a downslope movement of plants with a drying of the river valley following construction in 1967 of the Gardiner Dam, approximately 100 km upstream of Saskatoon. The significant alteration of the flood regime of the South Saskatchewan River has been cited as a factor affecting recruitment of riparian cottonwoods (*Populus deltoides*) (Begg et al., 1998). Flooding could have been a factor in limiting spread of European buckthorn south of Saskatoon prior to 1967. European buckthorn remains relatively rare in valley bottom sites, but it has established and likely will continue to spread. This raises some concern for the Bristle-leaved sedge (*Carex eburnea*), which is ranked by the Saskatchewan Conservation Data Centre as provincially rare (Bizecki Robson and Nelson, 1998), as it is found in the understory of these sites.

Thompson and Hansen (2001), categorize forest communities for the prairie region into four climax or habitat types with ten associated seral or disclimax for the prairie region into types. Of the three local poplar woodlands—cottonwood, balsam poplar and trembling aspen—only trembling aspen is recognized as a stable climax habitat type. Cottonwood and balsam poplar are considered to be seral stages, with cottonwood evolving into yellow willow (*Salix lutea*)/red-osier dogwood (*Cornus stolonifera*) shrub communities, and balsam poplar succeeding to Manitoba maple/choke cherry or green ash/choke cherry stands. In all cases the woodland types support a rich mix of understory shrubs and ground cover herbs, although species richness declines as they approach maturity; this suggests that disturbances, such as fire, has a positive effect on biodiversity. The ash and maple forests are particularly rich with up to 31 shrub species and 125 herb species. All three climax types are susceptible to impact from European buckthorn.

Two tributary streams with sizeable riparian zones feed into the South Saskatchewan River in the area where European buckthorn is found. To the south Beaver Creek is the main drainage system for the Dundurn sandhills, a locally important natural area characterized by numerous aspen groves. Fish Creek to the north lies within the aspen parkland, a mosaic of grassland and woodland plant communities that forms a transitional zone between the prairies and the boreal forest (Archibold, 1999). The aspen parkland is likely at highest risk in terms of loss of regional biodiversity from European buckthorn. In addition to loss of biodiversity, some reduction in fall colour is anticipated, especially following displacement of green ash, poplars and choke cherry. Likewise, berry picking opportunities will also be compromised due to reduction in saskatoon and choke cherry.

Impact on Local Fauna

Further studies indicate European buckthorn has the ability to elevate soil nitrogen, pH and water retention levels and rapidly deplete the leaf litter (Heneghan et al., 2002). Species in the buckthorn family produce varying concentrations of emodin (Izhaki, 2002) which could enter the soil by way of fallen leaves and berries. Emodin is believed important in regulating fruit dispersal by birds, as well as protecting the plant from herbivores (Izhaki, 2002), and could account for the altered soil invertebrate population associated with European buckthorn patches. Earthworms are reportedly more numerous under buckthorn and are responsible for rapid decomposition of the leaf litter, even when buckthorn and native leaves are intermixed. Decomposition is so rapid that it depletes the leaf litter layer prior to fall renewal. This causes an 'ecological trap' for invertebrates with insidious implications for the forest food web (Heneghan, 2003).

Change in habitat structure as a result of establishment of European buckthorn has been implicated in increased nest predation for American robins (*Turdus migratorius*) and wood thrushes (*Hylocichla mustelina*) (Schmidt and Whelan, 1999). Loss of tall shrub types can also impact bird diversity. Hawthorn (*Crataegus columbiana*) is known to provide excellent habitat for the Loggerhead Shrike (*Lanius ludovicianus*), a threatened bird (Leighton et al., 2002). The main use of European buckthorn is by foraging birds. European buckthorn provides a heavy and reliable source of fruits especially for wintering birds, and this is the principal mechanism of seed dispersal. Browsing by white-tailed deer (*Odocoileus virginianus*) is common but not heavy. Cutting of stems by beaver (*Castor canadensis*) has been observed only once. Feeding by porcupine (*Erethizon dorsatum*) has not been observed, although porcupines are common in the valley and commonly feed on choke cherry. Overall utilization of European buckthorn by wildlife other than for berries appears low.

Treatment Methods

The combination of heavy fruit production and high germination rates allows European buckthorn to rapidly dominate an understory, and its prolific nature makes it difficult to eradicate. Various methods can be used to remove European buckthorn, but repeated treatments are required to kill mature plants and eliminate the seed bank. Fire is effective in killing buckthorn seedlings and can top-kill mature shrubs, but the species re-sprouts vigorously, and burning usually needs to be repeated annually or biennially over several years before it is suppressed. Chemical control is the most common method of managing European buckthorn, but like the other methods, a number of several treatments invariably are required for complete eradication (Heidorn, 1991).

European buckthorn was recorded at 239 land parcels around Saskatoon, of which 118 have received one or more herbicide treatments. MVA has chosen to treat only fruiting stems in an attempt to eradicate seed production. Competition from male plants can help suppress the maturation of immature females, although removal of males would be required for complete restoration. The majority of plants was treated with Garlon 4 (active ingredient triclopyr) using a streamline basal bark spray method. The herbicide was mixed with a light mineral oil and applied at temperatures as low as –15°C. The effects were monitored in treated and untreated test plots located at Sutherland Beach, an area of river terraces located within the city limits of Saskatoon, which supports some of the densest buckthorn stands in the area. Total fruit loads on the buckthorn shrubs were counted in late fall, 2005 after the plots had received their second application of herbicide.

Cost and Effectiveness of Treatment

Three principal factors have affected treatment cost in this study; these are the number of fruiting stems requiring treatment, the average cost to treat each stem, and the spatial distribution of the stems.

To calculate the number of stems needing treatment at a particular site, current and future fruiting stem production must be considered. It is widely recognized that a single treatment of fruiting stems will not provide complete seed control (Pearson and Gillette, 1999). Some fruiting stems will escape treatment and some immature female stems, seedlings and seeds will mature in the future. In this study the number of stems treated in the first pass only dropped by about 20 per cent due to vigorous population recruitment. However, about 80 per cent of the shrubs were killed by the second treatment, although inevitably this will be augmented by further seedling emergence. For the Saskatoon population it is assumed that three extensive treatments will be required before the European buckthorn seed product population is reduced to a manageable level. If herbicide is applied in a third 'follow-up' treatment, the cumulative number of stems sprayed would be approximately double the original population.

The average cost of herbicide treatment was \$ 0.18 per stem of which \$ 0.16 was for labour and \$ 0.02 for chemical. Although this was relatively cost efficient in low density sites, costs would be prohibitive in denser thickets. For example, costs for three herbicide treatments at Sutherland Beach with fruiting stem densities of over 9,600 ha^{-1} would exceed \$ 3,200 ha^{-1}, and there is every likelihood that more than three treatments would be required to eliminate the female shrubs and the potential for in situ reproduction. Although stem densities are expected to drop exponentially with successive treatments, and the remnant population will consist of small, sparsely distributed individuals, labour costs are not expected to drop as quickly. Treating low density sites effectively increases labour costs per stem because of travel time to and from the sites and time required searching for European buckthorn stems. Records from particular sites treated in this study indicate as many as 375 stems can be treated per hour in very dense sites compared to as few as 2 stems per hour in the lowest density sites.

For dense sites total restoration to a native plant community will require substantial and long-term effort. Stopping seed production is a more laudable goal for the short-term and the 120-fold reduction after two treatments noted at the Sutherland Beach site attests to the feasibility of this approach. The comparative cost of treatments in low density stands is an important consideration in determining management strategies. Ultimately, it will be cheaper and more effective to target low density sites first, because vegetation response is much more positive with a strong shift towards native species. Such an approach is supported by Moody and Mack (1988) who showed that it is more advantageous to suppress small satellite colonies of invasive species than the main population centre.

The results of this study suggest that a five-year treatment rotation should be appropriate to control the European buckthorn population. This strategy would be particularly effective if treatments were carried out initially in two successive seasons in order to address the inherently high 'miss' rate. The following phases are envisioned for eventual control and restoration of native communities:

1. Remove fruiting stems using 3–5 chemical girdling passes on a five-year rotation. This is expected to result in a stabilized plant community void of European buckthorn seed.
2. Remove immature female seedlings and saplings. Commence after two chemical girdling passes; use burning, mechanical, chemical or grazing methods on an opportunistic basis to remove European buckthorn from the ground layer.
3. Remove male stems using chemical girdling after ground layer removal is complete; start first in sites where native vegetation is still intact.
4. Restore a native plant dominant community. This stage may or may not require active seeding or planting of native species. Ideally, most sites will retain self-restorative capability.

Disturbance of the tall-shrub layer probably will result in competitive release of the ground layer with associated development of fruiting stems. Although removing European buckthorn from the ground layer is an important initiative, it can be done over a longer time-frame, ideally as part of a larger effort designed to increase biodiversity of the local flora and fauna. The work can be performed as opportunities arise through small-scale maintenance projects and would allow for selection of preferred micro-sites. In the Saskatoon region, for example, ideal sites would exhibit the following characteristics:

1. Small numbers of buckthorn seedlings/saplings.
2. Close source of seed from native plants, such as ash and maple trees.

3. Remote from invasive plants, such as smooth brome (*Bromus inermis*) and Kentucky blue grass (*Poa pratensis*).
4. Presence of native shrubs, grasses or herbs that are adapted to fire, such as snowberry and fescue (*Festuca sp.*).

Controlling the ground layer will not always result in natural recruitment of native plants, although this may be enhanced if all European buckthorn stems are removed from an area. Heneghan et al., (2002) speculated that changes in soil chemistry are a key factor which eliminates native flora beneath stands of buckthorn; this residual effect retards re-establishment of native flora long after buckthorn is completely eradicated. However, Boudreau and Willson (1992), reported that autumn removal of European buckthorn in Minnesota resulted in natural establishment of 26 herb species in riparian woodland and 6 herb species in oak savanna/rock-outcrop prairie the following spring.

The presence of exotic species in the Saskatoon area has been linked to changes in the disturbance regime which have led to loss of aspen forest and conversion to tall native shrubs (Thorpe and Godwin, 1992). Fire and occasional grazing may be required to ensure the long-term success of this restoration project. The initial treatment work in this study relied on herbicides because it was the most practical, given the extensive distribution of the shrub population. Use of fire was precluded because of close proximity to the city, while age and density of the stands limited the effectiveness of grazing. Aesthetically, chemical treatment was less intrusive because it was carried out in the fall at the end of the growing season and the killed plants merged with the winter aspect of the vegetation in the valley.

If the goals of European buckthorn control are to improve levels of biodiversity, or to construct a historically authentic landscape, then a good understanding of original vegetation composition is needed. A reasonable model is available for woodland patches in the prairie region of North America (Thompson and Hansen, 2001). Ideally, and over the long-term, implementation of a control programme for European buckthorn should dovetail with strategies sympathetic with achieving historical authenticity in the remnant natural woodlands. Small-scale disturbance from fire and grazing is certainly a viable strategy in the later stages of the restoration programme and would be preferred to chemical maintenance once the European buckthorn population has been reduced to a manageable level.

Conclusion

Results from this study indicate good initial progress in limiting seed production in dense European buckthorn sites over a 6-year period. Chemical girdling using Garlon 4 in a basal bark application is cost effective for low density sites. However, chemical treatment is very costly for high density sites, exceeding \$ 3,200 ha^{1}, because of the need for repeat applications. Plants that were missed in the initial application and new seedlings arising from the seed bank can restore stand density to its original level over the course of a few years. Significant reduction occurs after the third application of herbicide, and if this is followed by additional treatments, complete eradication of European buckthorn buckthorn seed production should be possible with occasional maintenance. The success rate at the outset of a restoration programme may appear discouragingly low because of vigorous recruitment. For every 100 stems sprayed a similar number will need to be treated in two subsequent applications. However, over the long-term the possibility of seedling recruitment from in situ seed sources is projected to be effectively eliminated using 3–5 fruiting stem treatments on a five-year rotation period. Long-term control will necessitate removal of European buckthorn seedlings and stunted mature stems in the ground layer. Effective restoration to a native plant community could require removal of residual male European buckthorn shrubs from the tall shrub layer due to their ability to change site conditions, such as soil chemistry and light levels.

The spread of many exotic species including European buckthorn can be linked to causal factors, such as changes in the disturbance regime. Re-establishing this regime through periodic burning or grazing is advocated as a long-term proactive method to address the issue of restoring local native plant communities. Reactive measures such as chemical girdling can limit spread and buy time until a long-term programme can be established. Containing European buckthorn to a few dense stands is a reasonable expectation for a control programme with limited resources.

References

Archibold, O.W., (1999), "The Aspen Parkland of Canada", R.C. Anderson, J.S. Fralish and J.M. Baskin (eds.), *Savannas, Barrens and Rock Outcrop Plant Communities of North America*, Cambridge University Press, 406–420.

Begg, C.S., Archibold, O.W. and Delanoy, L., (1998), "Preliminary Investigation into the Effects of Water-Level Control on Seedling Recruitment in Riparian Cottonwoods, *Populus Deltoides*, on the South Saskatchewan River", *Canadian Field-Naturalist*, 112, 684–693.

Bizecki Robson, D., and Nelson, J.V., (1998), "Vegetation Inventory of Saskatoon and Area", Nelson Dynes and Associates Ltd., Saskatoon.

Boudreau, D., and Willson, G., (1992), "Buckthorn Research and Control at Pipestone National Monument (Minnesota)", *Restoration and Management, Notes* 10, 94–95.

Delanoy, L., (1996) "Vegetation Survey of Natural Riverbank Lands North and West of the Regional Psychiatric Centre", Unpublished, Meewasin Valley Authority.

Gale, S. W., (2001), "Control of the Invasive Exotic, *Rhamnus Cathartica* L. in Temperate North American Forests" http://www.hort.agri.edu/h5015/00papers/gale.htm.

Godwin, B. and Thorpe, J., (1992), "A Biophysical Survey of the Silverwood Riverbank Area", Askatchewan Reasearch Council, Saskatoon.

Hall, R.S., (1973), "Vegetation and River Regime of the South Saskatchewan River Near Saskatoon", *Master of Science Thesis,* University of Saskatchewan.

Harder, D.E. and Chong, J., (1983), "Virulence and Distribution of Puccinia Coronata in Canada in 1982", *Canadian Journal of Plant Pathology*, 5, 185–198.

Heidorn, R., (1991), "Vegetation Management Guideline: Exotic Buckthorns-common Buckthorn (*Rhamnus Cathartica* L.), glossy buckthorn (*Rhamnus Frangula* L.), Dahurian Buckthorn (*Rhamnus Davurica* Pall.)," *Natural Areas Journal* 11, 216–217.

Heneghan, L., (2003) "And When They Got Together ... Impacts of Eurasian Earthworms and Invasive Shrubs on Chicago Woodland Ecosystems", *C.W. Journal, Best Practices in Conservation and Restoration* 1, 27–31.

Heneghan, L., Clay, C., and Brundage, C., (2002) "Rapid Decomposition of Buckthorn Litter May Change Soil Nutrient Levels", *Ecological Restoration*, 20, 108–111.

Izhaki, I., (2002) "Emodin: A Secondary Metabolite with Multiple Ecological Functions in Higher Plants", *New Phytologist*, 155, 205–217.

Leighton, A., Hay, J., Houston, C.J., Roy, J.F. and Shadick, S. (eds.), (2002), *Birds of the Saskatoon Area*, Nature Saskatchewan/Saskatoon Nature Society, Regina.

Lineman, M., (1997) "A Survey of the Natural Vegetation and Flora Along the South Saskatchewan River Valley within and Near Saskatoon, Saskatchewan", *Master of Science Thesis,* University of Saskatchewan.

Miller, R.O., (2005), "Regenerating Quaking Aspen on A Site Dominated by Buckthorn in Upper Michigan: A Case Study", Michigan State University.

Moody, M.E. and Mack, R.N. (1988) "Controlling the Spread of Plant Invasions: The Importance of Nascent Foci", *Journal of Applied Ecology*, 25, 1009–1021.

Pearson, C.W. and Gillette, L.N., (1999) "Monitoring Various Buckthorn Control Methods in Maple-basswood and Oak Woods within The Historic Range of the Big Woods", Minnesota Department of Natural Resources.

Schmidt, K.A. and Whelan, C.J., (1999) "Effects of Exotic *Lonicera* and *Rhamnus* on Songbird Nest Predation", *Conservation Biology* 13,1502–1506.

Thompson, W.H., and Hansen, P.L., (2001), "Classification and Management of Riparian and Wetland Sites of the Saskatchewan Prairie Ecozone and Parts of Adjacent Subregions", Riparian and Wetland Research Programme Montana Forest and Conservation Experiment Station School of Forestry, The University of Montana Missoula, MT.

Thorpe, J. and Godwin, B., (1994), "Saskatoon Natural Grasslands Resource Management Plan Final Report", Delcan Western Ltd. in Association with Saskatchewan Research Council.

Thorpe, J., and Godwin, B., (1992) "Regional Management Plan for Douglas Provincial Park and Elbow PFRA Pasture", Saskatchewan Research Council.

USGS, (2000) "Species Abstracts of Highly Disruptive Exotic Plants", Northern Prairie Wildlife Research Center, USA.

22 Ecotourism and Biogeographic Resources in Himachal Pradesh

Anupriya Chatterjee

Concept and Basic Arena of Ecotourism

Ecotourism means ecological tourism, where ecology has both environmental and social connotations. The term 'Ecotourism' was initially coined in 1987 by Ceballos-Lascurain (Ceballos-Lascurain, 1996) who defined ecotourism as an experience of "travelling to relatively undisturbed areas with the specific objective of studying, admiring, and enjoying the scenery and its wild life". Ceballos-Lascurain preferred to confine the scope of ecotourism to the objective of travelling and the area travelled.

Today Ecotourism is distinguished from Mass Tourism or Resort Tourism by having a lower impact on the environment and by requiring less infrastructural development. Ecotourism is now the fastest growing sector in the tourist-based industry. It offers a sound alternative to mass tourism and attracts consumers who wish to enjoy more ethical and nature-based holidays. If ecotourism projects are to be ecologically sound they must recognize that most parts of the world have been modified, managed and in some cases improved by indigenous people for many centuries.

The World Tourism Organisation defined ecotourism as "tourism that involves travelling to relatively undisturbed natural areas with the specified object of studying, admiring and enjoying the scenery and its wild plants and animals as well as any existing cultural aspects (both of the past and the present) found in these areas" (WTO, 1993). As is obvious from the definition the focus is solely on visiting nature-based areas. In May, 2000, as part of the side events on the 8th session of the United Nations Commission on Sustainable Development (CDS, 8), a group of Indigenous Peoples Organizations, NGOs and other members of civil society provided a proposal on guidelines for ecotourism. Born in its current form in the late 1980s, ecotourism came of age in 2002, when the United Nations celebrated the 'International Year of Ecotourism'. The International Ecotourism Society defines ecotourism as "responsible travel to natural areas that conserves the environment and improves the well-being of local people".

International Consensus on Ecotourism

Ecotourism is becoming the most significant tourism market for many developed as well as developing countries. For many centuries ecotourism has not been seen so much as a marginal activity intended to finance protection of the environment than as a major sector of national economy and as means of attracting tourists. As for example in countries like Costa Rica, Nepal, Kenya, Madagascar and Antarctica ecotourism represents a significant portion of the GDP or as economic activity.

Tradition of Ecotourism and its Growth Pattern in Himachal Pradesh

Natural Environmental Treasures and The Popular Sites of Tourist Interest in Himachal Pradesh

Himachal Pradesh, a relatively small state in India nestled in the lap of the Himalaya, is truly considered a *Travelers' Paradise* for its unique canvas of lofty ice-capped mountain summits, glaciers, deep gorges, turbulent rivers, picturesque green and golden meadows with orchards, ethnic settlements and culture of ancient tribes. Having an area of 55,673 km^2 it has a legally classified forest area of 37,033 km^2, which is 66 per cent of its total geographical area. Out of the area designated as forest land, 12 per cent is under Protected Area Network. The per capita availability of forests in this state is 0.22 hectares against the national average of 0.06 hectare. A small state like Himachal Pradesh, which has only 1.7 per cent of the geographical area of the country, has contributed a remarkable 4.5 per cent of the net increase in the forest cover of the country. In addition the state has thirty two wildlife sanctuaries, two national parks and three game reserves covering over 5,500 km^2. This state is a treasure house both for plants as well as wildlife. It is also the home of some rare endangered species like snow leopard, musk deer, brown bear and pheasants like the Western horned Tragopan which has been bred in captivity for the first time in the world in Himachal Pradesh. Table 1 demonstrates the pattern of sectorwise distribution of traditional tourist places in Himachal Pradesh.

Table 1

Sectorwise Distribution of Traditional Tourist Places in Himachal Pradesh

Travel Sectors	*Names of the important Tourist Sectors*
Shimla Sector	Parwanoo, Kasauli, Barog, Kiarighat, Shimla, Chail, Darlaghat, Kufri, Fagu, Naldehra, Chindi, Paonta Sahib, Renukaji, Rohroo, Narkanda, Rampur, Sarahan, Khara Pathar.
Kullu-Manali Sector	Kullu, Manali, Swarghat, Bilaspur, Deothsidh, Mandi, Rewalsar, Raison, Naggar, Manikaran.
Kangra Sector	Kangra, Palampur, Baijnath, Jogindarnagar
Dharamshala-Chamba Sector	Dharamshala, Palampur, Jogindernagar, Hamirpur, Chamundaji, Jwalamukhi, Chintpurni, Shahtalai, Chamba, Dalhousie, Khajjiar, Mc Leodganj, Bharmour.
Kinnaur, Lahaul-Spiti Sector	Kalpa, Nichar, Ribba, Morang, Kanam, Nako, Sangla (Baspa) Valley, Kaza, Kibber, Tabo, Dhankar, Pin Valley, Lingti Valley, Triloknath, Kunzam La, Rothang La, Udaipur, Baralacha La, Kardang, Pattan Valley, Keylong.

Source: Himachal Tourism Department.

Recent Trend of Tourist Influx in Himachal Pradesh

Himachal Pradesh witnesses a heavy rush of tourists in every summer and autumn months. Current report shows that this state receives over 60 lakh tourists every year. The domestic tourist inflow in the state has recorded a 22 per cent increase in the past five years (2001–2005), while the foreign tourist inflow figure has gone up by around 30 per cent during the same period. According to the report received from the Himachal Pradesh Tourism and Development Corporation office 63, 45,069 domestic tourists visited the state in 2004 compared to 55, 44,414 in 2003.

National Tourism Policy 2002 and Tenth Plan exhibit a change in Himachal government's approach to tourism as a promoter of national integration and facilitator of international understanding. As mentioned elsewhere (e.g., www.himsuchana.org) Himachal Pradesh in 2002 received 1.44 lakh of international tourists (5.76% of total 25 lakh foreign tourists who visited India in that year) and 49 lakh domestic tourist i.e., a total of 50.44 lakh. With this trend, by 2010, Himachal Predesh is expected to receive 1.5 crore tourists including 15 lakh foreigners. The major countries contributing foreign tourists in Himachal Pradesh are US (9.8%), UK (16.1%), Germany (6.6%) and France (5.1%). The statistics show the growth pattern of tourist flow in Himachal Pradesh during 1997–2005 and the projected pattern for the ensuing year 2010 (Table 2).

Table 2

Account of Tourist Flow in Himachal Pradesh (in Lakhs)

Year	*Domestic tourist*	*International tourist*	*Total*
1997	38.30	0.63	38.93
1998	41.20	0.75	41.95
1999	43.52	0.91	44.43
2000	45.70	1.11	46.81
2004	58.5	1.5	60
2005	77.5	2.5	80
2006	95.5	4.5	100
2010*	135.0	15.0	150

* Projected figure.

Source: H.P. Tourist Economic Survey, 2002.

Characterization of Ecotourists

The characteristics of ecotourists visiting or would be visiting Himachal Pradesh can be classified in three broad categories as given in Table 3.

The Himachal Pradesh Government Tourist Economic Survey, conducted in 2002, presented a detailed account of the contemporary pattern of tourist pressure received in all the districts of the state, the report of which is presented in the Table 4.

Table 3

Major Types of Ecotourists

Types of Tourists	*Age groups*	*Traveller groups*	*Purpose of visit*	*Mode of Transport used*	*Accommodation preferred*	*Food preferred*
Adventurer	Young to Middle aged	Individual or small group	Sport or adventure	Avail local buses or mainly trek to destination	Cheap lodging in eco-villages, mostly locally owned or personal tents	Cheap local food
Nature lover	Middle aged	Travel in small groups or guided by travel agents	Spend leisure time in midst of nature	Hire private car or drive in by their own car or avail small buses of travel agents	Eco-hotels or eco-huts with good sanitation, electricity and other luxurious facilities	Lavishly prepared local food, often imported food and beverages consumed
Environmentalist/Researcher/Specialist	Young to old	Travel individually or with specialized educational groups	Scientific or specialized interest or governmental or Non-governmental rcscarch projects	Wide range of transport types	Wide range used but indigenous mode preferred	Wide range used but local preferred

Source: Himachal Pradesh Tourism Department.

Table 4
Trend of Districtwise Tourist Inflow

District	Percentage of tourist	District	Percentage of tourist
1. Bilaspur	3.6	7. Lahaul & Spiti	3.1
2. Chamba	7.3	8. Mandi	3.8
3. Hamirpur	4.6	9. Shimla	24.8
4. Kangra	14.5	10. Sirmour	5.9
5. Kinnaur	2.1	11. Solan	4.0
6. Kullu	24.5	12. Una	1.8

Source: H.P. Tourist Economic Survey, 2002.

The districtwise pattern of tourist pressure demonstrates that out of the 12 districts Kullu and Shimla alone account for more than 50 per cent of the total tourist traffic.

Potentials of the Development of Tourism Industry in Himachal Pradesh

In India foreign exchange earning from tourism is the third largest; therefore it is now becoming the largest export industry of the country. Himachal Pradesh is one of the top five tourist destinations (both domestic and international) in India with over 60 lakh tourists visiting the state every year, the number equaling its total population.

The reasons for recent tourist flow boom in Himachal Pradesh in current years may be assessed as follows:

- Himachal Pradesh, being a Himalayan state, is endowed with beautiful and comfortable climate; the best times for visit are summer (March to May) and autumn (September to November).
- Vast areas of scenic beauty for passive recreation comprise the picturesque valleys of Kullu, Kangra and Kinnaur with lash of colours, and by contrast the barren terrain of Lahaul and Spiti have stunning lunar landscape-like beauty presenting another dimension of aesthetic view.
- This has ecologically very rich areas with exotic flora, fauna and avifauna habitating the sanctuaries.
- This state has enough treasure to promote both adventure tourism and cultural tourism.
- Under the growing pressure of activists the huge number of mountain lover tourists who earlier preferred to visit Kashmir valley, now shifted their choice to the almost similar but more peaceful mountain environments of Himachal Pradesh.

A detail account of adventure tourism and cultural tourism in this state is listed in the Table 5.

Table 5

Potential Areas and Important Months for Adventure Tourism and Cultural Tourism

Month	Adventure Tourism		Cultural Tourism		
	Event	Place	Festivals/Fairs	Place	Legend
	National Snow Statue Competition	Kufri	Lohri/Maghi	Shimla, Chail, Manali, Dalhousie	Traditional mid-winter day and last sowing of Rabi crop accompanied by community folk songs & dance
January	Water Sports Regatta	Maharana Pratap Sagar – Kangra	Dechhang	Spiti Valley	Celebration of the height of the winter
			Phagli	Lahaul's Pattan Valley	Idols of Shiva Linga & Hidimba are made on a roof by packed snow and worshipped
			Sizi	Kinnaur	—as above—
			Guru Govind Singh's birthday	Paonta Sahib	Guru Govind Singh's birthday celebrated in different Gurudwaras
February	—	—	Gochi	Bhaga Valley	Celebrate birth of male children and token marriages of children below 6 years
			Baba Barbahag Singh Mela	Una	Honours the sages with magical powers
			Vasant Panchami	Every lower terrain	Marks the arrival of spring with flying kites
			Lossar	All Buddhist areas	Stylized Chham dance with elaborate costumes mark triumph of good over evil
March	Venue of India Rafting Cup	In selected place	Char	Lahaul	Celebrated around the Triloknath temple
			Shivratri fair	Mandi town	Accompanied by folk bands, song, dance and drama, people pay obeisance to Lord Shiva at Bhoothnath temple.
			Nalwari fair	Bilaspur	Cattle are traded along with wrestling bouts, aero and water sports.
			Balasundari fair during Navratras	Trilokpur near Nahan, Sirmour	This fair coincides with the sacred days of Navratras
			Chait	Kullu and Chamba	Celebrates the first month of lunar calendar by folk singing and dancing
			Holi	Palampur, Sujanpur, Paonta Sahib	Exuberant celebration of spring with riot of colours.

Cont'd...

Cont'd...

April	–	–	Rali	Kangra	Clay models are made in every house of rali, unmarried girls pray for grooms of their choice, newly wed for marital happiness.
			Sui mela	Chamba	Mela thronged by women and children
			Mahu Nag	Mandi	At the village of Taraur
			Markandaya fair	Bilaspur	Holy fair
			Rohru Jatar		Held in honour of sikh deity, Shikru
			Beesh	Kinnaur	Time when deities are taken out of temples
			Baisakhi	Shimla and Mandi mainly	Rooted in rural agrarian tradition, it bids farewell to winter.
May	Hand Gliding & Para Gliding	Bir Billing near Kangra	Banjar fair and Sarhi jatar	Various parts of Kullu	
	Summer Skiing	Rothang Pass			
	River rafting festivals and water sports	Throughout the states	Dhongri fair	Kullu	Focussd around goddess Hidimba temple
			Sipi fair	Shimla	Traditionally time for match making
June	Numerous Trekking, Hiking and Camping expeditions	Various parts of the state	Kangra festival	Kangra	
			Shoolini fair	Solan	In honour of the presiding goddess Shoolini
			Ghantal festival	Lahaul	Held at Ghantal Monastry on full moon night
			Saranhuli fair	Mandi	All offerings are made in a lake near temple of Dev Kamrunag in the Chachiot tehsil
July	Trekking	Lahaul and Spiti, Kinnaur	Ladarcha fair	Kaza	Old trade routes become alive as traders barter and sell a variety of goods
			Lahaul festival	Keylong	
			Haryali (Rhyali)	Entire state	Announces the advent of monsoon rain
			Sharavana Sankranti	Nahan	

Cont'd...

Cont'd...

			Sair fair	Arki	Conducted in honour of Banar Devta of Shari this fair is marked by buffalo fights
			Rampur Jatar	Shimla	Held at Jubbal
August	–	–	Minjar fair	Chamba	Celebrates the bounty of nature and prays for a good harvest
			Manimahesh Yatra	Chamba	Yatra to the sacred tarn Manimahesh after Janmashtami
			Chrewal, Badronjo or Patroru	Chamba and Kullu	Festival of fire and flower and time for purification of fields
			Gugga fair	Chamba, Bilaspur,Sirmour	Worship of Gugga, a manifestation of Nag Devta
			Dal fair	Dharamshala	
			Ashapuri fair	Kangra	
			Sharavan fair		Held in the holy shrine of Nanda Devi
September	–	–	Fullaich or Okhayang	Kinnaur	Festival of flowers
			Sair	Kangra valley and Shimla	Celebrated with stalls, singing and buffalo fights at Arki and Mashobra
			Nagini fair	Kangra	Bids the summer farewell
			Bawan Dawadashi fair	Nahan	
October	Regatta	Waters of Govind Sagar	Phool Yatra	Chamba's Pangi Valley	Killer and Panai witnesses neighbourly affection and Dehant Nag is worshipped
			Gutor	Spiti	Prayers and dances are held in monasteries of Ki, Tabo and Dhankar
			Dussehra and International Folk Festival	Kullu	More than 200 deities converge on Kulluin Dussehra, pay homage to Lord Raghunath along with folk music and trading of colourful wares
	Anglers vie for the largest catch of trout	Sangla Valley			
	Para Gliding	Billing			

Cont'd...

Cont'd...

November	Water Sports Competition	Maharana Pratap Sagar, Kangra & Govind Sagar Lake, Bilaspur	Lavi fair	Town Rampur, once major entrepot on old trade routes to Kinnaur, Tibet, Ladakh and Afghanistan	A fair graces the bank of Renuka Lake, idols of Renukaji and Lord Parshurama are ceremoniously dipped in the sacred lake and a lot of matching is done
	Pre-World Cup Gliding Championship	Billing, Kangra			
December	Ice Sketing Carnival and Competition	Shimla	International Himalayan festival	Kangra	Held in McLeodganj along with cultural programmes
			Christmas	Shimla and Dalhousie	Event packed season

Source H.P. Tourist Economic Survey, 2002

- Ecologically rich areas with exotic flora, fauna and avifauna habiting the sanctuaries are given in Table 6.

Table 6

Areas in Himachal Pradesh with Sanctuaries of Exctic Flora, Fauna and Avifauna

District	*Sanctuary/National Park*	*Area covered (km²)*
Bilaspur	Govind Sagar, Sri Naina Devi	223
Chamba	Sach-tuan Nala, Gamgul-Siyabehi, Tundah, Kugti, Kalatop-Khajjiar	724
Kangra	Pongdam lake, Dhauladhar	1251
Kinnaur	Rakchham—Chhitkul, Lipa Asrang, Rupi-Bhaba	838
Kullu	Sainj, Tirthan, Kias, Khokhan, Kanawar, Manali	272
Lahaul & Spiti	Kibber	1400
Mandi	Bandli, Nargu, Shikari Devi	391
Shimla	Daranghati 1 & 2, Talra, Water Supply Catchment	217
Sirmour	Churdhar, Simbalbara, Renuka	89
Solan	Chail, Shilli, Majathal, Darlaghat	157
Total area of Wildlife Sanctuaries		5,562
Kullu	Great Himalayan National Park	765
Lahaul & Spiti	Pin Valley National Park	675
Total area of National Parks		1,440
Total area of the Protected Area Network		7,002

Source: Himachal Pradesh Tourism Dept.

Dimensions of Ecotourism in Himachal Pradesh

A Paradigm Shift From Popular Tourism to Ecotourism

Conferring to the Indian Tourism slogan, 'Leave Only Footprints, Take Only Photographs', it becomes essential to promote a symbiotic relationship between tourism and environmental consciousness. Ecotourism is regarded as a positive dimension of tourism involving minimum population density and low impact activities taking place at natural sites endowed with sufficient biological, cultural and geographical resources. The major factors contributing to this paradigm shift in Himachal tourism are:

- The rise in awareness of both international and national tourists visiting here, regarding global ecological realities.
- The desire among a rapidly growing and relatively affluent segment of modern industrialized tourists to experience ecologically and culturally rich nature-based tourism of Himachal Pradesh.
- The growing conviction and know-how among the local people and businessmen, tour operators, agents and hoteliers that the unbound natural resources which they harness to earn their livelihood, is finite and should be conserved for their future generation.

Role Played by the Government and NGOs through Planning, Policy Making and Conducting Workshops

The strategies, policies and work methodologies of the government and non-governmental organizations have become well-knitted and organized under which "ecotourism is expected to become old tourism in new clothes". This process has contributed to the following:

1. To ensure regulated growth of ecotourism the National Tourism Policy, 1997 formulated a new draft of tourism policy in consultation with the state government, industry, associations and professionals. Its formulations are:
- Creating awareness about promotion of ecotourism and preservation of natural environment and heritage.
- Preservation and protection of natural resources, environment and ecology.
2. Government of Himachal Pradesh came out in May 2001 with a set of Policy for Ecotourism with special stress on local community involvement:
- Five societies have been formed for Great Himalayan National Park, Lake Renuka, Potters Hill near Shimla, Kufri, Khajjiar and two more are proposed to be formed near Mandi and Chelsham Pass. These societies are empowered to collect money from tourism services and use funds for management of government assets such as forest rest houses and development of ecotourism.
- Training to local youths as nature guides has been undertaken at Kufri, training 27 nature guides.
- The societies have developed agreement with tour operators to take tourists on designated ecotourism route.
- Various publications from the societies in form of leaflets and brochures to publicize ecotourism.
3. To promote ecotourism forest department of Himachal formulated:
- A State Level Ecotourism Advisory Committee under the Chairmanship of Secretary, Department of Forest.
- Kullu Ecotourism Initiative (KETI) has been launched by an NGO; they organized workshop in Delhi in April, 2002.
4. The revised policy of Himachal Pradesh Forest Department on Development of Ecotourism, 2005 has targeted to make Himachal Pradesh a leading ecotourism destination in the country by 2010. Its vision being, to preserve and protect the natural heritage, both flora and fauna, of Himachal Pradesh and attract discerning tourists and visitors to the state and in the bargain provide opportunities to enhance livelihood of the local people and alongside generate revenue for the sustainable development of its resources through the creation of a special purpose vehicle (SPV). The Eco-Tourism Policy would encourage a partnership between

the SPV on the one hand and the civil society (local communities, NGOs, academic institutions and private enterprises/businesses) and the state (Forest Tourism and Fisheries Department) on the other.

5. Policies for non-governmental organizations, scientific and research institutions were formulated as to:

- Create awareness among all concerned about the importance of sound eco-practices in tourism development.
- Organize training programmes to train the local people for taking up various vocations relating to ecotourism

Popular Ecotourism Units Enlisted by Himachal Pradesh Tourism Development

The following popular ecotourism units have been identified by the Himachal Pradesh Tourism Development:

Potter's Hill Van Vihar

Location: Khalini, Shimla district, at an altitude of 2,050 m.

The project: Located at about 7 km west of Shimla, adjacent to Himachal Pradesh University, Summer Hill, is popularly known as Potter's Hill once used by potters to obtain clay. It covers 100 hectares of prime western Himalayan temperate forest with extravagant floral diversity, waterfalls (Chadwick falls—2 km, Glen falls—3 km) as well as genetic diversity of several species. This nature park is barely 5 km from famous Mall Road and offers tenting site, children's park, horse riding, and walking trails through magnificent cedar and oak forest.

Great Himalayan National Park

Location: Shamshi, Kullu; at an altitude ranging from 1,500 to 6,000 m.

The project: This national park is a prime protected area of the country and provides varied opportunities to the eco-tourists to enjoy nature. The park has well developed eco-treks, forest bungalows, private guest rooms in traditional houses and tenting sites for night halts. To visit, prior permission of the park director is required, local guides and porters are also available.

Himalayan Nature Park

Location: Khalini, Shimla; at an altitude of 2,700 m.

The project: Located about 18 km from Shimla, on the Hindustan-Tibet Road, this nature park covers about 100 hectares of western Himalayan temperate forest with extensive floral and faunal diversities. The park offers 3 to 6 hours guided trek to Mahasu peak, offers opportunity to appreciate brown bear, musk deer, yak, bharal, sambhar, ghoral, monal pheasant and many other avifauna. It has limited lodging facilities.

Renuka Wildlife Sanctuary

Location: Khalini, Shimla.

The project: This sanctuary clothes the slopes around the picturesque Renuka Lake in the outer Himalayas with deciduous forest cover. Major attractions include temples of Rishi Jamdagni, Lord Parashurama and Goddess Renuka, trekking, a mini zoo, a lion safari, boating facility in the lake, but limited lodging facility.

Ecotourism Unit at Craignaino

Location: Craignaino, Shimla.

Estimated project cost: Rs 5 crores (excluding cost of land).

The project: The site is located about 15 km from Shimla in a forest area, controlled by the Municipal Corporation of Shimla. Presently this ecotourism project has a well developed park, apple orchard and nursery spreading over an area of 1 hectare. Remaining 7 hectares bears flora of temperate coniferous forest dominated by blue pine and deodar trees. Possible development options are as follows:

- Health resort
- Development of water bodies for recreation and sports
- Botanical garden
- Walking trails through forest, 4–5 km long
- Trekking activities through forest using portable tents or huts
- Using flat stretches of land for amusement park, building cottages or villas or chalets

But more such projects are required to cater to the already high, and ever increasing inflow of tourists.

A Holistic Assessment of Himachal Tourism

Ecotourism, the ecologically sustainable nature-based, tourism has contributed significantly in the present dimensions of tourism industry, which plays a very important role in the economy of Himachal Pradesh. The five important sectors (Shimla, Kullu-Manali, Kangra, Dharamsala-Chamba and Kinnaur-Lahul-Spiti) that gained traditional importance as tourist places have enormous potentials for ecotourism development. These potential natural resources are: (1) seasonally developed snow fields (for skiing and ice-skating), (2) forest covered mountain slopes with attractive biotic species and varied wildlife (for forest trekking and enjoyment of wildlife), (3) turbulent streams (for water sports including rafting), (4) unique antiquities (for anthropological observations) and (5) wide range of cultural and ritual practices by the tribal communities (for taking part in fairs and festivals). At present this state has 5,562 km^2 areas of wildlife sanctuaries and 1,140 km^2 areas of national parks. According to reports Himachal Pradesh receives about 80 lakhs domestic and foreign tourists in the current years, accounting for 22 per cent increase over the last 5 years (2001–2005). If this trend continues, by 2010 the figure of annual total tourist influx will reach 1.5 crore mark.

Ecotourism has gained importance in Himachal Pradesh particularly from the 1990s. This has been in conformity to the National Tourist Policy of 1997. From 2001 the Himachal Pradesh government has come out with a set of well formulated policy for ecotourism with special emphasis on community development. Subsequently the state forest department also formulated a State Level Ecotourism Advisory Committee and at the same time the NGOs as well, have taken active role in promoting ecotourism in the state.

Select References

Bhartari, R., (2003), "Ecotourism in Northern India (Chap. 1)", *Ecotourism and Sustainable Development* (ed.) by Ravishankar Kumar Singh, Abhijeet Publication, New Delhi.

Ceballos-Lascurain, H., (1996), *Tourism, Ecotourism and Protected Areas*, Gland, Switzerland: IUCN.

France, L., (1997), "Introduction to Sustainable Tourism (Chap. 1)", *Sustainable Tourism* (ed.), Lesley France, Earthscan Pub. Ltd., UK

Government of India, (2002), Planning Commission Report on Himachal Tourism (Chap. 19), Government of Himachal Pradesh Publication.

Jenkins, C.L., (1998), "Tourism in Third World Development: Fact or Fiction (Chap. 6)", *Tourism Management and Global Perspective* (ed.) by G.C. Batra and A.S. Chawla, Deep and Deep Publication, New Delhi.

Kumar, M., (1998), "Himachal Tourism in Retrospect and Prospect (Chap. 10)", *Tourism Management and Global Perspective* (ed.) by G.C. Batra and A.S. Chawla, Deep and Deep Publication, New Delhi.

Raj, A., (2003), "Indian Tourism: Sustainable Development (Chap. 7)", *Ecotourism and Sustainable Development* (ed.) by Ravishankar Kumar Singh. Abhijeet Publication, New Delhi.

Singh, R., (2003), "Ecotourism Guidelines for India (Chap. 4)", *Handbook of Environmental Guidelines for Indian Tourism*, Kanishka Publishers and Distributers, New Delhi.

Tisdell, C., (1999), "Ecotourism, Economics and Environment (Chap. 8)", *Biodiversity, Conservation and Sustainable Development: Principles and Practices with Asian Examples*, Edward Elgar Publications, UK.

Verma, G.C., (1998), "Tourism in Indian Context: Experiences and Challenges (Chap. 12)", *Tourism Management and Global Perspective* (ed.), by G.C. Batra and A.S. Chawla, Deep and Deep Publication, New Delhi.

Whilan, T., (2003), "Ecotourism and its Role in Sustainable Development (Chap. 4)", *Ecotourism and Sustainable Development* (ed.) by Ravishankar Kumar Singh, Abhijeet Publication, New Delhi.

World Tourism Organisation, (1993), *Sustainable Tourism Development: Guideline for Local Planners*, WTO, Madrid.

www.himsuchana.org

http://himachaltourism.nic.in

http://www.hptdc.nic.in

http://www.hptdc.gov.in

23 Conservation Atlas of China

Fan Zhengyi

The Importance and Urgency of Natural Conservation

Mankind seeks economic development and utilizes natural resources, which must agree with limited resources and support of ecosystem, while giving full consideration for later generations. This is the meaning of conservation. With development of society and technology, the increasing ability of creation, and construction of mankind directs the same force of destruction and ravage. In recent decades, as a result of the expanding scale and range of production, their influence on natural environment becomes more and more serious. The irrational exploitation of natural resource has ravaged environment much more seriously, than in any historical period before, affecting the existence of human being dangerously, as cried in "Save The World" written by R. Alan, the Adviser of Alliance of International Nature and Conservation.

The damage to environment and resources is a critical problem that all countries in the world have to face. It is a difficult task for all countries to solve the problem. The World Conservation Strategy (WCS) was promulgated by IUCN, UNEP and WWF, March 5, 1980. China is one of the countries that took part. The main goal of WCS is to spur development continuously by means of conservation—development depends on conservation and conservation depends on development—as written in the WCS. For the existence of ourselves, for later generations, it is our duty to save, and protect whole of nature when developing our economy.

China is a great country with plenty of resource. Chinese have been living and working in this land for thousands of years. The thought of conservation began in Chinese dynasty, about two thousand years ago. The Great Canal and Dujangyan are two examples of development in conservation by ancient Chinese people. Many improvements have been achieved since the establishment of Peoples Republic of China (PRC). Now we are making great efforts on the development of our country. We should do our best to make rational and permanent usage of natural resources, to combine development and conservation closely. But the population in China increased too fast since last 50 years. To sustain the population with a rising living level caused some serious problems, such as pollution, over exploitation, and imbalance of ecosystem. So, conservation is an important and urgent problem in China too.

For the needs in our country, for the goal of saving resources and protecting environment, the 'China Conservation Programme' (CCP) was set up in May 1983 and published in May 1987, which is an important step of conservation for both China and the world. In CCP it is emphasized that conservation is one of the most important tasks in socialist construction and is a basic policy of our country. It is obvious that to make the CCP integral, a series of background materials and a Conservation Atlas of China are needed, which is the reason and background for editing this atlas.

An Atlas Designed for Serving Conservation Work

Conservation atlas belongs to environment mapping which is a new active branch of thematic cartography. Editing "The Conservation Atlas of China" is a trial without perfect theory and mature experience to refer. In addition to the characteristics of general atlas, this atlas has special feature to serve conservation work.

This is a middle scale comprehensive thematic atlas with the format of octavo (1/8). The major idea of this atlas is the combination of environment and resource conserving as well as exploiting. It provides plentiful information of different aspects including natural environment/natural resources and conservation, and so on. One can easily find out that this atlas is neither a general physical atlas nor a social and economical one, but one really designed for conservation only.

Being an organ of the CCP, expressing the main goal of CCP by means of maps which give direct vivid and deep impression, the atlas fulfils the CCP with abundant information. It having its own system and integrity is an independent integral work and could be published separately. It shows wide territory, various landform, the ample natural resources and advantageous environment in China, as well as the achievements made and problems that existed in conservation of our country.

To provide a suitable living and working conditions for generations, sustaining continuous economic development and proper social improvement are the aims of conservation. It is realized by means of a series rational managements of controlling and protecting. Its foundation is ecology that is the link of natural and social sciences. The roles revealed in ecology are important direction of conservation so that they are the theoretical bases of this atlas too.

Natural environment where people live is an active integration, with a lot of interaction in it. There should be an integral consideration in the atlas so that it could serve for conservation. The most deliberated points in the atlas are the following important factors, i.e., environment, resources protection and rational exploitation as well as the relations among them. Therefore, the topics, contents and, expression of atlas were carefully selected in order to give a clear link from beginning to end. In addition to making the maps up-to-date, there are many figures showing variations of different times.

Being an indispensable part of CCP and according to the CCP, the atlas contains conservation as main topic adding some environment and resources sheets. The conservation as the most important part include achievements of environment research, success of territory renovation, establishments of environment managing organizations and sites of environment detecting stations, in order to show today's situation and the recent results of conservation research in the country.

So as to make the atlas practical and meaningful the terms which are being used on conservation programme or in essential need for conservation works are selected.

The Structure of the Atlas

Since the nature is the cradle of human being and all creatures, the environment protecting and resource exploiting should be combined in order to spur steady improvement in society and economy. The whole edition is based on the ecological roles and made parallel to the CCP in its order. All the problems are mentioned from total to part, from whole country to regions, from high level to low level and from general case to special ones.

With the principle of conservation, all maps are divided into three parts according to their different usage and content. The first part is preface map, expressing human factors, emphasizing the relationship between mankind and nature in order to reveal the influence of human activity on natural circumstances. The second part is the physical map with the chief objective of conservation, including environment and resources. This part contains many maps, figures and much information. It is a principal objective of resource investigation, circumstances analysis and existing problem inspection. The third part is the maps of regional conservation and conservation regions. It is one of the parts which were mainly emphasized in the atlas.

In part two, the maps were edited in sequence of climate, land, biology, water, geology and sea areas. In part three, the sequence of maps for regional conservation and conservation regions was based on the sequence of state unique number for easy use.

The atlas has its own complete system. There are levels in its structure. For instance, about topic selection, there are maps for the whole country to show the general figure of some factors and maps for certain region to show background of some factors, examples of rational exploitation, good reforming, or some regions existing serious problems. As far as the map type is concerned, there are synthesized or combined maps for multi-factors as well as analyses or distribution maps for certain factor. In degree, there are both professionally specialized topics for specialists and common ones easily understood by amateurs.

Maps in the atlas are divided into three levels: (1) Maps for the whole country; they mainly provide information about natural environment, resources, conservation and their variation as well as existing problems. (2) Regional maps; they show the achievements of reforming, accomplishments of nature protecting and existing problems in different regions. (3) Maps of conservation regions; they are the background and examples of conservation work.

Maps in part two are also divided into three levels. The first is synthetic maps that give general or overall characteristics and roles of some factors, such as comprehensive physico-geographical regionization of China and regionization of flora of China. The second is maps for analyzing factor that show characters, distributions variations and problems for certain natural conditions or resources, such as land use and land cover of China, distribution of forest of China and so on. The last is maps for typical regions or typical problems, such as comprehensive utilization of dike-pond regions (Zhujiang Delta) which is a good example of rational exploitation. The next to the map of classification of soil erosion is the map of soil erosion of Loess Plateau. This was specially drawn to show the extent and depth of soil erosion in Loess Plateau which is the region where soil erosion occurred much more seriously than any other place in China. All the three levels depend on each other, forming an organic integer.

While the structure of the atlas being designed, both space concept and time concept were considered. In other words, the variations of factors with time were considered as important as their space distributions evolutions and relations among them. Making use of time and space variation of certain factor, one can disclose if the condition become better or worse, to find necessary steps for protection. With multi-views, multi-aspects and multi-comparison this atlas is helpful for conservation work.

The General Survey of the Atlas

There are total 99 sheets of maps in the atlas, among them 66 sheets are folded and 6 sheets are single paged.

In part one, i.e., the human factors part, there are 8 sheets of maps in four aspects. The first shows administrative divisions of China. The following three maps show the density of population, population distribution of major cities and towns as well as the growth of population from 1933 to 1982 in China. Then two maps of major sceneries, cultural relics, historical sites and well-known historical cultural cities give deep impression of bright orient culture of China. The last two maps give locations of environmental protective management, environment monitory agencies and monitory stations in China.

There are 62 sheets in part two. Among them there are 2 general maps, 8 climate maps, 12 sheets about land resources, 14 sheets about plants, 8 sheets about animals, 6 sheets of water resources and 6 relevant geological maps and 6 maps about factors in sea regions. The first map shows the landform of the country. The following map shows the general natural figures and geographic conditions, present the physico-geographical regions of China. Then there are a set of climate maps exhibiting the relation between climate and environment or between climate and conservation. A diagram of temperature variation of China for the last 5,000 years and its comparison with Norway and Greenland.

The same historical period reveals the tendency of climate variation not only for China but also for the whole world. The annual radiation and the extreme minimum temperature are two factors being important to the growth of creatures. There are two maps for them. The yearly and monthly temperature and precipitation are most important factors for agriculture; there are diagrams presenting them correspondingly.

Land is one of the most important bases of living things. The map of land resource gives information about land property, quality and its usage in China. And the map of land use give more details about the land resource being used in agriculture nowadays in China. The following figure is one of the best examples in land use. It is a model of making rational use of water, soil and biological resources while developing planting, tending and fishing in balance in South China. The soil types, properties and distributions are shown in following map. Land desertization, soil erosion and soil salinization are serious problems in China. The information about these problems are given in the following maps. After the map of soil erosion, an extended figure showing very serious erosion in Loess Plateau is given. And then other figures show how to treat the problems.

The distribution of desert and desertization lands of China gives the type, origins and extent of desert in the country. The next figure, by means of comparison between the sandy area and covered area in 1958 and 1981, shows the enlarging danger of desert in Daqinggou region. But an inverse example as shown in next figure, by means of comparison too, shows that a typical treatment to the problem has been made in Shapetou region.

The salinization problem is also expressed in similar ways. The distributions of saline soil are drawn in a map to give salty and other information. Then an example is given for Fengqiou county where this problem is successfully dealt with.

The first map of plants is the regionalization of flora of China. Since there are variety of natural conditions existing in our wide country, this map is necessary. Vegetation has functions of keeping water and soil, reducing wind speed, filtering dust, sterilizing and cleaning air. A map is specially edited to represent the protecting and depolluting function of vegetation in Beijing and Tianjin region. Another map shows the types of vegetation and their distribution in China. Forest, the important and irreplaceable national resource, is systematically represented by 6 maps. First the distribution of forest is plotted, then is the rate of forest cover of the country. Jilin is a province with plenty of forest resource, the alteration about forest is particularly depicted by four maps in different periods. Nowadays governments in the world pay much attention to plant trees or grass in and near cities in order to improve urban circumstances. Two figures are plotted to present the change of green area in cities of China.

The 'Three North' shelter forest system engineering is a great project being carried on for territory renovation in China. A special map is drawn to give an understanding about it, the first and second stages as well as the key projects of second stage are plotted.

The conditions of grassland, its types and its horizontal and vertical distribution rules, are given as in map of grasslands of China. The next shows to what degree the human activities disturbed the natural environment and resources in four grasslands. Three following maps are given to exhibit distributions of economic, important medical, valuable and rare plants in the country.

In animal group, the first map is about the regionization of birds and mammals, then are the resources of animals and insects. Following figures show the distributions of valuable and rare animals in imminent danger as well as the variation of the distributions of these animals during history period in China. The last two maps of the group reflect the distributions of major economic freshwater fishes as well as valuable and rare freshwater animals of China.

On the water resource, the first is a map of river system and great water conservational projects in China, showing the general pattern, natural structure and density of water system, which reflects achievements obtained by Chinese people in the field of reforming natural condition and developing production. Following maps show time and space distributions of water resource and its varying rules. Lakes and swamps connect closely so they are plotted in one map with their diverse types and quirks. The historical alteration of lakes of Jianghan Plain is a good example for lakes in the country. In the map the shrinking of lakes in recent 50 years is shown and the measures for reform and protection nowadays are also shown. The last sheet of map gives the occurrence, development, distribution and type of glacier, snow and frozen earth as well as some kinds of observed data.

In the field of geology, the first sheet, distribution of underground water of China, shows not only the chemical type of underground water and its forming geological conditions but also some new information related to conservation. The map of hot mineral water distribution gives its type region, chemical type, and temperature as well as the exploitation of earth thermal. The map of Karst in China shows its distribution, its rock type and cover extent, providing information for ascertaining intensity of Karst and weakness in surrounding. A separated map shows earthquake calamities in China and the measures made to reduce destruction. The mud-rock flow is another calamity distributed in China as shown in the following map. And the last one of this group shows typical inspection, observation and managements for a small region of this problem.

In the connection of oceanology and conservation, the environment of offshore and neighbouring sea regions of China are shown first. Following figures show growth and distributions of sea production resources including fishes, algae and invertebrates offshore in China. The economical fishes are specially shown with their migration patterns and living surroundings. The valuable and rare aquatic animals are also shown with their distributions in another figure.

Part three contains 29 sheet of maps including 7 sheets of regional conservation, 22 sheets of conservation areas. The choice of regional conservation maps depend on its special features such as fruitful reforming natural conditions, successful controlling of natural calamities or peerless usage of resource or regions with plenty of natural resources waiting to be exploited in China.

The Yellow River and Yangzi River are two main lifelines of China. The first is represented by the conservation in Huang-Huai-Hai Plain and the latter is represented by the conservation in Taihu region. Huang-Huai-Hai Plain was formed mainly by the force of Yellow River which is a harmful river. The map does not show the disastrous historical background of the river, but only the, influences of human being portrayed nature. In three aspects, i.e., natural regionization, environment type and natural protection engineering, the achievements are shown. The Yangzi River Delta is the cream of Yangzi River Valley. However, the generalized Taihu Region is a representation of the Delta which is an important region for agriculture productions in China. The map emphasizes ample water resource and fertile land in the region. It is a model of continuous good circulation among agriculture lake and urban ecosystems, displaying deep influences of humans on natural condition. Besides, conservation in Sanjiang Plain, Dongtinghu Region, Tulufan Basin, Hainan Island and Taiwan Island are also the models of different types.

The map, distribution of conservation areas of China, introduces the distribution, types and grade of conservation areas comprehensively. Among total 21 sheets of map, 16 sheets belong to national areas, e.g., natural conservation of Dinghu Mt., Wolong, Changbai Mt. and so on. There are 5 sheets picturing special conservation areas such as Wudalianchi for volcano? Honghe for swamp, Xilinguole for grassland, Zhangjiajie National Park for forest and Bird Island for birds. Till now there are 6 conservation areas in China–Changbai Mt., Dinghu Mt. Wolong, Xilingoule, Wuyi Mt. and Fenjing Mt.—belonging to the international network of conservation areas, as called "Human and Biosphere". Since China is a great country with various natural conditions, the type, eco-circumstance, protecting objects, territory, management level as well as extent and depth of research work in different areas are strikingly various. It is attempted to give real, vivid impression from maps, but the uniformity is avoided.

Necessary caption is given for each map to provide more information. In addition, hundred colour-photos are given in groups to show characteristics of rare animals, plants, landscapes and conservation areas. In the end, there are catalog of conservation areas and catalog of valuable and rare animals and plants.

The atlas, with two printing plates of Chinese and English, will be published soon both in the country and abroad.

24 Role of Wildlife Corridors in Biodiversity Conservation

Looking Beyond

C.S. Rathore and Yogesh Dubey

The 21st century has brought many conservation challenges to the fore. One very important and significant challenge that has evoked considerable scientific interest is the fragmentation of wildlife habitat. With rapidly expanding human populations and other competing land uses, areas that used to be continuous habitat have become broken and fragmented, isolating plant and animal populations contained within them. Habitat fragmentation is usually a time driven process that innocuously sets in the garb of human habitation or habitat alteration and eventually accelerates with the passage of time resulting in complete isolation of originally contiguous habitat. Populations thus isolated face survival pressures through increased competition for food and space and obligated risks in relation to disease outbreaks and episodic calamities such as fire and flood. Over a larger time span, species inhabiting isolated habitats also face the risk of extinction through mechanisms such as excessive inbreeding. It is worthwhile to mention that studies have shown that extinction has been quite pronounced in island communities. Around 93 per cent of bird species whose extinction has been recorded since 1600 have been island species (UWI, 2006).

Habitat fragmentation and reduction at various spatial scales has been widely acknowledged as a primary cause of the decline of many species worldwide (Ehrlich 1986; Lovejoy et al., 1986; Harris, 1984; MacDonald, 2003). Habitat fragmentation generally results in smaller and more isolated animal populations (Noss, 1987, 1992; Beier 1993; Noss et al., 1996). Smaller populations are then faced with the threat of local extinction due to stochastic events and catastrophes (Shaffer, 1978; Gilpin and Soule, 1986), and they become more susceptible to the inbreeding depression and its effects on long-term viability of populations. Fragmentation has resulted in varied effects both on individual animals and ecological processes. Gibbs and Faarborg (1990) have reported that fragmentation and reduction in the size of habitats has resulted in decline of forest interior species such as ovenbird and Kentucky warblers showed. Kattan et al. (1994) have also reported the loss of 24 species

(31 per cent of the original avifauna) in San Antonio upland forest due to effects of fragmentation. A number of conservation biologists have recommended maintaining habitat connectivity at landscape level to reduce the isolation of fragmented habitats (Noss 1983, 1987; Noss and Harris, 1986; Craighead et al., 1997; Craighead and Vyse, 1995; Paetkau et al., 1997).

The habitat fragmentation issue is of particular relevance to developing countries where most of the biodiversity rich tropical ecosystems are located. Natural ecosystems in many developing nations are currently facing an unprecedented threat due to diverse competing pressures resulting from a burgeoning human population and unregulated economic growth. India is one of the twelve mega-biodiversity nations of the world. Out of a total of 12,28,153 recorded animal species, 89,451 are known to occur in India which is around 7.28 per cent of the total global faunal diversity. India is home to around 4,952 higher chordates (Table 1) of which around 175 animal species are in the International Union of Conservation of Nature and Natural Resources (IUCN) red list threat category. India is also home to over one billion people many of whom live proximate to forest areas depending on them for their livelihood and subsistence. Furthermore, urbanization, industrialization, infrastructure development projects, agriculture, grazing, deforestation, wildlife trade and poaching continue to create tremendous stress on wildlife and pristine natural environs. As habitats shrink and populations become more isolated on 'habitat islands' studded in a matrix of altered land use, serious questions on long-term survival of many key species is now being asked. The status of large cats located at the apex of the food pyramid, is a grim reminder of the intense pressure that these animals face due to habitat loss. In recent times, considerable scientific and media attention has been focused in India on large mammals—particularly large cats—and their conflict with man primarily attributed to dwindling habitat. The fierce conflict between man and leopard in Mumbai (Sanctuary, 2004) due to habitat loss, the small and vulnerable population of 359 Asiatic lions located in only 259 sq km of core forest habitat in Gir, Gujarat (Gujarat

Table 1

Endangered Higher Chordates in India (Alfred, 1998)

Group	*No of Species*		*Percentage in India*	*Threatened*
	In World	*In India*		
Mammals	4692	390	8.42	213
Birds	9062	1232	13.66	149
Reptiles	5817	456	7.84	33
Amphibia	5150	209	4.06	148
Pisces	21723	256	11.72	75
	46,444	2543	5.47	618

Forest Department, 2006); the highly endangered population of 3,600–4,000 tigers in India threatened by habitat loss and poaching (Project Tiger, 2006) are all indicative of the immense threat that these fiercely territorial apex predators face today. Shrinking habitats also mean genetic isolation for many species in the long run.

One solution to ameliorate the undesirable effects of habitat loss is to provide connectivity between isolated habitat patches through wildlife corridors. Creation of dispersal corridors has received much global attention during the last two decades. While the utility of wildlife corridors has been debated (Simberloff and Cox 1987; Simberloff et al., 1992; Beier and Noss 1998), conservationists largely believe that wildlife corridors facilitate animal dispersal from isolated habitats and help counter biological processes that lead to species extinction. In the recent years, there has been considerable advocacy for planning dispersal corridors for wildlife in the Indian subcontinent. Even government agencies have augmented research efforts in identifying corridor opportunities for tiger conservation between important tiger reserves in India. The declining status of wildlife has been largely attributed to extensive loss of habitat and the ensuing circumscription of wildlife to small and isolated protected areas in India eliciting conservationists to agree that connectivity between larger habitats may be the key to long-term survival of wildlife populations.

What are Wildlife Corridors?

A wildlife corridor has been defined as a linear landscape element which serves as a linkage between historically connected habitat/natural areas, and is meant to facilitate movement between these natural areas (McEuen, 1993). From a landscape ecology viewpoint, a corridor has been defined as a linear habitat that is embedded in a dissimilar matrix which connects two or more larger blocks of habitat and which is proposed for conservation on the grounds that it will enhance or maintain the viability of specific wildlife populations in the habitat blocks (Beier and Noss, 1998). At large spatial scales, providing connectivity between large patches of core wildlife habitat requires corridors—land managed for its function as routes for wildlife movement and dispersal (Saunders and Hobbs, 1991). The connectivity between wildlife habitats by means of corridors in fact implies a system of corridors and the core areas of habitat which they serve to link. At a conceptual level, various models of core areas, movement corridors and buffer zones have been proposed by several researchers as frameworks for long-term survival and conservation of wildlife (Noss 1992; Noss and Harris 1996). McEuen (1993) has summarized the following conservation benefits of corridors:

1. Enhanced immigration, which will support genetic flow, increase genetic diversity and enhance overall meta-population survival in connected patches.
2. Provide opportunity to avoid predation.

3. Accommodation of range shifts due to climate change.
4. Provision of a fire escape function.
5. Maintenance of ecological process connectivity.

The utility of wildlife corridors has been debated in ecological literature with contesting claims arguing that there is little evidence to suggest that corridors in fact deliver the benefits attributed to them (Simberloff and Abele, 1976). Simberloff et al. (1992) have argued that there is insufficient data on corridor use by animals to come to a conclusion on their benefits. Corridors, they argue, may set into motion undesirable outcomes like high rates of poaching and trapping in corridor belts, increased exposure of wildlife to domestic animals causing disease, entry routes for weeds and exotic species and corridors acting as genetic sinks.

It is worthwhile to point out here that while there is a significant body of research on the negative impacts of habitat fragmentation on forested landscape, scientific knowledge examining the thresholds of habitat connectivity is quite inadequate. In the past, for example, the concept of Population Viability Analysis (PVA) has been used to estimate Minimum Viable Populations (MVP) as well as the amount of habitat needed to support them (Shaffer 1978, 1981, 1987; Gilpin and Soule, 1986). However little work has been has been done to estimate minimum thresholds of connectivity that would be required for protection of these species. The scientific debate over the usefulness and necessity of species corridors is a reflection of this problem (Simberloff and Abele 1976, Simberloff et al., 1992). Despite of a lack of established scientific body of knowledge on the functions of corridors, a large number of corridor projects have been implemented in the US in 1980s. All these projects were heavily criticized by some of the researchers (Simberloff and Cox 1987, Simberloff et al., 1992). The general objection of the critics was against the glaring lack of evidence supporting the use and importance of corridors and extreme expenses involved.

However, many studies during the last few years have indicated that wildlife corridors have become a critical component in effective wildlife management and conservation strategy. Results from Parks Canada's (2006) effort to manage elk populations suggest that wildlife corridors have facilitated the connectivity between wolves (major predators) and elks restoring natural ecological processes and help reduce man animal conflict. In one of the most exhaustive studies done on corridors so far, researchers at the University of Florida have concluded that birds transfer more seeds in their droppings between habitats connected by corridors than between those that are unconnected (UFL, 2005). The study concludes that corridors not only help animal dispersal, they also help plant propagation. In another study of dispersing cougars, Beier (1995) observed use of three corridors by dispersers by radio-tagging juvenile cougars. The Coal Canyon corridor was interspersed by an 8-lane freeway with only one underpass, two golf courts and one horse stable. The habitat of the area was degraded. The Penchang corridor showed analogous conditions. The third corridor was near Arroyo Tabuco and had nearly undisturbed vegetation. Studies revealed that in the Coal Canyon corridor area, two out of three dispersers could successfully traverse the area while one was killed in an accident. Heavy use of culverts and underpasses was observed. The second corridor—the Penchang corridor—was rarely used by cougars. However all three cougars made a successful traverse through the Arroyo corridor, Beier found that lower quality habitat was not a problem for the dispersing cougar juveniles. The results of this study made Bier observe—"Any connection is always better than no connection in habitat patches".

John Singh et al. (1989) examined four corridors to study elephant movement between Rajaji and Corbett National Parks of India. They observed that three corridors—Binjrao, Malin river, and Kotdwar-Amsod corridors—were rarely used by elephants. The fourth corridor—the Chilla-Motichur corridor—was however frequently used, particularly during summers. The study found that the main reason why elephants avoided the three corridors was the rugged terrain. However, it was found that while elephants did not use these corridors, they were being used by other animals such as tigers and ungulates. The Chilla-Motichur corridor used by the elephants was found to be vulnerable to human disturbance, grazing, weed proliferation and lack of regeneration of natural species.

Hilty (2004) attempted to examine the use of riparian corridors and vineyards by native mammalian predators. Overall 21 corridors were identified which were classified into three categories—denuded, narrow, or wide-based on their width. Unbaited and remotely triggered cameras were used to determine the presence of predators. The investigators found that mammalian predator detection rate was 11 folds higher in riparian corridors than in vineyards. More native predator species were found in wide corridors than in narrow or denuded corridors. The study concluded that wide and well vegetated corridors—which are vital for native predators—were highly desirable for maintaining connectivity.

Population Dynamics in Fragmented Landscapes

Fragmentation of habitat and patchiness may affect various individual, population and community level processes. Endeavours of ecologists and conservationists to protect wildlife populations in such habitats have resulted in various conservation theories and management options. One early theory—the theory of island biogeography—was proposed by McArthur and Wilson (1967). The theory of island biogeography states that in a fragmented community, the species number remains constant, but that local turnover among islands changes the species composition of island communities. Levins (1970) furthered this concept for populations in fragmented habitats through the rates of meta-population extinction theory. A meta-population is a dynamic demographic unit composed of several smaller sub-populations of a species where extinction in one patch is balanced by recolonization in another. Levin's theory states that if recolonization rates match the extinction rates of sub-populations in a meta-population, then the persistence of a meta-population is possible. However, the ability of a sub-population to recolonize is dependent on how well the patches are connected. However, an important precondition is that the constituent sub-populations should be "isolated enough to constitute separate populations; and yet be interconnected enough to permit recolonization to balance extinction of thus isolated sub-populations" (Harrison 1992). It is worthwhile to mention in support of the above theories that as islands are isolated, and in many cases the species found on them are endemic, extinction has been particularly common on islands. About 93 per cent of the bird species whose extinction has been recorded since 1600 have been island species.

The Percolation theory deals with modelling irregular animal behaviour in a hostile matrix. The theory is important with respect to animal movements and use of resources. One other theory—the Source-Sink theory evolved from the demographic model suggested by Pulliam (1988, 1996). The theory defines 'Source' as a population in which births exceed deaths and immigration rates. On the other hand, a 'Sink' is defined as a population showing negative balance between offspring and deaths and in which juvenile population does not have the capability to replace adult mortality.

Merriam (1984) defines 'connectivity' as a parameter to measure the processes by which the sub-populations of a landscape are interconnected into a demographic functional unit. Connectivity can be seen as the inverse correlate of hostility of the inter-patch habitat. It is a functional parameter that varies according to specific organisms and is frequently not related to landscape structure (Farina 1998). However, connectivity itself stems from the complex interactions between species-specific movement patterns and landscape features. For example, woodland connectedness plays a fundamental role for species that need tree cover for their movements (ibid.). It is important to note that that 'connectedness' and 'connectivity' between habitats are two different terms. Thus, while connectedness—the structural connection between habitat patches—enhances the possibility of connectivity, connectivity itself may not be guaranteed by connectedness. Whatever the dynamics of a population in a fragmented landscape, Beier and Noss (1998) argue that connectivity "enhances population viability for many species".

Corridor Design

Designing wildlife corridors can be an arduous task. There is no set prescription on design parameters and adopting the right strategy depends on a multitude of factors which include species requirements, geographical setting, land availability, and human interference amongst other things. Literature in this area is scant and at best posits panoptic best practice guidelines. While there seems to be broad understanding and agreement on useful principles in designing wildlife corridors, there are no universally applicable rules that can be applied without modification to diverse corridor scenarios. A scrutiny of literature in this area brings out empiric guidelines and significant concerns while designing corridors. Bond (2003) lists the following broad guidelines with reference to design of wildlife dispersal corridors:

- The corridor should be as wide as possible. The corridor width may vary with habitat type or target species, but a rule of thumb is about a minimum of 1,000 feet wide (but larger if possible).
- Maintain as much natural open space as possible next to any culverts to encourage the use of the culverts.
- Maximize land uses adjacent to the corridor that reduce human impacts to the corridor (Beier and Loe, 1992). Isolation effects along corridors can be offset by having surrounding habitat similar to that found within corridors (Perault and Lomolino, 2000).
- Do not allow housing or other impacts to project into the corridor to form impediments to movement and increase harmful edge effects.
- If housing is to be permitted next to the corridor, put conservation easements on adjacent lots to prohibit structures nearest the corridor.
- Develop strict lighting restrictions for the houses adjacent to the corridor to prevent light pollution into the corridor. Lights must be directed downward and inward towards the home.

Stenseth and Lidicker (1992) refer to three types of movements and habitats in corridors (Table 2 and 3).

Table 2

Corridors Characterized by Movement

Movement	
Type	*Characteristic*
Dispersal	One way movement away from home site
Migration	Round trip movements
Home Range	Movements within home range

Table 3

Corridors Characterized by Habitat

Habitat	
Type	*Characteristic*
Transitional Habitat	Suitable only for the movement of a disperser
Marginal Habitat	Allows survival and sometimes reproduction
Survival Habitat	Good habitat which allows survival and reproduction

McKenzie and Bio (1995) suggest the following important parameters in designing corridors (other references below are as cited in McKenzie and Bio, 1995).

Habitat

The extent to which the corridor will be used by dispersers depends on the habitat within the landscape. The habitat within the corridor must be connected by 'high quality' habitat that provides for survival and reproduction. 'Suitable habitat' patches should be continuous and movement is not hindered by presence of gaps (Lovejoy et al., 1986; McEuen,1993). Studies on several species of rodents have shown that there is a strong tendency to remain within suitable habitat while dispersing.

Corridor Shape

Linear corridor shape was found to be superior than other shapes (Soule and Gilpin, 1991).

Corridor Width

Minimum corridor widths may be estimated using home range data of target species. If the corridor is to contain enough suitable habitat for the target species to permanently occupy the corridor, then the corridor width should be at least one home range wide and contain home ranges that are designed to be rectangular and twice as long as wide (Harrison, 1992).

Corridor Length

Corridors may be narrower than the minimum width as described above if they are less than one home range in length to facilitate passage to dispersers without foraging (ibid.).

Corridor Location

It is important to align corridors with other habitat suitable for the species (ibid.).

Landscape Context

Context may be important. In a study of arboreal marsupials, it was found that corridors that included variety of topographic positions—like gullies and ridges—supported a greater abundance of animals and higher species diversity (Lindenmayer and Nix, 1992).

Human Activities

Intensity of human activities will determine the effectiveness of the corridor. In this context type and extent of human activities—like livestock grazing, intrusion by dogs, human presence etc.—inside and adjacent to the corridor will have an important bearing (Harrison, 1992). Corridor design can include buffer zones to reduce or mitigate undesirable human activities.

Horn (2003) has designed corridors to connect forest patches in South Australia's green triangle plantations. Horn suggests that varying the width along longer corridors, in particular by creating 'nodes' may also improve their viability and reduce edge effects on target species. The value of suitable habitat has been again underlined in this study which proposes that corridor value decreases with length and as access to native vegetation decreases, potential use of corridor is also reduced. Corridors 40 m and 80 m wide were found to be suitable in this case. Plat (2002) suggests some broad wildlife corridor design principles (Table 4).

Table 4
Summary of Corridor Design Principles

Issue	*Principle (s)*
Patches	
Quality size and number	• Protect the best native vegetation first • The bigger the better • The more types of habitat, the better
Shape and edges	• The more compact, the better • Consider 'edge effects' • Include buffers
Position	• Include all land classes
Sites	
Local significance	• Include watercourses • Provide for threatened species (rare, vulnerable, endangered)
Linkages	
Connectivity and corridors	• The more connected, the better • Include corridors • Provide stepping stones
Matrix	
Mosaics	• Integrate nature conservation areas with surrounding land use

Some important points that need to be highlighted from the above principles include the stress on the need to conserve native vegetation, inclusion of buffers, consideration for edge effects, provision of 'stepping stones' and integration of nature conservation areas with surrounding land use.

Need for Studies on Corridor Use

Dispersal is a demographic characteristic of a population of some animal species in which young individuals move away from their maternal ranges to find and establish their own territories. In this process, individuals are exchanged between populations which are otherwise isolated. If there was a large contiguous patch of suitable habitat, the dispersers would move between areas of this habitat. However, the typical habitats of today represent more or less isolated patches sometimes connected by narrow linear habitat linkages. Thus, in such fragmented landscapes, dispersal is the process that brings about connectivity between otherwise isolated patches/populations. This connectivity helps in keeping the population viable by avoiding inbreeding depression.

To understand the role of corridors in establishing connectivity through dispersal, it is essential to understand what Revilla et al. (2004) call interplay between the behavioural dispersal characteristics of the species, the structure of the underlying landscape, and the demographic frame where it occurs.

The only way to study such behaviours in animals today is to analyze information stored in a dispersal track obtained using telemetry. Information can be decoded from these tracks to reveal the general patterns in dispersal movements (Revilla et al., 2004). Such tracks are best obtained for dispersing individuals using Global Positioning System (GPS) telemetry. In this, the animal is fitted with a collar containing a GPS receiver that calculates and records the animal's location, time and date at programmed interval. The benefits of GPS tracking outweigh its cost which at first looks prohibitory. The researcher can take locations as frequently as every minute, can reduce the frequency when the animal is not moving or during the general resting hours with a potential accuracy of 5m. A very informative and interesting comparison of conventional VHF radio telemetry, satellite telemetry and GPS telemetry is given by Mech and Barber (2002) in their report to the US National Parks Service.

Studies involving GPS tracking of animals will immensely advance our knowledge of their dispersal behaviour. This, if carried out in context of a fragmented landscape containing corridor elements as well as isolated habitat patches, would also reveal the value of corridors as agents of connectivity for wild populations. Of course, such studies have their own limitations of enormous expenses and procedural difficulties such as permissions to put GPS-collars on animals in statistically significant numbers which is necessary to get required data and time constraint in selectively getting juvenile males.

Conclusion

It is indeed extremely difficult to assess and quantify the benefits drawn from corridors in biodiversity conservation because of large number of species-specific factors like their preference for vegetation types, corridor dimensions and the biogeographic location. The use of corridors can also vary intraspecifically according the gender and age. However, a large number of studies indicate that corridors have significant value in maintaining biological diversity and are therefore an important factor in the conservation management system. Corridors help reduce isolation of spatially separated populations and at the same time potentially increase the total habitat available to a species. Dispersal across the connected patches may also enhance the meta-popualtion survival by increased opportunity through individual survival. Despite contrasting viewpoints on the utility functions and effectiveness of corridors, there is sufficient evidence to suggest that the linkages which connect larger habitats need to be preserved to maintain viable populations in the long run. As suggested by Beier and Noss (1998), the burden of proof against the value of corridors lies with their destructors.

While wildlife corridors may not provide immediate respite for loss of connectivity between key habitats, with careful planning, design and implementation, they can however help reduce overall negative effects of fragmentation by providing a dispersal medium across patches. More studies focusing on monitoring the use of corridors by various species need to be undertaken to fully understand and document the conservation value of corridors. This will greatly help in better planning and management of forested landscapes and species conservation in the future.

References

Alfred, J.R.B., (1998), "Faunal Diversity in India: An Overview", in *Faunal Diversity in India*, I–VIII, 1–485. (ed.), Alfred, J.R.B. et al. (1998), Envis Centre, Zoological Survey of India, Kolkata.

Beier, P. and R.F. Noss, (1998), "Do Habitat Corridors Provide Connectivity?", *Conservation Biology*, 12(6), 1241–52, Beier, P. (1993), "Determining Minimum Habitat Areas and Habitat Corridors for Cougars", *Conservation Biology* 7:94–108.

Beier, P. and S. Loe, (1992), "A Checklist for Evaluating Impacts to Wildlife Movement Corridors", *Wildlife*, Society Bulletin 20: 434–440.

Beier, P., (1995), "Dispersal of Juvenile Cougars in Fragmented Habitat", *Journal of Wildlife Management*, 59: 228–37.

Beier, P. and R.F. Noss, (1998), "Do Habitat Corridors Provide Connectivity?", *Conservation Biology* 12(6), 1241–52.

Bond, Monica, (2003), "Principles of Wildlife Corridor Design", http://www.biologicaldiversity.org/swcbd/Programs/sprawl/wild-corridors.pdf accessed April 2006.

Craighead, L., R., Walker, R. Noss, and K. Aune, (1997), "Applying Conceptual Models to Landscapes: Using Habitat Suitability Models of Selected Species to Define Core Areas and Buffer Zones for Wildlife Corridors", Oral presentation, 1997 Annual Meeting of the Society for Conservation Biology, Victoria.

Craighead, F.L. and E. Vyse, (1996), "Brown/grizzly Bear Metapopulations", D. McCullough (ed.), *Metapopulations and Wildlife Conservation Management*, Island Press, Washington DC and Covelo California, Chapter 14: pp. 325–51.

Ehrlich, P.R., (1986), "The Loss of Diversity", E.O. Wilson (ed.), *Biodiversity*, National Academy Press, Washington D.C. pp. 21–27.

Farina, A., (1998), *Principle and Methods in Landscape Ecology*, London: Chapman and Hall.

Gilpin, M.E., and M.E. Soulé, (1986), "Minimum Viable Populations: Processes of Species Extinction", M.E. Soulé (ed.), *Conservation Biology*, The Science of Scarcity and Diversity, pp. 19–34.

Gibbs, J.P., and J. Faaborg, (1990), "Estimating the Viability of Ovenbird and Kentucky Warbler Populations in Forest Fragments", *Conservation Biology*, 4: 193–196.

Gujarat Forest Department, (2006), GFD website at http://gujaratforest.gov.in/ accessed, April 2006.

Hilty, J.A., (2004), "Use of Riparine Corridors and Vineyards by Mammalian Predators in Northern California", *Conservation Biology*.

Harris, L., (1984), *The Fragmented Forest: Island Biogeography Theory and the Preservation of Biotic Diversity*, The University of Chicago Press, Chicago.

Harrison, R.L., (1992), "Toward a Theory of Inter-Refuge Corridor Design", *Conservation Biology*, 6(2).

Horn, T., (2003), "South East Biodiversity Corridor Strategy: Planning for Tomorrow's Biodiversity, Forestry SA", Government of South Australia.

ICE6, (1994), "ICE6 Tool Kit for Cumulative Effects Analysis. Systems for Environmental Management", Missoula, MT.

John Singh, A.J.T., Prasad, S.N and S.P. Goyal, (1990), "Conservation Status of the Chila-Motichur Corridor for Elephant Movement in Rajaji-Corbett National Parks Area, India", *Biological Conservation*, 51, 125–138.

Kattan, G.H., H. Alvarez-Lopez, and M. Giraldo, (1994), "Forest Fragmentation and Bird Extinctions: San Antonio Eighty Years Later", *Conservation Biology* 8: 138–146.

Levins, R., (1970), *Extinction*, pp. 77–107, M. Gerstenhaber (ed.), "Some Mathematical Problems in Biology", Providence, Rhode Island: American Mathematical Society.

Lindenmayer, D.B. and H.A. Nix, (1993), "Ecological Principles for the Design of Wildlife Corridors", *Conservation Biology* 7(3).

Lovejoy, T.E., Bierregaard, R.O. Jr. Rylands, A.B. Quintela, C.E. Harper, L.H. Brown, K.S. Jr. Powell and A.H. Powel, G.V.N., Schubart, H.O.R. and M.B. Hays, (1986), "Edge and Other Effects of Isolation on Amazon Forest Fragments", Soule, M. E. (ed.), *Conservation Biology*, The Science of Scarcity and Diversity, Sinauer and Associates, Sunderland, MA. pp. 251–285.

MacDonald, M.A., (2003), "The Role of Corridors in Biodiversity Conservation in Production Forest Landscapes: A Literature Review", *Tasforests*, Vol. 14, 41–52.

McAurthur. R.H. and E.O. Wilson, (1967), *Theory of Island Biogeography*, Princeton University Press, Princeton, New York, USA.

McCord, C.M., (1974), "Selection of Winter Habitat by Bobcats on the Quabbin Reservation", Massachusetts, *Journal of Mammalogy* 55: 428–437.

McEuen. A., (1993), "The Wildlife Corridor Controversy: A Review", *Endangered Species Update*, 10 (11 and 12).

Mech, L.D., and S.M. Barber, (2002), "A Critique of Wildlife Radio-tracking and its Use in National Parks", Report submitted to the US National Parks Service, Fort Collins, Colorado.

Merriam G., (1984), "Connectivity: A Fundamental Ecological Characteristic of Landscape Pattern", *Proceedings of the International Association for Landscape Ecology* 1: 5–15.

McKenzie, E. and R.P. Bio, (1995), "Important Criteria and Parameters of Wildlife Movement Corridors: A Partial Literature Review", Available online http://www.silvafor.org/publications/library/docs/Landscapepercent20 Corridors.pdf., Accessed February, 2007.

Noss, R.F., (1983), "A Regional Landscape Approach to Maintain Diversity", *BioScience* 33:700–706.

Noss, R.F., (1987), "Protecting Natural Areas in Fragmented Landscapes", *Natural Areas Journal*, 7:2–13.

Noss, R.F., (1992), "The Wildlands Project: Land Conservation Strategy", *Wild Earth* (special issue),10–25.

Noss, R.F. and L.D. Harris, (1986), "Nodes, Networks, and MUMs: Preserving Diversity at all Scales", *Environmental Management*, 10:299–309.

Noss, R.F., Quigley, H.B., Hornocker, M.G., Merrill, T. and P.C. Paquet, (1996), "Conservation Biology and Carnivore Conservation in the Rocky Mountains", *Conservation Biology*, 10(4), 949–963.

Paetkau, D., L. Waits, Clarkson, P. Craighead, L., Vyse, E., Ward, R., and C. Strobeck, (1997), "Dramatic Variation in Genetic Diversity Across the Range of North American Brown Bears".

Perault, D.R. and M.V. Lomolino, (2000), "Corridors and Mammal Community Structure Across a Fragmented, Old-growth Forest Landscape", *Ecological Monographs,* 70: 401–422.

Parks Canada, (2006), "Elk Management in Banff National Park", http://www.pc.gc.ca/pn-np/ab/banff/plan/plan23a_E.asp (April 2006).

Platt S.J., (2002), "How to Plan Wildlife Landscapes: A guide for Community Organisations", Department of Natural Resources and Environment, Melbourne, VIC. Project Tiger (2006). http://projecttiger.nic.in/ whyaretigerindanger.htm, Accessed, April, 2006.

Pulliam, H.R., (1988), "Sources, Sinks, and Population Regulation", *American Naturalist*, 132: 653–661.

Pulliam, H.R., (1996), "Sources and Sinks; Empirical Evidence and Population Consequences", pp. 45–69 in O.E. Rhodes, R.K. Chesser, and M.H. Smith, (ed.), *Population Dynamics in Ecological Space and Time*, University of Chicago Press, Chicago, Illinois, USA.

Revilla. E., Wiegand. T., Palomares F., Ferreras,P. and M. Delibes, (2004), "Effects of Matrix Heterogeneity on Animal Dispersal: From Individual Behaviour to Metapopulation-level Parameters", *The American Naturalist*, 164(5), 130–153.

Sanctuary Asia, (2004), "Borivili National Park, Mumbai, Man Vs Animal", July, 2004 available at http://www.sanctuaryasia.com/news/detailnews.php? fid=1001, April, 2006.

Saunders, D.A. and R.J. Hobbs, (eds.), (1991). *Nature Conservation 2: The Role of Corridors*, Beatty and Sons, Surrey UK, p. 442.

Shaffer, M., (1978), "Determining Minimum Viable Population Sizes: A Case Study of the Grizzly Bear (*Ursus arctos L.*)", Ph.D Dissertation. Duke University.

Shaffer, M., (1981), "Minimum Population Sizes for Species Conservation", *BioScience*, 31:131–134.

Shaffer, M., (1987), "Minimum Viable Populations: Coping with Uncertainty", M.E. Soulé (ed.), *Viable Populations for Conservation*, Cambridge: Cambridge University Press. pp. 69–86.

Simberloff, D.S, and L.G. Abele, (1976), "Island Biogeograhy Theory and Conservation Practice", *Science,* 191:285–286.

Simberloff, D. and J. Cox, (1987), "Consequences and Costs of Conservation Corridors", *Conservation Biology*, 1: 63–71.

Simberloff, D., Farr, J.A., Cox, J. and D.W. Mehlman, (1992), "Movement Corridors: Conservation Bargains or Poor Investments?", *Conservation Biology*, 6(4), 493–501.

Soule, M.E., and M.E. Gilpin, (1991), "The Theory of Wildlife Corridor Capability", D.A. Saunders and R.J. Hobbs (eds.), *Nature Conservation 2: The Role of Corridors,* Beatty and Sons, Surrey UK. pp. 3–8.

Stenseth, N.C. and Lidicker. W.Z., (1992), *Animal Dispersal: Small Mammals as a Model*, Chapman and Hall, New York, 365.

UFL, (2005) (online), "When Birds Use Wildlife Corridors, Plants Benefit", http://news.ufl.edu/2005/06/30/wildlife/Accessed April 2006.

USDA Forest Service, (1996), "Status of the Interior Columbia Basin: Summary of Scientific Findings", PNW-GTR-385. Portland, or. p. 144 .

USFWS, (1993), "Grizzly Bear Recovery Plan", US Fish and Wildlife Service, Missoula, Montana, p. 181.

UWI (online), "The Theory of Island Biogeography", (http://web2.uwindsor.ca/courses/biology/macisaac/55-437/lecture9.htm) Accessed April, 2006.

25 Sustainable Forestry for Mitigating Carbon Sequestration

Case of Delhi

R.B. Singh and Abhineety Goel

Forest ecosystem has potential to capture and retain large volumes of carbon over long periods as trees absorb carbon through photosynthesis process. A young forest, when growing rapidly, can sequester relatively large volumes of additional carbon roughly proportional to the forests growth in biomass. A mature forest acts as a reservoir, holding large volumes of carbon even if it is not experiencing net growth. Forest management can thus have an influence on carbon sequestration. Present estimates indicate that with appropriate policies the carbon pool in the terrestrial system could increase by up to 100 Gtc (Gigatones of Carbon) by the year 2050 compared to the level of carbon that would be sequestered without such policies (IPCC, 2001). This is equivalent to about 10 to 20 per cent of projected Greenhouse Gases emissions from fossil fuel consumption during the same period. Reducing deforestation, expanding forest cover, expanding forest biomass per unit area, and expanding the inventory of long-lived wood products are some of the activities that could help global community realizing the carbon sequestration potential of forest ecosystem.

The modification of atmospheric environment by land use of cities is the most apparent anthropogenic activity that impacts on city's microclimate. The city's land use pattern destroys existing microclimate and creates great complexity. With the onset of the industrialization era, the trend towards urbanization accelerated leading to significant changes in the climate in and around the most mega-cities (Singh, 2006).

The Study Region

The urban population of Delhi, which was only 0.21 million in 1901, has increased by about 60 times to 12.82 million in 2001. Population density of Delhi was 9,294 people per km^2 in 2001. The capital of India attracts a growing population and about 93 per cent of Delhi's population lives in urban part. The present migration rate is more than 1.6 lakh persons per annum. In 1991, Delhi's total 1483 km^2 area, out of it 46.2 per cent was urban settlement, with only 53.79 per cent remaining in rural areas, as compared to 39.9 per cent and 60.1 per cent, respectively in 1981. At present National Capital Territory(NCT) Delhi has 60 per cent urban and 40 per cent rural area (2001). Delhi with 28 industrial areas is a major centre for the whole of North India, with wholesale markets for nine different types of goods including fruits and vegetables, automotive parts, textiles etc. The total number of industries in Delhi was about 8,160 in the year 1951, which steeply increased to 1,29,363 by 1998 (approximately 18 times) while the employment in industry grew from 16.94 per cent in 1951 to 32.43 per cent in 1991 (ibid.).

Urban Development and Carbon Emission

The rapid growth of urbanization and industrialization inevitably led to higher consumption of energy usually reflected in increasing pollutant emission. The main sources to atmospheric gases and matters are due to burning of fossil fuels to generate heat and power. In the Indian urban cities, atmosphere constantly receives inputs from anthropogenic activities sources. Among the anthropogenic sources are pollutants from domestic heating and lighting, transport automobiles, railway, industrial activities, and incineration of municipal and industrial and domestic wastes. Some of these may also constitute fine particles from fossil fuel like coal. In Delhi and Mumbai, the major sources of atmospheric pollution are the vehicular source followed by industry and domestic sources (Figure 1), while in Kolkata major emission sources are industry and power generation plants. In these cities, the industrial activity is a major source of SO_2 and SPM emission whereas vehicular and domestic are the main source of CO_2 emission in the city's atmosphere (Kumar and Singh, 2003).

Figure 1

Registered Motor Vehicles in Selected Indian Metros

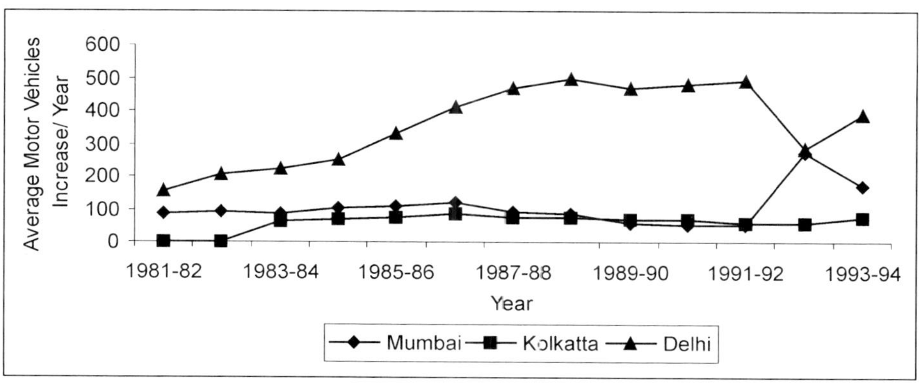

Source: Kumar and Singh, 2003.

The major gaseous pollutants include sulphur dioxide (SO_2), oxides of nitrogen (NO_x), and carbon dioxide (CO_2). The suspended particulate matter (SPM) levels were observed to be much above the permissible levels at all the monitoring stations. The vehicular and industrial sectors are the major sources of oxides of nitrogen. The major anthropogenic sources of SO_2 in the atmosphere are burning of fossil fuels for industrial and domestic purposes. For CO_2, emissions taken into account for the release is 15.9 million tonnes. These emissions can rise to 32.6 million tonnes if the present state of development continues without any major emission mitigation efforts.

Sources of Pollution

Delhi faces air pollution problems due to three major sources: transport, domestic, and industrial sectors. Vehicle population has increased from 2.4 million in 1993–94 to 4.2 million in 2003–04. Thus, an increase of about 87 per cent has been registered in a period of 10 years (Government of NCT of Delhi). Delhi has experienced an exponential growth in the number of personalized vehicles over the last two decades. Petrol driven vehicles account for more than 81 per cent of the pollution, whereas diesel vehicles account for only 19 per cent of the total pollution of the city.

The other two major sources of Delhi's air pollution loads are fuel combustion in industrial activities and domestic stoves. Among the large-scale industries, thermal power plants are the most prominent contributors to air pollution. Alone they are responsible for the release of about 302 tonnes of pollutants per day. They were responsible for as much as 10 per cent (approx.) of the air pollution load in 2001. Although the concentration of carbon monoxide had a gradual decline from 4,183 mg/m^3 in 2,001 to 3,258 mg/m^3 in 2002, still it continues to be above the danger mark of 2,000 mg/m^3.

According to Figure 2, emissions from the industrial sector have gone down from 56 per cent in 1970–71 to 40 per cent in 1980–81 and 29 per cent in 1990–91. It has been projected to further decrease to 20 per cent in 2000–01. The emission load from vehicular sources also shows a clear increase.

Figure 2

Sources of Pollution in Ambient Air Quality in Delhi

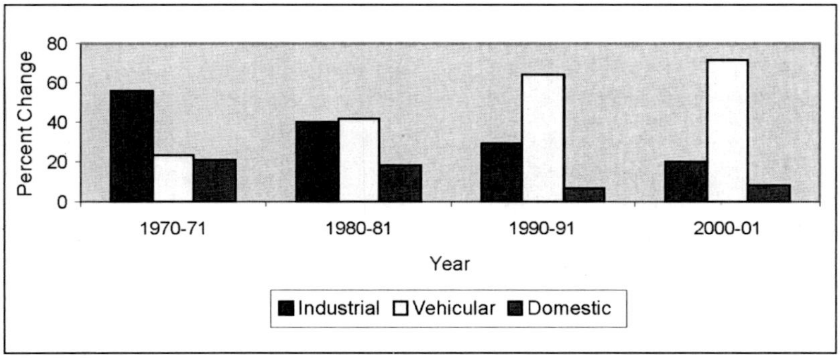

Source: TERI, 2003.

Presently 1,68,000 metric tonnes of pollutants are being emitted. This is about 50 per cent more than 87,000 metric tonnes emitted in 1987. In the last eight years, emission of carbon monoxide has gone up by 120 times, hydrocarbons by 204 times, sulphur dioxide by 5.8 per cent, oxides of nitrogen by 45 times and lead particles by 100 times (Malik, 1997). Every day 1,220 metric tonnes of carbon dioxide, 570 tonnes of hydrocarbons, 330 metric tonnes of nitrogen oxide and 253 metric tonnes of sulphur dioxide are released in Delhi's air.

Though every Delhite breathes nearly 16.5 kg of air loaded with vehicular emissions, the auto, bus and truck drivers get maximum exposure to the vehicular pollution due to their long hours (14–18 hours) of driving (ibid.).

Suspended particulate matter at all stations is far higher than the annual average concentration (Central Pollution Control Board standard for residential and industrial areas is 140 mg/m^3). The annual NO_x concentrations at all the stations are within Central Pollution Control Board standards. The standard for residential areas is 60 mg/m^3. The annual concentration of SO_2 at all stations is well within the Central Pollution Control Board standard for residential and industrial areas (60 mg/m^3) (Figure 3).

Figure 3

Gases Emissions in Industrial and Residential Sectors in Delhi from 1987 to 2003.

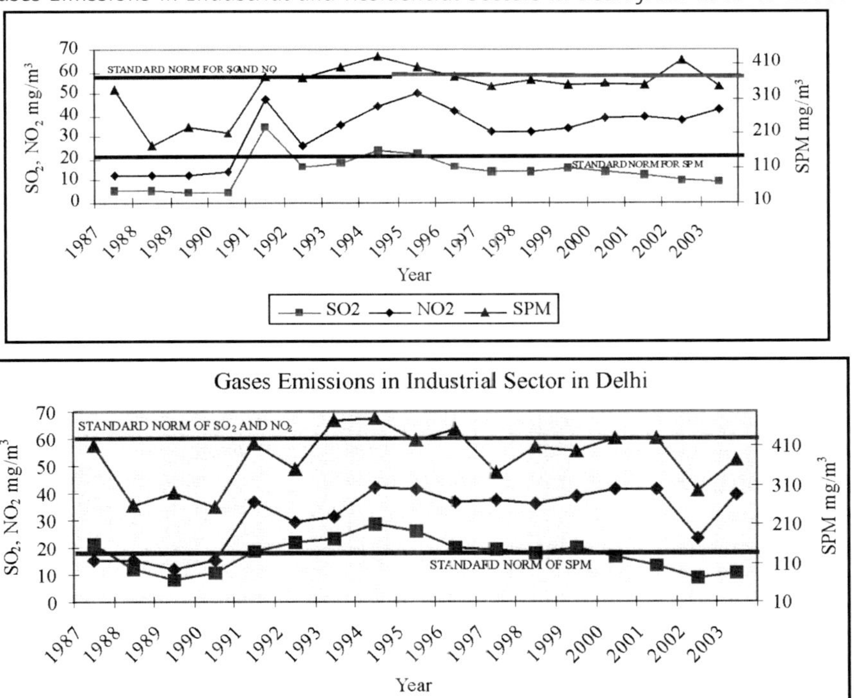

Source: http://www.cpcb.nic.in/cpcb/advance_ser/b.php

As such SO_2 level has decreased by 53.39 per cent in 2003 as compared to 1998 (i.e., year of CNG initiative). This shows tremendous achievement after conversion of all buses/taxis/autos in CNG mode. Annual average value of NO_2 has decreased significantly (4.82 per cent) in 2003 as compared to previous year. It was 45mg/m^3 in 2003 as against 47.28 mg/m^3 2002. Thus ambient air quality has improved significantly which can be gauged from the fact that as compared to 1997 the concentration of carbon monoxide has fallen by 41.14 per cent. Sulphur dioxide level has reduced by 49.20 per cent. However, not much change was observed in case of NO_2. There was tremendous improvement in concentration of particulate matters (SPM and RSPM) in the ambient air in 2003 as compared to 2002 (Table 1).

Table 1

Yearwise Annual Mean Ambient Air Quality Levels in Delhi: Year Concentration in Ambient Air (mg/m^3)

Year	*SO_2*	*NO_x*	*CO*	*SPM*	*RSPM*
1997	18.68	44.85	4810	362.58	—
1998	20.37	42.17	5450	377.92	—
1999	19.46	40.11	4241	374.92	—
2000	18.03	41.83	4686	430.83	191.00
2001	14.10	41.75	4183	381.67	150.08
2002	11.33	47.28	3258	455.92	192.25
2003	9.49	45.00	2831	352.30	148.86

Source: Economic Survey of Delhi, 2003–04.

Major Driving Forces

Population Growth

Promoted to capital of British India in 1911, Delhi became the capital of independent India in 1947. After 1947, Delhi had to accommodate 4,70,000 refugees from west Punjab and Sindh, while 3,20,000 Muslims left the city for Pakistan (Guilmoto and Vaguet, 2000). Delhi has experienced the highest demographic growth in the last few decades. Being the administrative capital, it attracts people from all over India for political, economic, social reasons.

According to Guilmoto and Vaguet, since 1961, Delhi has been the third largest Indian metropolis, behind Mumbai and Kolkata. Delhi has experienced the highest demographic growth of the last decades: 5.1 per cent annually from 1951 to 1961, 4.5 to 4.6 from 1961 to 1981, and 3.9 per cent annually from 1981 to 1991. Its population increased from 1.4 million in 1951 to 8.4 million in 1991, and 13.78 million inhabitants in 2001 (Figure 4).

Figure 4

Percentage Change in Urban Population

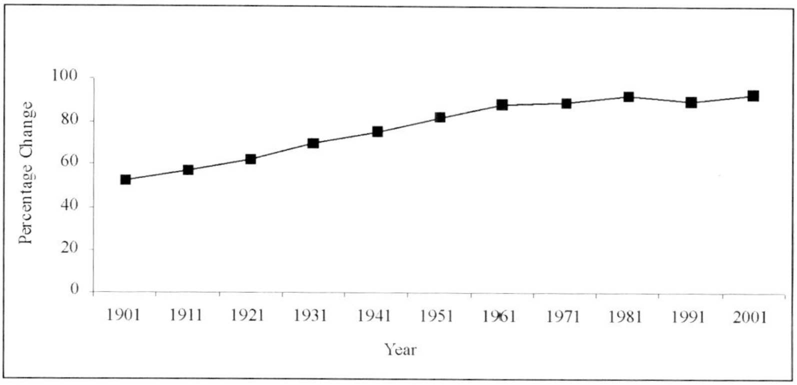

Source: http://delhiplanning.nic.in/Economic % 20Survey/Ecosur2001-02/PDF/chapter3.pdf

Industrialization in Delhi

Out of a total of 1,26,000 industries, there are 98,000 industries in non-conforming areas as per the Master Plan of Delhi, 2001 (Figure 5). The Supreme Court has directed a number of polluting industries in Delhi to close down and ordered certain categories of industries to shift from non-conforming to conforming areas. In July 1996, closure and shifting of 168 industrial units took place in Delhi. In September 2000, 27 'polluting' types of industries were shut down. Industries were shifted to outskirt villages of Narela and Bawana.

Figure 5

Growth of Industries in Delhi from 1951–2001

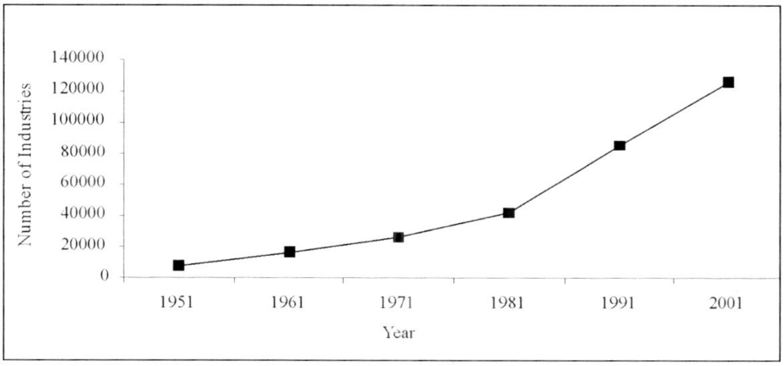

Source: Kumar and Singh, 2003

Developmental Activities

The Metro Rail Transit System has been a very well planned network from Rithala to Shahdara. The trees cut and transplanted for the construction of the metro corridor at the various locations in the north and central Delhi has tremendously changed the scenario of Delhi. Total, 3,548 trees were cut and against these only 1,749 trees were planted. Thereby making a total deficit of 1,799 (Delhi Metro Rail Corporation, 2003). More trees have been cut than the trees planted. The plantations have been done in areas where it is not required. It should have been planted at the locations of the construction of the Metro and Rail corridors where it is urgently required as the existing trees have been cut leading to the locations becoming prone to the environmental degradation in form of pollution etc. This is a case of faulty implementation.

Forest as an Instrument of Carbon Sequestration

The Ridge forest in Delhi is one of the world's largest city forests and this natural vegetation of Delhi is botanically described as a 'Scrub forest' (Singh, 2002). In 2003, Delhi's total forest and tree cover constituted about 268 sq km, which is about 18.07 per cent of the state's geographic area (Forest Survey of India, 2003) (Table 2). This is a remarkable increase in the forest cover from 15 sq km in 1987 to 170 sq km in 2003.

Table 2

Forest and Tree Cover in Delhi

Total Forest and Tree cover	268 sq km
Of State's Geographic Area	18.07 per cent
Per-capita Forest and Tree Cover	0.002 ha

Source: State of Forest Report, 2003.

Role of Vegetation

Plants have been seen to play an important role in both reducing the environmental pollution load as well as serve as pollution indicators without sustaining damage or retardation. It acts as an effective carbon sink and natural air conditioner lowering temperatures. Trees reduce thermal pollution by checking heat build up in the atmosphere by preventing the radiation of heat from base surfaces.

Vegetation cover acts as a pollution scavenger or pollution sink, as it absorbs gases and gathers particulate matter through leaves having large surface areas. The green and leafy portion of the trees and plants has the capacity to filter dust, smoke and other pollutants in the air. Some species like *Ficus, Mangifera,* etc., also act as good dust collectors. One hectare of woodland (1,000 trees) absorbs as much as 3.7 tonnes of CO_2 from the atmosphere and gives out 2.5 tonnes of life sustaining oxygen. Studies show over a distance of 200 sq m or less, a green cover can reduce the concentration of pollutants by 120gm/cubic metre. A full-grown Peepal (*Ficus religiosa*) tree is, for instance, estimated to give out 600 kg of oxygen in 24 hours. Herein lies the importance of the 151 sq km of forest and tree cover in Delhi, though this constitutes 10.2 per cent of the Delhi's geographical area of 1,483 sq km.

According to the Vatavaran report, when a patch of wilderness is converted to a park the following changes are noticed:

- Temperature goes by 2–4°C.
- Noise level goes up by 40–50 decibels.
- 1 sq km of forest releases 3 tonnes of oxygen whereas a park of same size releases one tonne of oxygen. Oxygen generation goes down.
- 1 sq km of forest absorbs 4.2 tonnes of carbon dioxide whereas a park absorbs 1.7 tonnes of carbon dioxide. Absorption of carbon dioxide goes down.
- Dust absorption capacity goes down. A 10 sq m forest area can absorb 11 gms of dust in 24 hours whereas a park of same size at the same time absorbs 6 gms of dust.

There is a great need to identify tree and plant species that can withstand and combat the pollution of the city effectively. For instance, it was found that Jamun (*Eugenia jambolana*), was the most suitable to plant along roadsides as it was the least affected by lead exhaust. Tall plants are less susceptible to pollutants as revealed from more than the 90 per cent survival percentage of plantation along Shahdara Link drain running parallel to a road from Delhi-Uttar Pradesh border (Report, TCPO, 2003). Along the air pollution tolerant tree species, those suitable for Delhi are Peepal (*Ficus religiosa*), Siris (*Albizzia lebbek*), Amaltas (*Cassia Fistula*), Ber (*Zizyphus jujuba*), Neem (*Azadirachta indica*), Imli (*Tamarindus indica*), Shisham (*Dalbergia sisso*), Dhak (*Butea monosperma*), Gular (*Ficus glomerata*), Pilkhan (*Ficus infectoria*), Sataun (*Alstonia Scholaris*), Desi kikar (*Acacia nilotica*).

Species suitable for area where gaseous pollutants are dominant are—Desi Kikar (*Acacia nilotica*), Bel (*Aegle marmelos*), Ulloo Neem (*Ailanthus excelsa*), Siris (*Albizzia lebbek*), Sataun (*Alstonia Scholaris*), Neem (*Azadirachta indica*), hisham (*Dalbergia sisso*), Peepal (*Ficus religiosa*), Pilkhan (*Ficus infectoria*), Lagerstromia flosreginae, Maulsiri (*Mimusops elengi*), Kaner (*Nerium indicum*) (Greening Delhi Action Plan, 2001).

Species suitable for abatement of noise and dust pollution are—Neem (*Azadirachta indica*), Amaltas (*Cassia Fistula*), Peepal (*Ficus religiosa*), Pilkhan (*Ficus infectoria*), Maulsiri (*Mimusops elengi*), Gular (*Ficus glomerata*), Arjun (*Terminalia arjuna*), Kigelia pinnata, Jamun (*syzygium cumini*), Bauhinia variegata, Semul (*Bombax ceiba*), *Cassia nodusa, Cassia javanica, Jacaranda mimosaefolia*.

Afforestation as Carbon Mitigation Measures

The forestry sector provides a number of carbon sequestration options such as increasing number of 'green belts' of pollutant resistant trees and shrubs. One hectare of woodland absorbs 3.7 tonnes of carbon dioxide from the atmosphere and gives out 2.5 tonnes of oxygen. Herein lies the importance of the 151 sq km of forest and tree cover in Delhi, though this constitutes 10.2 per cent of the total geographical area of 1,483 sq km The current rate of afforestation in Delhi is about 1.1 per cent, which has only marginal carbon sequestration potential. In Delhi, various afforestation programmes has been started by different governmental and non-governmental organizations along with the efforts taken by its citizens.

According to one such programme started by the Government of India in 2003–04, the various Residents Welfare Associations, particularly those who are associated with the Bhagidari Programme (a programme started by the state government in 1999 to involve people of Delhi in improving the social and environmental situation of Delhi) will act as the basic co-coordinating unit in the programme. Their work will include—informing the residents about the programme, gathering volunteers for planting, identifying places for planting, assessing seedling requirement, identify types of species desired, identify nursery from where desired seedling will be procured, planning labour input for digging the pits, coordinating with the non-government organization and government agencies and chalking out programme of plantation in their respective colonies so that planting can be done by government agencies in association with Resident Welfare Association (Greening Delhi Action Plan, 2003).Several organizations such as environmental groups, educational groups, Rotary Clubs, Lions Club, governmental organizations like Municipal Corporation of Delhi, Central Public Works Department, Department of Horticulture, Government of Delhi, Public Works Department (Horticulture), New Delhi Municipal Committee, Delhi Development Authority etc., will be identified and involved closely in the programme as motivators and for conducting nodal activities. Their work will include: coordination with Resident Welfare Associations and other bodies like market associations for plantation activities, guiding people to plant trees properly.

The corporate sector can play decisive role in the programme by sponsoring and arranging logistic and other requirements. This can include sponsorship of advertisements for awareness; provide vehicles for transportation of seedling; provide tree guards for the plants. Besides, various corporate bodies can adopt schools for planting and other environment improvement programmes. The forest department would coordinate the programme at various levels. It will decide upon distribution points for free distribution of saplings as also the species to be distributed for planting. The involved stakeholders have already started contributing towards the goals of the programme by distributing more than 3 lakh saplings free of cost.

Conclusion

The rapid industrialization and fast urbanization in Delhi has resulted in its environmental degradation in various ways such as population growth, developmental activities such as metros, flyovers, etc. Forest management can thus have an influence on carbon sequestration. Reducing deforestation, expanding forest cover, expanding forest biomass per unit area, and expanding the inventory of long-lived wood products are some of the activities that could help global community realizing the carbon sequestration potential of forest ecosystem. As can be seen in the case of Delhi, the forest cover over the years has been increasing, from 15 sq km in 1987 to 170 sq km in 2003.

References

Delhi Metro Rail Corporation, (2003), Unpublished Data.

Forest Survey of India, (2003), "State of Forest Report, 2003", Forest Survey of India, Dehradun.

Government of India, "Greening Delhi Action Plan, 2003–2004", Department of Forest and Wildlife, Delhi.

Government of NCT of Delhi, (2003), "Economic Survey of Delhi 2003–04", Planning Department, Government of NCT of Delhi.

Guilmoto, C.Z. and Vaguet, A., (2000), *Essays on Population and Space in India*, French Institute of Pondichery, Pondichery, pp. 133–151.

Intergovernmental Panel on Climate Change, (2001), *Climate Change 2001: Mitigation*, Cambridge University Press, Cambridge, UK.

Kumar, Bhuwan and Singh R.B., (2003), *Urban Development and Anthropogenic Climate Change: Experience in Indian Metropolitan Cities*, Manak Publications Private Limited, Delhi pp. 136 to 173.

Malik, Iqbal, (1997), "Environmental Landmarks of Delhi and their Impact on Biodiversity", in *Fauna in Delhi*, Zoological Survey of India.

Malik, Iqbal, (1998), "Static Ridge-Vatavaran Report", Vatavaran, Delhi.

Singh, Prithipal, (2002), "Plants in Delhi's Environment" in Nair, K.R.G. et al., (2002), *Environment and Sustainable Development: Indo-Canadian Perspectives*, Kaveri Books, Delhi.

Singh, R.B., (2006), (ed.), *Sustainable Urban Development*, New Delhi: Concept Pub.

TERI, (2003), "Work Stream 4: Air Quality Management", New Delhi: Tata Energy Research Institute, 96 pp.

Town and Country Planning Organization, (2003), "Plants for Abatement of Vehicular Pollution: A Case Study of Delhi", Town and Country Planning Organization, Delhi.

Internet references:

http://www.cpcb.nic.in/cpcb/advance_ser/b.php

http://delhiplanning.nic.in/Economicpercent20Survey/Ecosur2001-02/PDF/chapter3.pdf

26 Indigenous Knowledge for Sustainable Biodiversity and Food Security in the Kollihills, Tamil Nadu

T. Vasantha Kumaran

For long years, tribal people have been the custodians of biodiversity. They have been more so in the hill areas where they are confined even today. Unfortunately, however, their basic requirements sometimes force them into activities such as deforestation for monetary gains or for extending agriculture that leads to a loss of biodiversity (Kumaran, 1983; 1991; 1998; 2003). This has happened through what we now believe is a resource process in the hills. It is now a common understanding that indigenous or local knowledge can be used to fulfil socio-economic needs as well as conserve biodiversity (Rajasekaran, 1994; Mitchell, 1997). This study explores the diversity of the indigenous agroforestry systems of the Kollihills, identifies the factors which force the tribal people to engage in activities that erode those systems, and also the farming systems in practice, and formulates policy interventions designed to make effective use of indigenous knowledge in biodiversity building and agricultural development through a modified resource process.

Although material has been drawn from earlier research by author (Kumaran, 1983; 1991; 1993; 2003), field work for the study reported was conducted during 1995, 1998 and 2003, with interviews and participant observation, and the method of data collection was in the form of book-keeping and diary method for indigenous or local knowledge and human time-budgeting. Elder tribals were closely consulted for their indigenous knowledge on agroforestry, tree cropping, and crops such as pineapple, jack, banana, rice and tapioca. Technologies, techniques and practices were studied in terms of crops, seasons and hamlet/village communities. Alternatives were considered in terms of resources and constraints analyzed through conventional methods.

Relevance to Present Day Problems and Needs of Tribal Society

The study is relevant to the present day for it takes on the 'biodiversity and indigenous knowledge' paradigm, in geographical research. The researchers recognize that the tribals have systems of knowledge which could solve the problem of biodiversity better than any 'imposed systems'. The study thus forms a model research for similar efforts, especially with the scattered tribes in the state of Tamil Nadu. The policy options developed here will have wider applications. Food security and biodiversity are the two very aspects of human survival and the human experience with them in traditional or tribal society could be a welcome addition.

Indigenous knowledge and biodiversity are increasingly being talked about in geographical research (Mitchell, 1997, for example), there has not been any major efforts at research on the two areas. In Indian geography, this focus would hopefully usher in 'new perspectives and new methodologies' besides enriching the knowledge on human experience in space and time, as we take both a spatial and a temporal view of indigenous knowledge and biodiversity and the agro-forestry systems and food security system in the Kollihills.

Indigenous or Local Knowledge System

As indigenous or local knowledge is central to the study, it is important that it is defined and elaborated in its use. The concept of indigenous or local knowledge systems has its roots in the traditional knowledge and resources and environmental management systems (Mitchell, 1997: 179). Indigenous or native or tribal people are found on every continent and in many countries. In the course of human history, there have been human groups whose interests were linked to the prudent use of their resource base and that such groups evolved approariate resources use practices, based on simple rules of thumb, that ensured the long-term sustainability of the resource base. These practices and the knowledge about them were arrived at through a process of trial and error. Such knowledge is now known as the traditional or local knowledge (Gadgil and Berkes, 1991: 136 and see Box 1).

Box 1

Indigenous or Local Knowledge: Definition

Indigenous or local Knowledge can be defined (Kumaran, 1998:1) as:

- "the knowledge developed by a given culture to classify the objects, activities, and events of its universe".
- Indigenous or local knowledge system is a cumulative body of knowledge and beliefs, handed down through generations by cultural transmission, about the relationship of living beings (including humans) with one another and with their environment.
- Further, indigenous knowledge system is an attribute of societies with historical continuity in resource use practices; by and large, these are non-industrial or less technologically advanced societies, many of them indigenous or tribal.

Any knowledge is a process. It is a useful process. Indigenous or local knowledge cannot be understood independently from the ways in which it changes. It evolves through a process of experimentation, carried out by indigenous, traditional people (Box 2 for some key words and their meanings), involving a diverse set of variables and objectives. This leads to innovations and adaptations. Observation is an important part of the process (Shah, 1993: 38).

Box 2

Key Words and their Meanings

Indigenous	Something which is indigenous is originally from the country in which it is found, rather than coming or being brought there from some other country.
Traditional	1. Traditional is something—like dress and customs—that has existed in a place without changing for a long time. 2. Traditional are the beliefs and attitudes that have existed in a place for a long time and are not exposed to change. 3. Traditional is an organization or institution in which older methods are used in preference to modern ones.
Knowledge	Knowledge is information and understanding about a subject, which a person has in his or her mind or which is shared by all human beings.
System	1. A system is a way or working, organizing, or doing something in which you follow a fixed plan or set of rules. 2. A system is also the way that a whole institution or aspect of society has been organized and arranged. 3. If a situation or activity has some system, it has a sense of orderliness or good organization. 4. The system of something is the way in which it is arranged so that all its parts fit together or work together.

Source: *Collins Cobuild English Language Dictionary*, 1988, London and Glasgow: Collins, pages in order: 741, 1551, 803, and 1484-1485.

Indigenous knowledge is not generated or acquired equally throughout a community; nor is the existing knowledge about products and processes equally distributed in a society. Differences in knowledge exist in and among individuals, in terms of their ability, opportunity or wish to observe and experiment, according to gender, age, social status or personal ability. Mathen (1997: 175) tells about a ten year old tribal boy who could identify 275 varieties of vegetation in and around his village, including 14 different types of grass. He is illiterate; but his knowledge about the plants, their characteristics, and their uses is more than worthy of a doctorate in agriculture. His is not of course a solitary example. There are several hundreds, may be thousands among the indigenous, traditional peoples who are equally knowledgeable. It must be mentioned here that indigenous or local knowledge and technology are quite often supplanted by modern ones. In traditional cultures, life was, and is, organized around a highly refined awareness of the environment. Indigenous knowledge system, which also goes by other names such as local knowledge systems, traditional ecological knowledge, and indigenous technical knowledge, represent experience acquired over several millennia of direct human contact with the environment (Box 3).

Although the term 'indigenous knowledge system' came into wider use only in the 1980s, its practice is as old as ancient hunter-gatherer cultures.

There has been a growing recognition of the value of IKS in several areas, most importantly in matters relating to environment building, improving

Box 3
Knowledge: Source, Generation and Use

Source	*Generation*	*Use*
Local	Locally generated	Locally used
	Externally generated	Locally used
External	Locally generated	Externally used
	Externally generated	Externally used

biodiversity and sustainable development. Indigenous knowledge system represents also an intellectual process of creating order out of disorder, of practising art and science.

The quantity and quality of indigenous knowledge varies among community members, depending upon gender, age, social status, intellectual capability and occupation. Indigenous ecclogical knowledge takes the following forms:

- *Knowledge of Biotic Materials:* an intimate and detailed knowledge of the environment, including plants, animals and natural phenomena.
- *Technical Knowledge*: development and use of appropriate technologies for primary resource utilization, uses of biotic materials, and humane environmental conservation practices.
- *Cultural Knowledge*: cultural practices and beliefs, a holistic worldview that parallels the scientific discipline of ecology.

The Kollihills: A Brief Profile

The Kollihills, an area of about 282.93 sq km, lies in Namakkal and Rasipuram taluks of Salem district. It has 14 Panchayats, 16 revenue villages and 263 hamlets. These hills are forested. The Kollihills are more than 1,100 m above mean sea level. The typical village is called 'Nadu', seven of the 14 Nadus are in Rasipuram taluk while the other seven are in Namakkal taluk, of Salem district. The villages under the taluks are:

Namakkal taluk: Valappur Nadu, Gundur Nadu, Ariyur Nadu, Valavanthi Nadu, Selur Nadu, Thinnanur Nadu and Devanur Nadu.

Rasipuram taluk: Bail Nadu, Edapuli Nadu, Chittoor Nadu, Tirupuli Nadu, Alathur Nadu, Gundani Nadu and Perakkarai Nadu.

Valavanthi Nadu is the headquarters of the Kollihills Tribal Development Block, with Semmedu, a hamlet of Valavanthi Nadu being the administrative headquarters of the block. The only weekly market centre on the hills is at Solakkadu.

The Tribal Population

According to 2001 Census, the population of the Kollihills was nearly 40,000. The Kollihills are populated by the Malayalis (= hill humans) living in the 14 Nadus. People here practice a clanish, kinship based social intercations.

The historical account of the people of the Kollihills, which the elders of the hill people remember well, suggest that they were, until they moved into the hills in the historical times, the peasants of the coastal areas near about Kanchipuram. People belonging to a lineage of three brothers were driven out of their lands and they took to heels to reach the Kollihills, the Pachamalai and the Kalrayan Hills.

In fact, each of the three lineages of the brothers took to these hills such that the youngest landed here. It is widely claimed by the tribals themselves that they belonged to the Kongu Vellala caste, while they settled down in the hills in the 16th century. In the centuries after their occupation of the hills, they acquired tribal characteristics. They are today known as the 'Malayalis', meaning the 'hill people'. They are distinctly different from the Malayalis of Kerala, as the hill tribes speak Tamil, which is their mother tongue. However, they speak Tamil with a slang, which often makes Tamil they speak sound more like Malayalam. They were given the status of the tribe and they are so treated by the government for all administrative purposes.

Education of the tribes is taken care of by a high school which is located in Valavanthi Nadu Panchayat and a number of elementary, middle and high schools in and around the hills. The children generally continue their education up to standard V and drop out heavily at standard VI, as this was the age at which they become farm help to their parents.

The Focus

The study focuses on three important areas of the Kollihills milieu:

1. Resource Process (RP) towards socio-economic development;
2. Indigenous Agro-forestry system (IAFS) meeting needs for a variety of goods and services; and
3. Indigenous Food Security Systems (IFSS), where crops fulfil caloric energy requirements.

Connected to all of these are: (a) biodiversity, and (b) indigeneous knowledge systems. Therefore, the study is an attempt at connecting the three focuses of analysis to the two aspects above.

The significance of the study lies in the fact that the interrelationships between indigenous knowledge, biodiversity and food security system and deforestation are looked at for the first time in the context of the Kollihills. The study also recognizes the fact that the tribes have the ability to solve their problems in food security through the use of indigenous knowledge, if only their socio-economic conditions change and the compulsions of the times are overcome. It is important to note that the proposal envisages also a careful look at the sustainable policy options for sustainable biodiversity and agricultural development. It is moreover a natural extension of the researches conducted by Kumaran (1983) and Kumaran and Aruchamy (1989).

Resource Process (RP)

A Resource Process is defined by its utility humans perceive in the substance and by the technology of transforming the potential of the substance into an actuality of satisfaction.

It is a function of utility, substance and technology.

That is, RP = f{ust}.

The substance for all practical purposes is fixed, finite and limited. But the utility factor and technology factor are not limited at all. Natural resources as functions rather than inventories, are not the least bit limited (Rosell, 1979: 3).

It is with the premise above that the study looks at the resource process and agricultural development in the Kollihills of Tamil Nadu. Agricultural development in tribal areas is seen to be related to decision making on that part of the human environment which comprises of resources (for agriculture). Current insights on and the present situation of agricultural development in tribal areas (for example, Kumaran, 1983; 1987; 1991) help us to arrive at policy interventions designed to make use of the resource process and the indigenous knowledge towards biodiversity building. In this process, it would be useful to take into account the traditional habits and usages peculiar to tribal people concerned, the Malayalis of the Kollihills, and also pay attention to use of plants and crops by them. This naturally involves, in addition, a study of the natural resources that surround the people and form an integral part of their ecology.

Indigenous Agroforestry System

Indigenous agroforestry system involves an integrated approach to land use, characterized by deliberate maintenance of trees and other woody perennials in fields and pastures. The agroforestry of the Kollihills are found on bench terraces (pine apple and tapioca), moderately sloping lands (citrus fruits, citron, guava and pomegranate) and steeply sloping lands (wild trees). Banana suckers are planted randomly inside the forest gardens and around the edges. In some places, the banana is intercropped with tapioca and pepper.

Indigenous Food Security Systems

Indigenous food security systems in the Kollihills form the basis of food and nutritional security. Cereal grains harvested from semi-irrigated agricultural lands form the bulk of the food consumed by the people of the Kollihills. Local landraces of beans and puchased pulses and oils provide the protein they require. The money used in the purchases come from the sale of fruits in the local market, mainly at Solakkadu. Changing socio-cultural values, population growth, poor revenues from the fruit markets and extravagance in cultural and religious activities have resulted in greater need for monetary income. Although the people are aware of the consequences of deforestation, the compulsions of their socio-economic priorities promote the cutting of trees thus leading to increasing disappearance of the indigeneous agroforestry system. It is also clear that the destruction of the forest ecosystem has placed people in a vulnerable situation. These situations lead to the need for an understanding of the interrelationships between (a) indigeneous knowledge, (b) biodiversity, (c) food security and deforestation.

Agricultural Development in the Tribal Kollihills.

Agriculture is a resource process. Agricultural development in tribal Kollihills is related to decision-making on that part of the human environment which comprises of resources (land, water, land utilization types, crop combinations and mixes, crop protection and management and a whole gamut of related matters). Expert opinion on decision-making leading to agricultural development, reveals conflicting features which, by gross simplification, may be designated as clashes between 'the biological and the economic spheres of interest' (Kumaran, 1991; Weitz, Pelley and Applebaum, 1980; Norman, 1974; Weitz, 1971). This situation necessitates a search for a compromise satisfying either aspect. Current insights on and the present situation of agricultural development in tribal areas helps us also to arrive at such a compromise. For example, agroforestry as an agricultural system comes to our assistance, where there is a 'building and strengthening of new and old ecosystems' in balance with nature and the tribal culture. Because the tribe depends on agroforestry, it has a stake to upholding its own interest by support activities that compromise on the biological and the economic. In this process, it would be useful to take into account the traditional habits and usages peculiar to the Malayalis of the Kollihills and also pay attention to plant uses by them. This naturally involves, in addition, a study of the natural resources that surround the people and form an integral part of their ecology.

In recent years, we have become aware of the importance of every step in agricultural development and what each of them holds for the more distant future: food, especially subsistence, security and the fears about 'it' being hard to achieve. And yet it is here we find how little we know of the future, especially, of ecological pressures and their interaction. For this reason, we have to pay equal attention to both the general principles of ecology and the analysis of all points of departure including a framework for future oriented decision-making for agricultural development.

It is pertinent here to note that, in the not too distant past, the economic approach to man and society has been a clearcut winner in disputes with ecology. But confirmed by the test of time, in spite of warnings of biologists, two strategies offer themselves invariably: expansion in space and expansion in time. In the Kollihills, for obvious reasons of the 'limits of the hills' expansion in space can be achieved only in terms of intensity of use. Inter-cropping systems (3, 4, 5 and 7-crop systems in the Kollihills, see Kumaran, 1983; 1993) very widely practised is the testimony to expansion in space. Expansion in time, however, occurs in the very 'characteristics' of agriculture, from traditional to modern, and from purely economic to both biologically and ecologically sound. These strategies are a survival mechanism that the tribe has been dependent upon, and in tune with the changing times and needs. Experience in the hills points out to limitations of short-range possibilities in the strategies. What is important, the course of social progress in time is gaining momentum. (It is inevitable to both economists and biologists to gain an understanding of the very substratum of their conflict. It is where geographers, as human ecologists, can help. The study of the Kollihills is best suited for such an understanding and it is this understanding that the present study on the Kollihills attempts at).

The total area of the Kollihills is 28,293 ha. Of this, the cultivated land is of 14,609 ha (51.63%). The net sown area is 12,009 ha (42.44% of the total area). Cultivable waste is 1,223 ha. Fallow is of the order of 2,669 ha while land put to other uses is 3,830 ha. Forest occupies 12,454 ha (44%). Grassland is of the order of 985 ha.

The crops grown are paddy, ragi, tapioca, millets, pepper, coffee, cardomon, mango, guava, pine apple and jack. Agricultural Officer of the Kollihills has provided the latest data on the land area under different crops and they are given below:

> Tapioca 4,453 ha, millets 1,727 ha, paddy 1,502 ha, ragi 1,337 ha, pine apple 1,000 ha, mochai 599 ha, guava 381 ha, banana 300 ha, soya beans 279 ha, coffee 234 ha, pepper 100 ha, silver oak 332 ha and mangoes 107 ha. Crops such as cardamom, clover and turmeric are grown in areas less than 100 ha. Thus millets including ragi and pine apple have greater proportion of area under them, with plantation crops and tree crops occupying considerable area under cultivation.

The 7-System Management and Terraced Agriculture

The agriculture of the tribes is not only hill agriculture but also terraced agriculture. The tribes are capable of land management given the difficulties of slope cultivation. They have greater skills at land and water management by managing the slopes very well. They have developed the technique of bunding, and planting trees to avert soil erosion and landslides. The farmers of the Kollihills have also developed inter-cropping which benefit them most and helps them with balancing family labour schedule given the assets and assignments. The *ayacut* of the streams, which is irrigated by the waters from the springs along them, is very fertile. The area under well irrigation is only 28 ha, whereas land under irrigation along stream bed is 1,873 ha.

Family farm is the basic unit of agriculture, exception being the plantation agriculture (cardamom, coffee). Family farm is characterized by family labour. Tribal farmers we are concerned with are essentially owner-cultivators, who take occasional hired help, including women, especially during the weeding and harvest operations. Women's contribution is higher in terms of family labour contributions rather than through wage labour. Tribal agriculture as practised here is no distinct from the plains as the modern inputs (seed, irrigation and chemical technologies) are very much the same.

Terraced agriculture brings their traditional managerial capabilities to the fore and the tribe have mastered a system of management, involving soil, slope, water, micro-climate, plant-vegetation, animal and space, the 7-system management of terraced agriculture (Table 1).

Table 1
The 7-System Management:
Indigenous Knowledge System and Traditional Managerial Practices of the Kollihills

Management Category	Forms of Practices
1. Slope	Terracing, levelling, hedging, timber and rock heaping.
2. Soil	Bunding, application of green and cattle manure, tillage and field forms.
3. Water	*Ayacut* principle of irrigation, storage in checkdams, traditional lifts (manual, animal drawn), diversion devices and traditional delivery systems.
4. Micro-climate	Shade management (plantation and homeyard bushes of coffee), surface geometry management, tillage and wind management (hedging to keep off winds).
5. Plant-Vegetation	Crop selection, mixtures, inter-cropping, planting methods, agroforestry practices, and predation management.
6. Animal	Domestication, and preventive measures against destruction of crops by animals and wildlife.
7. Space	Spatial arrangement of fields and crops (land utilization types) in relation to homes, distance minimization and crop care maximization.

Source: Kumaran, 1983; 1991; 1993.

What is more is that the system has been in practice for more than a century. The Table in effect is an elaboration of the internal, community support as no activity or practice is possible of the 7-system management, without the active involvement of the entire community.

The spatial mosaic that emerges from the terraced agriculture can only be appreciated upon careful observation of the nuances of the 7-system management. It must be mentioned that the 7-system management is part of the indigenous knowledge system of the tribal Kollihills, on a broader scale in agriculture than the ones that follow in other sections of this paper.

The tribe possesses the expertise in executive and managerial aspects of agriculture and agroforestry to an extent that is beneficial in the long run to the ecology and the society, in achieving certain production and welfare levels both for the tribe and for the ecology. The ills of hill ecology may thus be placed squarely on the externally influenced development efforts (by the government) and the ill-equipped bureaucracy of the tribal development administration. What is needed is the appreciation and encouragement of the wisdom of the tribe in formulating the 7-system management strategy for agriculture and agroforestry.

Indigenous Agroforestry System of the Kollihills

The agroforestry system of the Kollihills consists of semi-irrigated agriculture and forest gardens. Semi-irrigated agricultural lands are flat lands adjacent to streams. There is an intensive application of indigenous knowledge systems in this system of tribal economy. For example, the tribal farmers classify their soils in these areas into three types: red soil (*sem mann*), clayey soil (*kali mann*) and ash soil (*sambal mann*). Table 2 summarizes the diversity, status and uses of trees in the forests of the Kollihills.

Table 2
Tree Diversity and Uses of Forests

English Name	Local Name	Status	Uses (in Tamil)
Trees			
Acid lime Banana	Elumichai Vazhai	rare available	fruit; pickle flowers food; fruit for cooking; leaves for social occasions and ceremonies
Citron Eucalyptus	Narathai Thylam	available introduced	fruit; pickle firewood;charcoal; medicinal;posts; ornamental; windbreaks
Fig	Athi	rare	fruit; (fresh,dried) medicinal
Guava	Goiyya	available	fruit; firewood and tool handles
Jack	Pala	endangered	timber;groundwater recharge
Mandarin tangerine	Kamala	rare	fruit
Moringa	Kattu murungai	rare	leaves as plant nutrients; avenues; firewood
Jambulana	Naval	endangered	fruit; firewood; avenues
Orange	Sathu kudi	available	fruit
Pomegranate	Madulai	available	fruit
Sissoo	Visiri	endangered	firewood; timber
	Perikka	endangered	avenues
	Kola	endangered	avenues
	Enuga	endangered	groundwater recharge
	Konnai	rare	leaves as plant nutrients; avenues; firewood
Spices and Beverage			
Cardamom	Elakkai	rare	spice
Clove	Kirambu	rare	spice
Nutmeg	Jathikkai	rare	spice
Pepper	Milagu	available	spice
Coffee	Coffee	rare	beverage

Source: Fieldwork 1983; 1995; 2004. See Rajasekaran and Warren 1994: 14.

Semi-Irrigated Agricultural System

In clayey soils, they cultivate paddy (dry) and ragi. Ragi is also grown in red soils. Samai and thenai (a minor millet) are grown in ash soils. The soils are enriched by the application of the leaves of the forest trees such as konnai, karu vembu and kattu murungai. The availability of water which depends on the rainfall determines the type of crops to be grown in the semi-irrigated lands. In the case of early or late monsoon and failures, the tribal farmers cultivate drought resistant crops such as legumes, cowpea, sesamum and mustard.

Forest Garden System

This is known locally as the *kattu kazhani*. It is of three types, depending on whether the crops are planted on terraces, moderately sloping lands and steeply sloping lands. The soils of the terraces are enriched by the wild tree leaves, while lines of stones are placed at the bench edges to prevent soil erosion and to conserve moisture (see Reij, 1993). Pineapple and tapioca are cultivated mainly on the terraces, while citrus fruits, citron, guava and pomegranate are cultivated on the moderately sloping lands above the terraces.

Wild trees such as konnai, karvembu, naval and kattu murungai are found on the steeply sloping lands above the moderately sloping lands. Banana suckers are planted randomly amidst the forest gardens and also around the edges. In several hamlets, banana is intercropped with tapioca and pepper.

Indigenous Knowledge Systems in Agroforestry

Tapioca, it must be mentioned, was introduced to the tribal farmers by the agroindustry, while the inter-cropping of tapioca with banana was innovated by the people through their informal experiments. The farmers have shown that the banana grows in any type of soil in the garden. The farmers of the hills also grow local landraces of beans (karuppu mochai and avarai) and different types of root and tuber vegetable crops (senai and karunai kizhangu) as inter-crops in forest gardens.

The Malayalis of the Kollihills are knowledgeable in the harvest of fruits, methods of ripening, classification of fruits for sale in the market and preservation of fruits. They know a great deal about the fruit trees and their adaptability to altitudes of the slopes. Traditionally, the methods of cultivation are different for different tree crops and the seasons. Jack trees, the fruit of which is known for its flavour and taste, are grown in the higher altitudes of the terraced lands. In the tribe's knowledge, the jack trees are believed to significantly recharge groundwater, and the recharging of the groundwater by jack trees is locally referred to as 'sunaippu'. The farmers of the Kollihills claim that the jack trees encourage monsoon rains. The tribe has laid down laws and regulations towards minimizing the cutting of trees and these are underpinned by religious and spiritual belief systems. They have a sacred forest site, known locally as the kovil kaadu (temple forest), which has a variety of fruit and spice species (Rajasekaran and Warren, 1994).

Indigenous Food Security System of the Kollihills

Agricultural and agroforestry systems of the Kollihills are a basis of food and nutritional security in the tribal enclave. Cereals from the semi-irrigated agriculture form the bulk of the food while local landraces of beans play an important role in providing protein and the tuber crops satisfy the caloric energy requirements. The tribe often deny themselves the luxury of eating their own home grown fruits as they sell them in the markets. But they provide for their protein and fat requirements from the pulses and oil they buy using the income from the sale of the fruits. The portion the people set aside for home consumption is limited and thus it meets their vitamins and mineral salts requirements in a limited way.

Disappearance of the Indigenous Agroforestry System

The disappearance of the indigenous agroforestry system in the Kollihills over the decades, was due primarily to changing socio-cultural values, population increases, poor income from the marketed agricultural produce and extravagance of the tribal people in their social and cultural activities and celebrations. These have resulted in the depletion of forests and self-denial of the food requirements of the people. All through the events, however, the tribal people have remained aware of the consequences of deforestation and self-denial of food security. Besides, their socio-economic priorities have resulted in the cutting of trees and sale of vital agricultural produce in the markets around the hills.

The problem is further aggravated by the middlemen of the markets who lend money to the tribal farmers with the view to take their products at the harvest in repayment for the loan and interest. Their prices are the marginal which make the ventures unprofitable. Now that the agro-industries (sago industry, for example) have moved in with the large scale changes in the cropping pattern (especially of tapioca), they have persuaded or forced the tribal farmers to grow tapioca instead of fruit and wild trees. True, tapioca offers immediate income but has a devastating effect on the indigeneous forest systems. There has been a large scale cutting of the wild trees and forests to make way for the tapioca cultivation in recent years. This is fast endangering the agroforestry of the Kollihills. Apart from the socio-cultural constraints, there are some external factors which are unfavourable for the sustainability of the agroforestry systems, even while the Forest Department has moved with a large scale investment on social forestry. But the social forestry schemes which favour eucalyptus species deplete groundwater at an alarming rate.

Undoubtedly, the deterioration in the agroforestry systems has placed the people and the natural ecosystems (plant, animal) in a most vulnerable position. It has been found that the deforestation has already resulted in the drying up of some forest streams, which in turn has reduced the yield of cereal crops such as the paddy and the millets. These conditions also inhibit the growth of the spice crops such as cardamom, coffee, cloves and pepper which require shade and cool environments for vegetative growth and reproductive functions.

The rate of deforestation has had a direct impact on the food security system and nutrition availability. Inter-crops have suffered by the removal of shade trees. The declining fruit production in the recent years is blamed on the increasing area under tapioca, catering to the industry. Importantly, there has been a change in the rythm of monsoons which has reduced the availability of water in the streams and the recharge of groundwater from the surface run-off. This has particularly affected the yield of paddy which is the staple food of the Malayalis. In sum, the climatic change, deforestation and deterioration of the agro-forestry systems have resulted in a threat to the food security. What is more worse is that the tribe known for their 'altruistic tendencies' are slowly turning themselves into people who have no compassion for others, because of the hard times (see Kumaran, 1987). With the youngsters finding jobs in the towns and cities of the country, the cultural values have been changing very fast.

It is now a common sight that the children of the hills suffer from malnutrition, due to the combined effects of the deforestation and deterioration of forest systems, monsoon failure, inadequate water, loss of food production and inadequate revenues from the markets.

What Needs to be Done in the Changing Resource Process

It is imperative that the changing resource processes, in agroforestry, forest garden systems, cropping patterns of the modern agriculture and agro-industrial development, have spelt, as Table 2 indicates, disaster for some species of trees which are endangered and depletion of the agroforestry and crop systems to the detriment of food security of the Malayalis. While the tribe appear to "climb a foot along the pole of development, they seem to slide down by three feet": the path of development is indeed slippery and the slide in the health of the ecosystems and the people gives cause for worry. It is the contention of this study that the resource process in action is faulty and has to be drastically corrected with the participation of the tribal community. Commitment on the part of the community to a balanced, ecosystem supporting resource process is the need of the hour.

It is important that the efforts at such participatory development of the hills take into account the relationships between the indigenous or local knowledge system and indigenous agroforestry system on the one hand and the indigenous or local knowledge system and the biodiversity on the other. If these can be taken advantage of, in the interests of the Malayalis and sustainable development of the Kollihills, then indigenous food security system can be shored up and sustained. "There is a need for a large scale awareness campaign, stimulating action at the community level" needs no mention. But the planners and the policy makers of the state must rethink their inputs to a socio-economic development and formulate schemes with participatory approaches which will benefit the tribe in the long run while sustaining their indigenous knowledge system, indigenous agroforestry system and indigeous food security system through maintenance and management of biodiversity.

Some suggestions that emerge from the study are:

1. Allowing indigenous or local knowledge system full scale sway over new and modern systems of agriculture, including indigenous agroforestry system and forest garden systems. NGO initiative in this regard would be much welcome, provided the NGO is hill-based and not an outsider to the hills.
2. It is a fact that unless the tribals satisfy their food needs, fuel supplies required and curbing their population at the same time, deforestation cannot be stopped. NGO initiative and government cooperation could accomplish this provided joint management of the forest cover and alternatives to satisfying food and fuel needs could be found. Such initiative and cooperation should take into account, and clearly understand, the interrelationships between indigenous knowledge system, biodiversity, food security and deforestation.

3. There is need to increase the income of the people from their current sources of income, as poor income from fruit harvests marketed is a rule rather than exception. Efforts must be made, in a participatory way, to explore the realistic possibilities for increasing farmers' income through fruits and forest produce.
4. At the regional level, it would be worthwhile to set up a structured market in tune with the demands of the tribe, by making the tribe responsible for running it so that middlemen, who gives loan at exhorbitant interests, could be eliminated. What is more important is that the tribal farmers must be organized into an 'Association' so that they could take on the responsibility of running this market as a collective and not individual owners. This is quite possible and can be done in the hills, if an NGO could be identified and assigned the job.
5. There is need to create an awareness among the population about the need to keep their indigenous agroforestry system intact so that their indigenous food security system could be taken care of. This would be a challenge to the tribal population and not easy to accomplish unless they realize the worth of the forests and stay away from them. In fact, they should be helped to develop forests both as a community and as individual farmers.
6. It has been found that much of the indigenous knowledge system remains dormant and must be revived towards developing the agroforestry systems, forest gardens and cropping systems. NGO could not only do it but also assure community participation by developing dedicated village groups.
7. Agroindustry must be encouraged to set up industries but be made responsive to ecosystems conservation and upholding biodiversity of the hills.
8. Considering the agro-ecological and biophysical conditions of the hills, suitable indigenous landraces or hybrid varieties which were developed using traditional varieties may be advocated for cropping. As for forests, the rate of reforestation should be high.

References

Gadgil, M. and F. Berkes, (1991), "Traditional Resource Management Systems", *Resource Management and Optimization*, 8(3/4), 127–141.

Kumaran, T.V., (1983), "Agricultural Development in the Kollihills, A Project funded by Indian Council of Social Science Research", Department of Geography, University of Madras, Madras.

——, (1987), "Agricultural Systems and Foodwebs in the Kollihills of Salem District", *The Journal of Madras University*, LIX(1), 1–11.

——, (1991), "Community, Supporting Services and Crop Diversification in Tribal Agriculture", Proceedings of the Eighth Afro-Asian Regional Conference, Bangkok, International Commission on Irrigation and Drainage.

——, (1993), "Is Externally Influenced Tribal Development Viable? A Case of the Kollihills", *The Geography Teachers*, India (new series), 4(1), 17–32.

——, (1998), "Indigenous Knowledge Systems: Concepts, Cases and Assessment", A Workshop, Department of Geography, University of Madras, Chennai and Centre for Alternative Energy and Rural Technology, Chengalpattu.

——, (2003), "Sustainable Bio-diversity and Food Security through Traditional Ecological Knowledge Systems in the Kollihills of Tamil Nadu", India, A Major Research project Funded by the University Grants Commission, Department of Geography, University of Madras, Chennai.

Kumaran, T.V. and S. Aruchamy, (1989), "Human Time Budgets in Agriculture: A Case of Kambam Valley", L.Vidyanath and R.M. Rao (eds.), *Patterns of Resources Use in India*, New Delhi, BH Publishers.

Mathen, K., (1997), "The Burden of Literacy: A Letter from India", *The Ecologist*, 27(5), 175–177 (September-October).

Mitchell, B., (1997), *Resource and Environmental Management*, Essex, England: Addison Wesley Longman.

Norman, D.W., (1974), "Rationalising Mixed Cropping Under Indigenous Conditions: The Example of Nigeria", *The Journal of Development Studies*, 11: 3–21.

Rajasekaran, B., (1994), "A Framework for Incorporating Indigenous Knowledge Systems into Agricultural Research and Extension Organisations for Sustainable Agricultural Development", *Technology and Social Change Series* No. 22, Ames, Iowa: Iowa State University.

Rajasekaran. B. and D.M. Warren, (1994), "IK for Socio-Economic Development and Bio-Diversity Conservation: The Kollihills", Indigenous Knowledge and Development Monitor, 2(2), 13–17.

Reij, C., (1993), "Improving Indigenous Soil and Water Conservation Techniques: Does it Work?", Indigenous Knowledge and Development Monitor, 1(1), 11–13.

Rosell, D.Z., (1979), "Appropriate Technology: The Philippine Setting", *The Philippine Geographical Journal*, 23: 1–10.

Shah, P., (1993), "Participatory Watershed Management Programmes in India: Reversing our Roles and Revising our Theories", in IIED: Rural People's Knowledge, Agricultural Research and Extension Practice, *Asia Papers, Research Series*, 1(3), 38–67.

Weitz, R., (1971), *From Peasant to Farmer: A Revolutionary Strategy for Development*, New York, Columbia University Press.

Weitz, R., D. Pelley and L. Applebaum, (1980), "A Model for the Planning of New Settlement Projects", *World Development*, 8: 705–723.

Contributors

Abha Lakshmi Singh, Department of Geography, Aligarh Muslim University, Aligarh, India.

Abhineety Goel, Department of Geography, Delhi School of Economics, University of Delhi, Delhi, India.

Anu, Faculty of Science, University of Plymarth, Plymarth, UK.

Anupriya Chatterjee, Department of Geography, Visva-Bharati, Santiniketan, India.

Anuradha Sharma, Department of Geography, University of Jammu, Jammu Tawi, India.

Bat-Ochir Bold, Nordic Institute of Asian Studies, Leifsgade 33, DK 2300, Copenhagen S, Denmark.

Bulti Roy, Department of Geography and Disaster Management, Tripura University, Suryamaninagar, Tripura (West), India.

C.S.Rathore, Indian Institute of Forest Management, Bhopal, India.

Durgadas Mukhopadhyay, A21, Sector- 31, Noida, India.

Fan Zhengyi, Changchun Institute of Geography, Academia Sinica, Beijing, P.R. of China.

Huiping Lin, Department of Geography, Peking University, Beijing 100871, China.

J.S. Bali, Retd., Head of GoI Division of Soil Conservation, FAO Consultant, Dehradun, India.

Jennifer L. Hill, School of Geography and Environmental Management, University of the West of England, Bristol BS16 1QY, UK.

Jimmi Debbarma, Department of Geography and Disaster Management,Tripura University, Suryamaninagar-799130, Tripura (West), India.

Kailash Choubey, Department of Geography, Dr. H.S.Gaur University, Sagar, India.

Kaoru Takara, Disaster Prevention Research Institute, Kyoto University, Uji, Kyoto, Japan.

L. Delanoy, Meewasin Valley Authority, Saskatoon, SK, S7K 3G5, Canada.

M. Bortamuly, Department of Geography, Delhi School of Economics, University of Delhi, Delhi, India.

Md. Bahar-Ud-Din Shah, Department of Geography, Aligarh Muslim Univesity, Aligarth, India.

O.W. Archibold, Daprtment of Geography, University of Saskatchewan, Saskatoon, SK, 57N, 5A5, Canada.

Paul J. Curran, Bournemouth University, Talbot Campus, Fern Barrow, Poole, Dorset BH12 5BB, UK.

Roshan Shrestha, Disaster Prevention Research Institute, Kyoto University, Uji, Kyoto, Japan.

R.B. Singh, Department of Geography, Delhi School of Economics, University of Delhi, Delhi, India.

Rina Debbarma, Department of Geography and Disaster Management,Tripura University, Suryamaninagar, Tripura (West), India.

Rubeena T.A., Department of Geography, University of Madras, Chennai, India.

S.P. Singh, Faculty Areas of Forest Resource Economics and Management, Indian Institute of Forest Management, Bhopal, India.

Shailesh R. Shukla, Natural Resources Institute, The University of Manitoba, Winnipeg, Manitoba, R3T 2N2, Canada.

Sudip Dey, Department of Geography and Disaster Management, Tripura University, Suryamaninagar, Tripura (West), India.

Suraj Mal, Department of Geography, Delhi School of Economics, Univeristy of Delhi, Delhi.

Susmita Choudhuri, Department of Geography and Disaster Management,Tripura University, Suryamaninagar, Tripura (West), India.

Toshiharu Kojima, River Basin Research Center, Gifu University, Gifu, 501-1193, Japan.

T. Vasantha Kumaran, Department of Geography, University of Madras, Chennai, India.

Udo Schickhoff, Institute of Geography, University of Hamburg, Bundesstr, 55, D-20146 Hamburg, Germany.

Ved Prakash Meena, Department of Geography, M.S.J. (Government) P. G. College, Bharatpur, India.

Vishwambhar Prasad Sati, Department of Geography, Government P. G. College, Shivpuri, MP, India.

Xuegong Xu, Department of Geography, Peking University, Beijing, China.

Yasuto Tachikawa, Disaster Prevention Research Institute, Kyoto University, Uji, Kyoto, Japan.

Yogesh Dubey, Indian Institute of Forest Management, Bhopal, India.

Zaiyi Fu, Department of Geography, Peking University, Beijing, China.